Studies in Computational Intelligence

Volume 1061

The series "Studies in Computational Intelligence" (SCI) publishes new developments and advances in the various areas of computational intelligence—quickly and with a high quality. The intent is to cover the theory, applications, and design methods of computational intelligence, as embedded in the fields of engineering, computer science, physics and life sciences, as well as the methodologies behind them. The series contains monographs, lecture notes and edited volumes in computational intelligence spanning the areas of neural networks, connectionist systems, genetic algorithms, evolutionary computation, artificial intelligence, cellular automata, self-organizing systems, soft computing, fuzzy systems, and hybrid intelligent systems. Of particular value to both the contributors and the readership are the short publication timeframe and the world-wide distribution, which enable both wide and rapid dissemination of research output.

Indexed by SCOPUS, DBLP, WTI Frankfurt eG, zbMATH, SCImago.

All books published in the series are submitted for consideration in Web of Science.

Oscar Castillo · Patricia Melin
Editors

Fuzzy Logic and Neural Networks for Hybrid Intelligent System Design

Springer

Editors
Oscar Castillo
Division of Graduate Studies and Research
Tijuana Institute of Technology
Tijuana, Baja California, Mexico

Patricia Melin
Division of Graduate Studies and Research
Tijuana Institute of Technology
Tijuana, Baja California, Mexico

ISSN 1860-949X ISSN 1860-9503 (electronic)
Studies in Computational Intelligence
ISBN 978-3-031-22041-8 ISBN 978-3-031-22042-5 (eBook)
https://doi.org/10.1007/978-3-031-22042-5

This Springer imprint is published by the registered company Springer Nature Switzerland AG
The registered company address is: Gewerbestrasse 11, 6330 Cham, Switzerland

Preface

We describe in this book recent developments on fuzzy logic and neural networks, as well as their hybrid combinations, and their application in areas such as, intelligent control and robotics, pattern recognition, medical diagnosis, time series prediction and optimization of complex problems. There are papers with the main topics of type-1, type-2 and type-3 fuzzy systems, which basically consists of papers that propose new concepts and algorithms based on type-1, type-2 and type-3 fuzzy logic theory and their applications. There are also papers that present theory and practice of meta-heuristics in diverse areas of application. There are interesting papers on different applications of fuzzy logic, neural networks and hybrid intelligent systems in medical problems. In addition, we can find papers describing applications of fuzzy logic, neural networks and meta-heuristics in robotics problems. There are a total of 14 papers forming the book in the above-mentioned topics.

In conclusion, the edited book comprises papers on diverse aspects of fuzzy logic, neural networks and nature-inspired optimization meta-heuristics for designing and implementing hybrid intelligent systems and their application in areas , such as intelligent control and robotics, pattern recognition, time series prediction and optimization of complex problems. We expect that the book will serve as reference for researchers and graduate students working in the theory and applications of the computational intelligence area.

Tijuana, Mexico — Oscar Castillo
Tijuana, Mexico — Patricia Melin
May 2022

About This Book

This book covers recent developments on fuzzy logic, neural networks and optimization algorithms, as well as their hybrid combinations. In addition, the above-mentioned methods are applied to areas such as intelligent control and robotics, pattern recognition, medical diagnosis, time series prediction and optimization of complex problems. Nowadays, the main topic of the book is highly relevant, as most current intelligent systems and devices in use utilize some form of intelligent feature to enhance their performance. In addition, on the theoretical side, new and advanced models and algorithms of type-2 and type-3 fuzzy logic are presented, which will be of great interest to researchers in these areas. Also, new nature-inspired optimization algorithms and innovative neural models are put forward in the manuscript, that are very popular subjects, at this moment. There are contributions on theoretical aspects as well as applications, which make the book very appealing to a wide audience, ranging from researchers to professors and graduate students.

Contents

On Decision Making Applications via Distance Measures 1
Feride Tuğrul and Mehmet Çitil

On Intuitionistic Fuzzy Abstract Algebras 23
Gökhan Çuvalcıoğlu and Sinem Tarsuslu

Generalization of Intuitionistic Fuzzy Submodules of a Module by Using Triangular Norms and Conorms and (T,S)-L Subrings 51
Ümit Deniz

Fuzzy Dynamic Parameter Adaptation in the Mayfly Algorithm: Implementation of Fuzzy Adaptation and Tests on Benchmark Functions and Neural Networks 69
Enrique Lizarraga, Fevrier Valdez, Oscar Castillo, and Patricia Melin

Fuzzy Classifier Using the Particle Swarm Optimization Algorithm for the Diagnosis of Arterial Hypertension 85
Martha Pulido and Patricia Melin

A Survey of Models and Solution Methods for the Internet Shopping Optimization Problem 105
Miguel Ángel García Morales, Hector Joaquín Fraire Huacuja, Juan Frausto Solís, Laura Cruz Reyes, and Claudia Guadalupe Gómez Santillán

A Comparison Between MFCC and MSE Features for Text-Independent Speaker Recognition Using Machine Learning Algorithms 123
Joseph Isaac Ramírez-Hernández, Alain Manzo-Martínez, Fernando Gaxiola, Luis C. González-Gurrola, Vania C. Álvarez-Oliva, and Roberto López-Santillán

Forecasting Based on Fuzzy Logic of the Level of Epidemiological Risk for the Mexican State of Tamaulipas 141
Paula Hernández-Hernández and Norberto Castillo-García

Bio-inspired Flower Pollination Algorithm for the Optimization of a Monolithic Neural Network 155
Hector Carreon-Ortiz, Patricia Melin, and Fevrier Valdez

Rendezvous and Docking Control of Satellites Using Chaos Synchronization Method with Intuitionistic Fuzzy Sliding Mode Control 177
Onur Silahtar, Fatih Kutlu, Özkan Atan, and Oscar Castillo

Optimizing a Convolutional Neural Network with a Hierarchical Genetic Algorithm for Diabetic Retinopathy Detection 199
Rodrigo Cordero-Martínez, Daniela Sánchez, and Patricia Melin

Interval Type-3 Fuzzy Systems: A Natural Evolution from Type-1 and Type-2 Fuzzy Systems 209
Oscar Castillo, Juan R. Castro, and Patricia Melin

A Comparative Study Between Bird Swarm Algorithm and Artificial Gorilla Troops Optimizer 223
Ivette Miramontes and Patricia Melin

Particle Swarm Optimization of Convolutional Neural Networks for Diabetic Retinopathy Classification 237
Patricia Melin, Daniela Sánchez, and Rodrigo Cordero-Martínez

On Decision Making Applications via Distance Measures

Feride Tuğrul and Mehmet Çitil

Abstract Intuitionistic fuzzy sets are an area that has attracted the attention of many researchers recently and is used in many application areas. Researchers have made use of intuitionistic fuzzy sets because of their usefulness in all applications involving decision making. In decision making problems including criteria and alternatives, intuitive fuzzy sets are evaluated with distance measures defined, and more sensitive results are obtained in application areas than many methods.

Keywords Fuzzy logic · Intuitionistic fuzzy set · Decision making · Distance measure

1 Introduction

The notion of fuzzy logic was firstly defined by Zadeh in 1965 [1]. Then, Intuitionistic fuzzy sets (shortly IFS) were defined by Atanassov [2, 3]. Intuitionistic fuzzy sets form a generalization of the notion of fuzzy sets. The intuitionistic fuzzy set theory is useful in various areas, such as artificial intelligence, engineering, education, algebraic structures, topologic spaces, algebraic structures, control systems, agriculture areas, computer, economy and various engineering fields, real life situations [4–14]. Various applications of intuitionistic fuzzy set have been carried out through distance measures approach [15–20].

Education has many components and is an important issue for the continuity of societies. Each occupational group must complete the training process related to their field. The talents of an individual belonging to a profession and the field in which he/she specializes should be compatible with the needs of the society. Although the basis of success is human, the most important factor affecting success

---

F. Tuğrul (✉) · M. Çitil
Department of Mathematics, Kahramanmaraş Sütcü İmam University, Kahramanmaraş, Turkey
e-mail: feridetugrul@gmail.com

M. Çitil
e-mail: citil@ksu.edu.tr

O. Castillo and P. Melin (eds.), *Fuzzy Logic and Neural Networks for Hybrid Intelligent System Design*, Studies in Computational Intelligence 1061,
https://doi.org/10.1007/978-3-031-22042-5_1

is education. Education has a wide range of applications in many fields such as fuzzy logic, intuitionistic fuzzy logic, decision making, multi-criteria decision making, intercriteria analysis [21–23].

Choosing the most accurate and rational system while performing selection, evaluation and ranking in education; it is very important to achieve our goal. We used the distance measure while determining the goals and enroll in the training. By expressing the exam scores in intuitionistic fuzzy clusters instead of just a single grade, we can observe the concepts of membership, non-membership and hesitancy together. Therefore, it will give the most accurate result to be expressed with the intuitionistic fuzzy sets given in education. Also, it is very useful to use the distance measure when listing the intuitionistic fuzzy alternatives [14]. For this paper; high schools in any province in Turkey have been researched. If it is desired to work with these methods in the field of education, data can be obtained from the desired country and the desired province. The data we used in our study were taken from Turkey. For these papers, Approximately 42,000 students have been researched according to official data from the Ministry of Education.

In our studies, we preferred the field of education as the application areas of intuitionistic fuzzy logic. The methods we will use are useful methods that can be applied in all areas. Thanks to these methods, many goals such as selection, ranking, decision making, placement, and element selection can be achieved in the most accurate way.

2 Preliminaries

Definition 1 : [1] Let $X \neq \emptyset$. A fuzzy set A in X defined as:

$$A = \{\langle x, \mu_A(x)\rangle | x \in X\},$$

where $\mu_A(x) : X \to [0, 1]$ is the membership function of the fuzzy set A.

Definition 2 : [2, 3] Let $X \neq \emptyset$. An intuitionistic fuzzy set A in X;

$$A = \{\langle x, \mu_A(x), \nu_A(x)\rangle | x \in X\}$$

$$\mu_A(x), \nu_A(x), \pi_A(x) : X \to [0, 1]$$

defined membership degree, nonmembership degree and hesitation degree of the element $x \in X$ respectively. In which,

$$\mu_A(x) + \nu_A(x) + \pi_A(x) = 1$$

Definition 3 : Let $X \neq \emptyset$. Intuitionistic fuzzy sets $A, B, C \in X$. The distance measure d between intuitionistic fuzzy sets A and B mapping $d : X \times X \rightarrow [0, 1]$; if $d(A, B)$ sstisfies the following axioms:

A1. $0 \leq d(A, B) \leq 1$

A2. $d(A, B)$ if and only if $A = B$

A3. $d(A, B) = d(B, A)$

A4. If $A \subseteq B \subseteq C$ then $d(A, C) \geq d(A, B)$ and $d(A, C) \geq d(B, C)$

Distance measure is a term that describes the difference between intuitionistic fuzzy sets and can be considered as a dual concept of similarity measure.

Definition 4 : [24, 25] Let $A = \{\langle x, \mu_A(x), \nu_A(x), \pi_A(x)\rangle | x \in X\}$ and $B = \{x, \mu_B(x), \nu_B(x), \pi_B(x) | x \in X\}$ be two intuitionistic fuzzy sets in $X = x_1, x_2, \ldots, x_n$; $i = 1, 2, \ldots, n$. Based on geometric interpretation of intuitionistic fuzzy set proposed the following four distance measures between A and B:

The Hamming Distance:

$$d_H(A, B) = \frac{1}{2}\sum_{i=1}^{n}(|\mu_A(x_i) - \mu_B(x_i)| + |\nu_A(x_i) - \nu_B(x_i)| + |\pi_A(x_i) - \pi_B(x_i)|)$$

The Euclidean Distance:

$$d_E(A, B) = \sqrt{\frac{1}{2}\sum\nolimits_{i=1}^{n}\left[(\mu_A(x_i) - \mu_B(x_i))^2 + (\nu_A(x_i) - \nu_B(x_i))^2 + (\pi_A(x_i) - \pi_B(x_i))^2\right]}$$

The Normalized Hamming Distance:

$$d_{n-H}(A, B) = \frac{1}{2n}\sum_{i=1}^{n}(|\mu_A(x_i) - \mu_B(x_i)| + |\nu_A(x_i) - \nu_B(x_i)| + |\pi_A(x_i) - \pi_B(x_i)|)$$

The Normalized Euclidean Distance:

$$d_{n-E}(A, B) = \sqrt{\frac{1}{2n}\sum\nolimits_{i=1}^{n}\left[(\mu_A(x_i) - \mu_B(x_i))^2 + (\nu_A(x_i) - \nu_B(x_i))^2 + (\pi_A(x_i) - \pi_B(x_i))^2\right]}$$

Definition 5 : [26] M is set of options and C is a set of criteria. $M = \{M_1, M_2, \ldots, M_m\}$, $C = \{C_1, C_2, \ldots, C_n\}$ where each option M_i is expressed via intuitionistic fuzzy description, namely

$$M_i = \{(C_1, \mu_{i1}, \nu_{i1}), (C_2, \mu_{i2}, \nu_{i2}), \ldots, (C_n, \mu_{in}, \nu_{in})\},\ i = 1, 2, \ldots, m.$$

where μ_{ij} indicates the degree to which option M_i satisfies criterion C_j, ν_{ij} indicates the degree to which option M_i does not satisfy criterion C_j. Our goal is to point out the best option (to rank the considered options). The options should satisfy the criteria $C_j, C_k, \ldots, C_p$ or criterion C_s, i.e.:

$$\left(C_j \text{and} C_k \text{and}, \ldots, \text{ and } C_p\right) \text{or } C_s \tag{1}$$

Definition 6 : [13]

$$Sim(C_i, A) = \frac{l_{IFS}(C_i, A)}{l_{IFS}(C_i, B)} \tag{2}$$

where $l_{IFS}(C_i, A)$ is a distance from $C_i\left(\mu_{C_i}, \nu_{C_i}, \pi_{C_i}\right)$ to $A(1, 0, 0)$, $l_{IFS}(C_i, B)$ is a distance from $C_i\left(\mu_{C_i}, \nu_{C_i}, \pi_{C_i}\right)$ to $B(0, 1, 0)$. The distances $l_{IFS}(C_i, A)$ and $l_{IFS}(C_i, B)$ are calculated from:

$$l_{IFS}(C_i, A) = \frac{1}{2}\sum_{i=1}^{n}\left(\left|1 - \mu_{C_i}\right| + \left|0 - \nu_{C_i}\right| + \left|0 - \pi_{C_i}\right|\right) \tag{3}$$

$$l_{IFS}(C_i, B) = \frac{1}{2}\sum_{i=1}^{n}\left(\left|0 - \mu_{C_i}\right| + \left|1 - \nu_{C_i}\right| + \left|0 - \pi_{C_i}\right|\right) \tag{4}$$

For, $0 \le Sim(C_i, A) \le \infty$.

The problem of finding an option M_i satisfying in the best way condition (1) can be solved by evaluating each option M_i

$$\begin{aligned} E(M_i) =& Sim(A, M_i) \\ =& \min\left\{\left[Sim\left(A, C_j\right), Sim(A, C_k), \ldots\ldots, Sim\left(A, C_p\right)\right], Sim(A, C_s)\right\} \end{aligned} \tag{5}$$

Condition (5) means that for each M_i we look for the worst satisfied criterion W_i among $C_j, C_k, \ldots,$ and C_p and next we look for the better criterion between W_i and C_s. The worst means the least similar and the least similar and the best means the most similar.

The smallest value among $E(M_i), i = 1, 2, \ldots, m$ in (5) points out the option which best satisfies condition in (1).

3 An Application of Intuitionistic Fuzzy Logic with Distance Measure

The objective of this application is to determine the relationship between students' official tests and the pilot tests by means of distance measures in intuitionistic fuzzy sets [27]. The solution has been accepted by measuring the shortest distance between each student and each school by using distance measure. Owing to this application; distance measures have been compared. The sequence of validity and reliability of the distance measures has been determined. Also; this application has been carried out on students who studied in two different years.

3.1 Comparison of Distance Measures with the Application of Intuitionistic Fuzzy Sets

The official exam results of the students who were arbitrarily selected from one of those years have been investigated. Let $H = \{H_1, H_2, H_3, H_4, H_5\}$ be set of high schools. L ={Turkish, Mathematics, Science, Social, English, Religion} be set of lessons,

$S = \{S_1, S_2, S_3, S_4, S_5, S_6, S_7, S_8, S_9, S_{10}\}$ be set of students.

High schools' reference points have been calculated in one of those years for each lesson in L, in Table 1.

Students' official test scores in one of those years academic year have been indicated in Table 2.

Shortest distance between each student (i.e. Table 2) and each high school (i.e. Table 1) has been calculated using Euclidean distance method, in Table 3.

Shortest distance between each student (i.e. Table 2) and each high school (i.e. Table 1) has been calculated using normalized Euclidean distance method, in Table 4.

Shortest distance between each student (i.e. Table 2) and each high school (i.e. Table 1) has been calculated using hamming distance method, in Table 5.

Shortest distance between each student (i.e. Table 2) and each high school (i.e. Table 1) has been calculated using normalized hamming distance method, in Table 6.

Distance between each student and each school has been calculated by four different distance measures. Comparison of distance measures is in the following table (Table 7).

If we want to interpret these tables, we can achieve the following results:

- The results of all distance measurements are consistent with each other.
- The most accurate distance measurement is the normalized measure of hamming.

According to Table 7; the sequence of validity and reliability of the distance measures is this;

$$d_{n-H} < d_{n-E} < d_E < d_H.$$

Table 1 High schools' reference points

	Turkish	Mathematics	Science	Social	English	Religion
H_1	(0.965,0.028,0.007)	(0.985,0.012,0.003)	(0.995,0.004,0.001)	(0.990,0.008,0.002)	(0.975,0.02,0.005)	(0.995,0.004,0.001)
H_2	(0.91,0.08,0.01)	(0.945,0.044,0.011)	(0.945,0.044,0.011)	(0.92,0.064,0.016)	(0.9,0.08,0.02)	(0.99,0.008,0.02)
H_3	(0.845,0.124,0.031)	(0.8,0.18,0.02)	(0.825,0.14,0.035)	(0.85,0.12,0.03)	(0.855,0.126,0.019)	(0.95,0.04,0.01)
H_4	(0.71,0.261,0.029)	(0.57,0.387,0.043)	(0.75,0.2,0.05)	(0.825,0.14,0.035)	(0.65,0.28,0.07)	(0.69,0.248,0.062)
H_5	(0.64,0.324,0.036)	(0.4,0.48,0.12)	(0.55,0.405,0.045)	(0.61,0.351,0.039)	(0.57,0.345,0.085)	(0.83,0.153,0.017)

Table 2 Students' official test scores

	Turkish	Mathematics	Science	Social	English	Religion
S_1	(0.95,0.04,0.01)	(0.98,0.01,0.01)	(0.99,[illegible].005,0.005)	(0.95,0.03,0.02)	(0.98,0.01,0.01)	(0.95,0.04,0.01)
S_2	(0.9,0.08,0.02)	(0.98,0.01,0.01)	(0.95,[illegible]03,0.02)	(0.99,0.008,0.002)	(0.98,0.015,0.005)	(0.99,0.006,0.004)
S_3	(0.65,0.28,0.07)	(0.45,0.44,0.11)	(0.6,0[illegible]2,0.08)	(0.4,0.48,0.12)	(0.35,0.52,0.13)	(0.65,0.3,0.05)
S_4	(0.6,0.32,0.08)	(0.45,0.44,0.11)	(0.7,0[illegible]4,0.06)	(0.8,0.16,0.04)	(0.8,0.17,0.03)	(0.8,0.16,0.04)
S_5	(0.7,0.24,0.06)	(0.8,0.16,0.04)	(0.85,[illegible]12,0.03)	(0.75,0.2,0.05)	(0.9,0.08,0.02)	(0.95,0.04,0.01)
S_6	(0.85,0.12,0.03)	(0.95,0.04,0.01)	(0.95,[illegible]02,0.03)	(0.95,0.05,0)	(0.85,0.11,0.04)	(0.95,0.02,0.03)
S_7	(0.9,0.08,0.02)	(0.95,0.04,0.01)	(0.95,[illegible]02,0.03)	(0.95,0.015,0.035)	(0.9,0.09,0.01)	(0.95,0.04,0.01)
S_8	(0.75,0.2,0.05)	(0.61,0.35,0.04)	(0.72,[illegible]24,0.04)	(0.82,0.15,0.03)	(0.7,0.25,0.05)	(0.7,0.2,0.1)
S_9	(0.65,0.3,0.05)	(0.68,0.26,0.06)	(0.72,[illegible]24,0.04)	(0.79,0.15,0.06)	(0.75,0.18,0.07)	(0.78,0.15,0.07)
S_{10}	(0.6,0.3,0.1)	(0.55,0.4,0.05)	(0.8,0.[illegible]5,0.05)	(0.74,0.22,0.04)	(0.9,0.07,0.03)	(0.4,0.51,0.09)

Table 3 Values calculated using Euclidean distance

E	H_1	H_2	H_3	H_4	H_5
S_1	0.0568	0.1107	0.2949	0.6491	0.9039
S_2	0.0757	0.0987	0.697	0.6729	0.8889
S_3	0.9633	0.8184	0.61004	0.4247	0.4119
S_4	0.7075	0.6225	0.4371	0.2585	0.3649
S_5	0.4002	0.2949	0.1704	0.4187	0.5935
S_6	0.5424	0.0775	0.2141	0.5296	0.7993
S_7	0.5507	0.0583	0.2142	0.6131	0.8261
S_8	0.6737	0.5254	0.3389	0.1012	0.3927
S_9	0.5839	0.4768	0.2972	0.1915	0.43001
S_{10}	0.8226	0.71707	0.6187	0.5204	0.53906

Table 4 Values calculated using normalized Euclidean distance

$n - E$	H_1	H_2	H_3	H_4	H_5
S_1	0.0231	0.045	0.1204	0.2675	0.36903
S_2	0.0308	0.04279	0.27589	0.27471	0.365
S_3	0.43412	0.391	0.3375	0.20901	0.13595
S_4	0.29438	0.37369	0.17615	0.17789	0.14159
S_5	0.22332	0.1204	0.06946	0.17097	0.24231
S_6	0.27333	0.0333	0.0815	0.22134	0.33569
S_7	0.05085	0.03291	0.09209	0.2341	0.33903
S_8	0.27506	0.2168	0.13836	0.0415	0.16034
S_9	0.2383	0.1946	0.1213	0.07818	0.17555
S_{10}	0.33582	0.29274	0.25261	0.21248	0.2266

Table 5 Values calculated using Hamming distance

H	H_1	H_2	H_3	H_4	H_5
S_1	0.0261	0.0458	0.1125	0.2675	0.3623
S_2	0.0257	0.0363	0.107	0.2658	0.365
S_3	0.434	0.4053	0.3375	0.1845	0.1315
S_4	0.2872	0.2418	0.1625	0.0941	0.1183
S_5	0.1465	0.1101	0.056	0.1576	0.2308
S_6	0.0692	0.0333	0.0701	0.2176	0.3184
S_7	0.0539	0.0163	0.08	0.2341	0.333
S_8	0.2519	0.2168	0.1375	0.0415	0.156
S_9	0.2539	0.2051	0.1258	0.0766	0.1513
S_{10}	0.3139	0.2701	0.206	0.1341	0.2266

Table 6 Values calculated using normalized Hamming distance

$n-H$	H_1	H_2	H_3	H_4	H_5
S_1	0.0043	0.0076	0.0187	0.0445	0.0603
S_2	0.0042	0.006	0.0178	0.0443	0.0608
S_3	0.0723	0.0675	0.0562	0.0307	0.0219
S_4	0.0478	0.0403	0.027	0.0156	0.0197
S_5	0.0244	0.0183	0.0093	0.0262	0.0384
S_6	0.0115	0.0055	0.0116	0.0362	0.053
S_7	0.0089	0.0027	0.0133	0.039	0.0555
S_8	0.0419	0.0361	0.0229	0.0069	0.026
S_9	0.0423	0.0341	0.0209	0.0127	0.0252
S_{10}	0.0523	0.045	0.0343	0.0223	0.0377

3.2 *Determination of the High School Where the Students Will Enroll Using the Normalized Hamming Distance Measure*

This section was implemented using the normalized hamming distance measurement in the light of the above results. The exams that are applied at regular intervals in the educational institutions where students work outside of the school are determined and their results are investigated [27]. At the end of the year, the official exam results of the students were obtained from the institution. The results of the formal exam and the exams in the education institution other than the school have been compared and interpreted. Let $H = \{H_1, H_2, H_3, H_4, H_5\}$ be set of high schools. $L = \{$Turkish, Mathematics, Science, Social, English, Religion$\}$ be set of lessons, $S = \{S_1, S_2, S_3, S_4, S_5, S_6, S_7, S_8, S_9, S_{10}\}$ be set of students.

High school base point has been calculated for each lesson in L, in Table 8.

Average of students' who were randomly selected pilot tests score has been determined in Table 9.

Distance between each student (i.e. Table 9) and each high school (i.e. Table 8) has been calculated using normalized hamming distance method depending upon average of students' pilot tests, in Table 10.

Table 10 depicts that the shortest distance between each student and each high school has given that the student will enroll in the high school depending upon average of students' pilot tests. According to Table 10; the student S_1 is to enroll in H_2 high school, the student S_2 is to enroll in H_3 high school, the student S_3 is to enroll in H_3 high school, etc. Official test scores of the students in 2016–2017 academic year has been determined in Table 11.

Distance between each student (i.e. Table 11) and each high school (i.e. Table 8) has been calculated using normalized hamming distance method depending upon official test scores of the students', in Table 12.

Table 7 Comparison of distance measures values

		H_1	H_2	H_3	H_4	H_5
S_1	d_E	0.0568	0.1107	0.2949	0.6491	0.9039
	d_H	0.0261	0.0458	0.1125	0.2675	0.3623
	d_{n-E}	0.0231	0.045	0.1204	0.2675	0.36903
	d_{n-H}	0.0043	0.0076	0.0187	0.0445	0.0603
S_2	d_E	0.0757	0.0987	0.697	0.6729	0.8889
	d_H	0.0257	0.0363	0.107	0.2658	0.365
	d_{n-E}	0.0308	0.04279	0.27589	0.27471	0.365
	d_{n-H}	0.0042	0.006	0.0178	0.0443	0.0608
S_3	d_E	0.9633	0.8184	0.61004	0.4247	0.4119
	d_H	0.434	0.4053	0.3375	0.1845	0.1315
	d_{n-E}	0.43412	0.391	0.3375	0.20901	0.13595
	d_{n-H}	0.0723	0.0675	0.0562	0.0307	0.0219
S_4	d_E	0.7075	0.6225	0.4371	0.2585	0.3649
	d_H	0.2872	0.2418	0.1625	0.0941	0.1183
	d_{n-E}	0.29438	0.37369	0.17615	0.17789	0.14159
	d_{n-H}	0.0478	0.0403	0.027	0.0156	0.0197
S_5	d_E	0.4002	0.2949	0.1704	0.4187	0.5935
	d_H	0.1465	0.1101	0.056	0.1576	0.2308
	d_{n-E}	0.22332	0.1204	0.06946	0.17097	0.24231
	d_{n-H}	0.0244	0.0183	0.0093	0.0262	0.0384
S_6	d_E	0.5424	0.0775	0.2141	0.5296	0.7993
	d_H	0.0692	0.0333	0.0701	0.2176	0.3184
	d_{n-E}	0.27333	0.0333	0.0815	0.22134	0.33569
	d_{n-H}	0.0115	0.0055	0.0116	0.0362	0.053
S_7	d_E	0.5507	0.0583	0.2142	0.6131	0.8261
	d_H	0.0539	0.0163	0.08	0.2341	0.333
	d_{n-E}	0.05085	0.03291	0.09209	0.2341	0.33903
	d_{n-H}	0.0089	0.0027	0.0133	0.039	0.0555
S_8	d_E	0.6737	0.5254	0.3389	0.1012	0.3927
	d_H	0.2519	0.2168	0.1375	0.0415	0.156
	d_{n-E}	0.27506	0.2168	0.13836	0.0415	0.16034
	d_{n-H}	0.0419	0.0361	0.0229	0.0069	0.026
S_9	d_E	0.5839	0.4768	0.2972	0.1915	0.43001
	d_H	0.2539	0.2051	0.1258	0.0766	0.1513
	d_{n-E}	0.2383	0.1946	0.1213	0.07818	0.17555
	d_{n-H}	0.0423	0.0341	0.0209	0.0127	0.0252

(continued)

Table 7 (continued)

		H_1	H_2	H_3	H_4	H_5
S_{10}	d_E	0.8226	0.71707	0.6187	0.5204	0.53906
	d_H	0.3139	0.2701	0.206	0.1341	0.2266
	d_{n-E}	0.33582	0.29274	0.25261	0.21248	0.2266
	d_{n-H}	0.0523	0.045	0.0343	0.0223	0.0377

Table 12 depicts that the shortest distance between each student and each high school has given that the student will enroll in the high school depending upon official test scores of the students'. According to Table 12; the student S_1 is to enroll in H_2 high school, the student S_2 is to enroll in H_3 high school, etc. When Tables 10 and 12 are researched, coherent results have been obtained. But the results of two students are different. According to the tables; it has been determined that two students (S_4 and S_7) went to a different high school than expected. Comments on inconsistent results may include:

- When school scores are too close together, even the slightest change in student scores can change the school.
- Every student cannot be expected to be as comfortable in the real exam as in normal life. There can be many factors that affect the day of the exam (health, psychology, etc.).

4 An Application of Intuitionistic Fuzzy Logic with Similarity Measure in Decision Making

In this section, we have implemented an application of decision making in the success ranking of middle schools. For this application, similarity measures defined in intuitionistic fuzzy sets are used [28]. Each middle school point has been calculated depending on the average student examination score.

The important thing in decision making is to determine the criteria. It is obvious that the criteria affecting school success are courses. While ranking the success of the schools, decisions were made according to these criteria. In this application, which is based on the basic courses, both the ranking was made and the most successful school was determined. For this application have been profited from similarity measures that proposed new solution by Szmidt and Kacprzyk [29]. The utility of this method; options have been compared to the positive-ideal solution and negative-ideal solution. The best considered option should be as close as possible to the positive-ideal solution and as far as possible to the negative-ideal solution. The best option taking into account only positive-ideal solution can be misleading.

$S = \{S_1, S_2, S_3, S_4, S_5, S_6, S_7, S_8, S_9, S_{10}\}$ be set of high schools.

$L = \{L_1, L_2, L_3, L_4, L_5, L_6\}$ be set of criteria. Set of criteria respectively are $L = $\{Turkish, Mathematics, Science, Social, English, Religion\}.

Table 8 High school base points

	Turkish	Mathematics	Science	Social	English	Religion
H_1	(0.96,0.03,0.01)	(0.95,0.04,0.01)	(0.995,0.004,0.001)	(0.97,0.02,0.01)	(0.95,0.03,0.02)	(0.997,0.002,0.001)
H_2	(0.92,0.06,0.02)	(0.93,0.06,0.01)	(0.93,0.05,0.02)	(0.9,0.05,0.05)	(0.91,0.05,0.04)	(0.98,0.01,0.01)
H_3	(0.85,0.07,0.08)	(0.82,0.12,0.06)	(0.75,0.20,0.05)	(0.85,0.12,0.03)	(0.81,0.18,0.01)	(0.9,0.05,0.05)
H_4	(0.65,0.3,0.05)	(0.55,0.44,0.01)	(0.65,0.3,0.05)	(0.81,0.17,0.02)	(0.63,0.3,0.07)	(0.55,0.42,0.03)
H_5	(0.5,0.45,0.05)	(0.3,0.65,0.05)	(0.63,0.33,0.04)	(0.66,0.25,0.09)	(0.65,0.32,0.03)	(0.81,0.15,0.04)

Table 9 Average of students' pilot tests

	Turkish	Mathematics	Science	Social	English	Religion
S_1	(0.9,0.05,0.05)	(0.92,0.07,0.01)	(0.9[illegible],0.03,0.02)	(0.87,0.1,0.03)	(0.95,0.05,0)	(0.92,0.04,0.04)
S_2	(0.86,0.1,0.04)	(0.83,0.15,0.02)	(0.8[illegible],0.14,0.04)	(0.89,0.1,0.01)	(0.68,0.25,0.07)	(0.89,0.09,0.02)
S_3	(0.82,0.1,0.08)	(0.77,0.15,0.08)	(0.6[illegible],0.25,0.09)	(0.77,0.2,0.03)	(0.63,0.3,0.07)	(0.9,0.05,0.05)
S_4	(0.76,0.2,0.04)	(0.59,0.36,0.05)	(0.5[illegible],0.45,0.04)	(0.68,0.24,0.08)	(0.54,0.4,0.06)	(0.89,0.09,0.02)
S_5	(0.95,0.04,0.01)	(0.9,0.05,0.05)	(0.[illegible],0.03,0.05)	(0.95,0.04,0.01)	(0.87,0.1,0.03)	(0.95,0.05,0)
S_6	(0.36,0.55,0.09)	(0.35,0.6,0.05)	(0.[illegible],0.6,0.11)	(0.375,0.5,0.125)	(0.265,0.65,0.085)	(0.4,0.5,0.1)
S_7	(0.9,0.05,0.05)	(0.95,0.03,0.02)	(0.9[illegible].06,0.04)	(0.95,0.01,0.04)	(0.95,0.02,0.03)	(0.95,0.05,0)
S_8	(0.95,0.04,0.01)	(1,0,0)	(0.9[illegible].05,0.05)	(1,0,0)	(1,0,0)	(1,0,0)
S_9	(0.7,0.2,0.1)	(0.8,0.15,0.05)	(0.9[illegible].05,0.05)	(0.95,0.03,0.02)	(0.9,0.07,0.03)	(0.9,0.08,0.02)
S_{10}	(0.6,0.3,0.1)	(0.8,0.18,0.02)	(0.4[illegible].35,0.25)	(0.9,0.07,0.03)	(0.5,0.42,0.08)	(0.8,0.11,0.09)

Table 10 Distance between students' pilot tests and high school base points

	H_1	H_2	H_3	H_4	H_5
S_1	0.0561	0.0358	0.0875	0.28	0.3258
S_2	0.136	0.10667	0.06833	0.19	0.24333
S_3	0.212	0.15667	0.0651	0.17166	0.195833
S_4	0.30033	0.24592	0.2	0.14167	0.14667
S_5	0.0471	0.0391	0.0958	0.2608	0.3333
S_6	0.63075	0.58833	0.49167	0.30667	0.29541
S_7	0.04216	0.03583	0.13	0.29	0.3333
S_8	0.03966	0.075	0.13667	0.335	0.385
S_9	0.112	0.08833	0.1075	0.2175	0.27833
S_{10}	0.30367	0.26333	0.18667	0.18333	0.235

The points of schools in 2014, 2015, 2016, 2017 years and the calculations of each school are given in Tables 13, 14, 15, 16, 17, 18, 19, 20, 21, 22, 23, 24, 25, 26, 27, 28, 29, 30, 31 and 32.

The smallest value among $E(M_i)$ points out the option which best satisfies condition. According to the above calculations, each school's from year to year ranking of school success is as follows:

For S_1 : $2016_{S_1}, 2017_{S_1}, 2014_{S_1}, 2015_{S_1}$
For S_2 : $2016_{S_2}, 2015_{S_2}, 2017_{S_2}, 2014_{S_2}$
For S_3 : $2015_{S_3}, 2016_{S_3}, 2017_{S_3}, 2014_{S_3}$
For S_4 : $2015_{S_4}, 2016_{S_4}, 2017_{S_4}, 2014_{S_4}$
For S_5 : $2017_{S_5}, 2015_{S_5}, 2016_{S_5}, 2014_{S_5}$
For S_6 : $2016_{S_6}, 2015_{S_6}, 2017_{S_6}, 2014_{S_6}$
For S_7 : $2015_{S_7}, 2016_{S_7}, 2017_{S_7}, 2014_{S_7}$
For S_8 : $2015_{S_8}, 2016_{S_8}, 2017_{S_8}, 2014_{S_8}$
For S_9 : $2015_{S_9}, 2017_{S_9}, 2016_{S_9}, 2014_{S_9}$
For S_{10} : $2015_{S_{10}}, 2017_{S_{10}}, 2016_{S_{10}}, 2014_{S_{10}}$

According to the above calculations, for each year ranking of school success is as follows:

For 2014: $S_2, S_4, S_3, S_5, S_7, S_6, S_8, S_9, S_1, S_{10}$
For 2015: $S_2, S_4, S_3, S_5, S_6, S_7, S_8, S_{10}, S_9, S_1$
For 2016: $S_2, S_4, S_5, S_6, S_3, S_8, S_7, S_1, S_{10}, S_9$
For 2017: $S_2, S_5, S_4, S_3, S_6, S_7, S_8, S_9, S_{10}, S_1$

In this paper; each year ranking of school success and each school's from year to year ranking of school success have been made separately. For each year ranking of school success has been varied. This situation has different causes: student change at school, teacher change at school, difficulty or simplicity of examination, socio-economic status and psychology of students.

Table 11 Official test scores of the students

	Turkish	Mathematics	Science	Social	English	Religion
S_1	(0.8,0.15,0.05)	(0.85,0.12,0.03)	(1,0,0)	(1,0,0)	(0.9,0.05,0.05)	(0.9,0.07,0.03)
S_2	(0.75,0.2,0.05)	(0.8,0.15,0.05)	(1,0,0)	(0.95,0.03,0.02)	(0.7,0.25,0.05)	(0.95,0.05,0)
S_3	(0.85,0.1,0.05)	(0.75,0.15,0.1)	(0.95,0.05,0)	(0.9,0.07,0.03)	(0.75,0.2,0.05)	(0.9,0.07,0.03)
S_4	(0.75,0.15,0.1)	(0.55,0.4,0.05)	(0.[illegible],0.35,0.05)	(0.6,0.3,0.1)	(0.5,0.45,0.05)	(0.85,0.12,0.03)
S_5	(0.75,0.2,0.05)	(0.75,0.35,0)	(0.[illegible]5,0.12,0.03)	(0.95,0.03,0.02)	(0.9,0.08,0.02)	(0.95,0.04,0.01)
S_6	(0.25,0.7,0.05)	(0.35,0.6,0.05)	(0.[illegible],0.4,0.1)	(0.45,0.5,0.05)	(0.45,0.55,0)	(0.7,0.25,0.05)
S_7	(0.8,0.15,0.05)	(0.95,0.03,0.02)	(1,0,0)	(0.95,0.02,0.03)	(0.85,0.12,0.03)	(0.9,0.05,0.05)
S_8	(1,0,0)	(1,0,0)	(0.[illegible],0.1,0.1)	(1,0,0)	(1,0,0)	(1,0,0)
S_9	(0.5,0.4,0.1)	(0.95,0.05,0)	(0.[illegible],0.08,0.02)	(0.8,0.15,0.05)	(0.9,0.05,0.05)	(0.85,0.1,0.05)
S_{10}	(0.61,0.3,0.09)	(0.5,0.45,0.05)	(0.[illegible]4,0.32,0.04)	(0.84,0.1,0.06)	(0.65,0.3,0.05)	(0.52,0.42,0.06)

Table 12 Distance between students' official tests and high school base points

	H_1	H_2	H_3	H_4	H_5
S_1	0.07367	0.06833	0.11	0.27166	0.36833
S_2	0.19466	0.18333	0.11166	0.22916	0.285
S_3	0.1205	0.16333	0.07166	0.22666	0.27
S_4	0.325	0.2891	0.2008	0.15	0.1441
S_5	0.10017	0.1	0.10916	0.21833	0.255
S_6	0.62142	0.58833	0.38	0.255	0.16167
S_7	0.067	0.07	0.10333	0.2833	0.31833
S_8	0.06133	0.07667	0.145	0.32667	0.385
S_9	0.15533	0.11833	0.14333	0.23667	0.2375
S_{10}	0.341	0.3	0.234	0.0375	0.1416

Table 13 The points of S_1 middle school

S_1	2014	2015	2016	2017
L_1	(0.461,0.486,0.053)	(0.535,0.419,0.046)	(0.519,0.433,0.048)	(0.41,0.531,0.059)
L_2	(0.311,0.621,0.068)	(0.338,0.596,0.066)	(0.317,0.615,0.068)	(0.3,0.63,0.07)
L_3	(0.42,0.522,0.058)	(0.484,0.465,0.051)	(0.502,0.449,0.049)	(0.54,0.414,0.046)
L_4	(0.448,0.497,0.055)	(0.566,0.391,0.043)	(0.512,0.44,0.048)	(0.5,0.45,0.05)
L_5	(0.374,0.564,0.062)	(0.349,0.586,0.065)	(0.488,0.461,0.051)	(0.38,0.558,0.062)
L_6	(0.509,0.442,0.049)	(0.762,0.785,0.023)	(0.685,0.284,0.031)	(0.54,0.414,0.046)

Table 14 Calculations for S_1

S_1	2014	$E(S_1) = 0.879$
	2015	$E(S_1) = 0.956$
	2016	$E(S_1) = 0.439$
	2017	$E(S_1) = 0.784$

Table 15 The points of S_2 middle school

S_2	2014	2015	2016	2017
L_1	(0.979,0.019,0.002)	(0.929,0.064,0.007)	(0.93,0.063,0.007)	(0.87,0.117,0.013)
L_2	(0.915,0.077,0.008)	(0.871,0.119,0.01)	(0.948,0.047,0.005)	(0.89,0.099,0.011)
L_3	(0.908,0.083,0.009)	(0.931,0.063,0.006)	(0.941,0.054,0.005)	(0.98,0.018,0.002)
L_4	(0.922,0.071,0.007)	(0.907,0.084,0.009)	(0.932,0.062,0.006)	(0.95,0.045,0.005)
L_5	(0.895,0.095,0.01)	(0.861,0.126,0.013)	(0.938,0.056,0.006)	(0.97,0.027,0.003)
L_6	(0.901,0.09,0.009)	(0.971,0.027,0.002)	(0.973,0.025,0.002)	(0.95,0.045,0.005)

Table 16 Calculations for S_2

S_2	2014	$\mathrm{E}(S_2) = 0.108$
	2015	$\mathrm{E}(S_2) = 0.029$
	2016	$\mathrm{E}(S_2) = 0.027$
	2017	$\mathrm{E}(S_2) = 0.052$

Table 17 The points of S_3 middle school

S_3	2014	2015	2016	2017
L_1	(0.739,0.241,0.02)	(0.72,0.252,0.028)	(0.673,0.295,0.032)	(0.752,0.223,0.02)
L_2	(0.509,0.442,0.049)	(0.488,0.461,0.051)	(0.489,0.46,0.051)	(0.559,0.396,0.04)
L_3	(0.608,0.353,0.039)	(0.628,0.665,0.037)	(0.651,0.315,0.034)	(0.65,0.315,0.035)
L_4	(0.653,0.313,0.034)	(0.658,0.308,0.034)	(0.646,0.319,0.035)	(0.745,0.229,0.02)
L_5	(0.493,0.457,0.05)	(0.482,0.467,0.051)	(0.608,0.353,0.039)	(0.636,0.327,0.03)
L_6	(0.732,0.242,0.026)	(0.873,0.115,0.012)	(0.818,0.164,0.018)	(0.805,0.175,0.01)

Table 18 Calculations for S_3

S_3	2014	$\mathrm{E}(S_3) = 0.353$
	2015	$\mathrm{E}(S_3) = 0.143$
	2016	$\mathrm{E}(S_3) = 0.217$
	2017	$\mathrm{E}(S_3) = 0.236$

Table 19 The points of S_4 middle school

S_4	2014	2015	2016	2017
L_1	(0.788,0.191,0.021)	(0.768,0.208,[illegible])	(0.736,0.238,0.026)	(0.69,0.279,0.031)
L_2	(0.59,0.369,0.041)	(0.546,0.409,0.045)	(0.608,0.353,0.039)	(0.67,0.297,0.033)
L_3	(0.682,0.287,0.031)	(0.698,0.272,0.03)	(0.711,0.261,0.028)	(0.83,0.153,0.017)
L_4	(0.722,0.251,0.027)	(0.728,0.245,0.027)	(0.737,0.237,0.026)	(0.77,0.207,0.023)
L_5	(0.544,0.411,0.045)	(0.552,0.404,0.044)	(0.688,0.281,0.031)	(0.68,0.288,0.032)
L_6	(0.756,0.22,0.024)	(0.881,0.108,0.011)	(0.875,0.113,0.012)	(0.82,0.162,0.018)

Table 20 Calculations for S_4

S_4	2014	$\mathrm{E}(S_4) = 0.312$
	2015	$\mathrm{E}(S_4) = 0.133$
	2016	$\mathrm{E}(S_4) = 0.14$
	2017	$\mathrm{E}(S_4) = 0.214$

Table 21 The points of S_5 middle school

S_5	2014	2015	2016	2017
L_1	(0.717,0.255,0.028)	(0.711,0.261,0.028)	(0.71,0.261,0.029)	(0.81,0.171,0.019)
L_2	(0.47,0.477,0.053)	(0.494,0.456,0.05)	(0.522,0.431,0.047)	(0.63,0.333,0.037)
L_3	(0.591,0.369,0.04)	(0.615,0.347,0.038)	(0.682,0.287,0.031)	(0.74,0.234,0.026)
L_4	(0.623,0.34,0.037)	(0.677,0.291,0.032)	(0.682,0.287,0.031)	(0.65,0.315,0.035)
L_5	(0.469,0.478,0.053)	(0.502,0.449,0.049)	(0.662,0.305,0.033)	(0.79,0.189,0.021)
L_6	(0.728,0.245,0.027)	(0.854,0.132,0.014)	(0.85,0.135,0.015)	(0.88,0.108,0.012)

Table 22 Calculations for S_5

S_5	2014	$E(S_5) = 0.36$
	2015	$E(S_5) = 0.168$
	2016	$E(S_5) = 0.173$
	2017	$E(S_5) = 0.134$

Table 23 The points of S_6 middle school

S_6	2014	2015	2016	2017
L_1	(0.603,0.358,0.039)	(0.583,0.376,0.041)	(0.603,0.358,0.039)	(0.5,0.45,0.05)
L_2	(0.342,0.593,0.065)	(0.328,0.605,0.067)	(0.406,0.535,0.059)	(0.4,0.54,0.06)
L_3	(0.482,0.467,0.051)	(0.517,0.435,0.048)	(0.528,0.425,0.047)	(0.6,0.36,0.04)
L_4	(0.528,0.425,0.047)	(0.566,0.391,0.043)	(0.671,0.297,0.032)	(0.55,0.405,0.045)
L_5	(0.464,0.483,0.053)	(0.403,0.538,0.059)	(0.553,0.403,0.044)	(0.5,0.45,0.05)
L_6	(0.603,0.358,0.039)	(0.739,0.235,0.026)	(0.837,0.147,0.016)	(0.71,0.261,0.029)

Table 24 Calculations for S_6

S_6	2014	$E(S_6) = 0.618$
	2015	$E(S_6) = 0.341$
	2016	$E(S_6) = 0.191$
	2017	$E(S_6) = 0.392$

Table 25 The points of S_7 middle school

S_7	2014	2015	2016	2017
L_1	(0.539,0.415,0.046)	(0.55,0.405,0.045)	(0.51,0.441,0.049)	(0.49,0.459,0.051)
L_2	(0.36,0.576,0.064)	(0.336,0.598,0.066)	(0.367,0.57,0.063)	(0.46,0.486,0.054)
L_3	(0.481,0.468,0.051)	(0.487,0.462,0.051)	(0.529,0.424,0.047)	(0.65,0.315,0.035)
L_4	(0.464,0.483,0.053)	(0.498,0.452,0.05)	(0.493,0.457,0.05)	(0.59,0.369,0.041)
L_5	(0.368,0.569,0.063)	(0.365,0.572,0.063)	(0.485,0.464,0.051)	(0.45,0.495,0.055)
L_6	(0.609,0.352,0.039)	(0.736,0.238,0.026)	(0.707,0.264,0.029)	(0.69,0.279,0.031)

Table 26 Calculations for S_7

S_7	2014	$E(S_7) = 0.603$
	2015	$E(S_7) = 0.346$
	2016	$E(S_7) = 0.398$
	2017	$E(S_7) = 0.429$

Table 27 The points of S_8 middle school

S_8	2014	2015	2016	2017
L_1	(0.56,0.396,0.044)	(0.552,0.404,0.044)	(0.521,0.432,0.047)	(0.49,0.459,0.051)
L_2	(0.355,0.581,0.064)	(0.314,0.618,0.068)	(0.337,0.597,0.066)	(0.45,0.505,0.055)
L_3	(0.479,0.469,0.052)	(0.512,0.44,0.048)	(0.52,0.432,0.048)	(0.69,0.279,0.031)
L_4	(0.526,0.427,0.047)	(0.503,0.448,0.049)	(0.499,0.451,0.05)	(0.52,0.432,0.048)
L_5	(0.368,0.569,0.063)	(0.36,0.576,0.064)	(0.409,0.532,0.059)	(0.46,0.486,0.054)
L_6	(0.6,0.36,0.04)	(0.714,0.258,0.028)	(0.709,0.262,0.029)	(0.66,0.306,0.034)

Table 28 Calculations for S_8

S_8	2014	$E(S_8) = 0.625$
	2015	$E(S_8) = 0.385$
	2016	$E(S_8) = 0.394$
	2017	$E(S_8) = 0.489$

Table 29 The points of S_9 middle school

S_9	2014	2015	2016	2017
L_1	[illegible]	(0.524,0.429,0.047)	(0.341,0.594,0.065)	(0.41,0.531,0.059)
L_2	(0.278,0.65,0.072)	(0.261,0.666,0.073)	(0.25,0.675,0.075)	(0.38,0.558,0.062)
L_3	(0.393,0.547,0.06)	(0.432,0.512,0.056)	(0.426,0.517,0.057)	(0.5,0.45,0.05)
L_4	(0.403,0.538,0.059)	(0.422,0.521,0.057)	(0.343,0.592,0.065)	(0.5,0.45,0.05)
L_5	(0.334,0.6,0.066)	(0.323,0.61,0.067)	(0.33,0.603,0.067)	(0.32,0.612,0.068)
L_6	(0.518,0.434,0.048)	(0.657,0.309,0.034)	(0.528,0.425,0.047)	(0.59,0.369,0.041)

Table 30 Calculations for S_9

S_9	2014	$E(S_9) = 0.851$
	2015	$E(S_9) = 0.496$
	2016	$E(S_9) = 0.82$
	2017	$E(S_9) = 0.649$

Table 31 The points of S_{10} middle school

S_{10}	2014	2015	2016	2017
L_1	(0.427,0.516,0.057)	(0.392,0.548,0.06)	(0.297,0.633,0.07)	(0.27,0.657,0.073)
L_2	(0.336,0.598,0.066)	(0.21,0.711,0.079)	(0.226,0.697,0.077)	(0.25,0.675,0.075)
L_3	(0.413,0.529,0.058)	(0.322,0.611,0.067)	(0.415,0.527,0.058)	(0.41,0.531,0.059)
L_4	(0.329,0.604,0.067)	(0.309,0.622,0.069)	(0.365,0.572,0.063)	(0.5,0.45,0.05)
L_5	(0.313,0.619,0.068)	(0.268,0.659,0.073)	(0.287,0.642,0.071)	(0.29,0.639,0.071)
L_6	(0.501,0.444,0.055)	(0.663,0.304,0.033)	(0.53,0.423,0.047)	(0.549,0.39,0.061)

Table 32 Calculations for S_{10}

S_{10}	2014	$E(S_{10}) = 0.897$
	2015	$E(S_{10}) = 0.484$
	2016	$E(S_{10}) = 0.814$
	2017	$E(S_{10}) = 0.739$

5 Conclusion

In all these studies, applications of decision making were made by using distance measures defined in heuristic fuzzy logic. By comparing the distance measurements, the most sensitive distance measurement was determined. As a result of the comparison, the most sensitive distance measurement is the normalized hamming distance measure. The applications made are applications made in the field of decision making in education. Thanks to these methods that we have chosen and hope to develop, selection, placement, determination and decision-making in education will be made quite regular. The application areas of these methods are not only education but also many fields. There are and will continue to be applications in engineering, computers, agriculture, economy, medicine and many more. The methods we use in our work have application areas in many areas. We preferred to apply these methods in the field of education. Researchers who want to take actions such as sequencing, preferences, and decision-making can use these methods in other application areas. In many application areas of intuitionistic fuzzy logic, with these methods will be reached the right goal. These are studies that shed light on many new application areas.

References

1. L.A. Zadeh, Fuzzy sets. Inf. Control **8**, 338–353 (1965)
2. K.T. Atanassov, Intuitionistic fuzzy sets, VII ITKR Session, Sofia, 20–23 June 1983 (Deposed in Centr. Sci.-Techn. Library of the Bulg. Acad. of Sci., 1697/84) (in Bulgarian), Reprinted: Int. J. Bioautomation **20**(S1) (2016)
3. K.T. Atanassov, Intuitionistic fuzzy sets. Fuzzy Sets Syst. **20**(1), 87–96 (1986)

4. G. Çuvalcıoğlu, S. Yılmaz, Some properties of OTMOs on IFSs. Adv. Stud. Contemp. Math. **20**(4), 621–628 (2010)
5. G. Çuvalcıoğlu, E. Aykut, An application of the intuitionistic fuzzy modal operator $E_{\alpha,\beta}$. NIFS **20**(5), 57–61 (2014)
6. G. Çuvalcıoğlu, S. Yılmaz, K.T. Atanassov, Matrix representation of the second type of intuitionistic fuzzy modal operators. Notes on Intuitionistic Fuzzy Sets **20**(5), 9–16 (2014)
7. G. Çuvalcıoğlu, S. Yılmaz, Some properties of intuitionistic fuzzy equivalence relations and class trees w.r.t. intuitionistic fuzzy equivalence relations. Adv. Stud. Contemp. Math. **24**(1), 77–86 (2014)
8. G. Çuvalcıoğlu, E. Aykut, An application of some intuitionistic fuzzy modal operators to agriculture. Notes on Intuitionistic Fuzzy Sets **21**(2), 140–149 (2015)
9. G. Çuvalcıoğlu, S. Tarsuslu (Ylmaz), Universal algebra in intuitionistic fuzzy set theory. Notes on Intuitionistic Fuzzy Sets **23**(1), 1–5 (2017)
10. P.A. Ejegwa, A.J. Akubo, O.M. Joshua, Intuitionistic fuzzy sets in career determination. Glob. J. Sci. Front. Res.: F Math. Decis. Sci. **14**(3) (2014)
11. S. Melliani, O. Castillo, in *Recent Advances in Intuitionistic Fuzzy Logic Systems: Theoretical Aspects and Applications*. Studies in Fuzziness and Soft Computing (Springer, 2019)
12. S. Melliani, O. Castillo, in *Recent Advances in Intuitionistic Fuzzy Logic Systems and Mathematics*. Studies in Fuzziness and Soft Computing (Springer, 2021)
13. E. Szmidt, J. Kacprzyk, Intuitionistic fuzzy sets in some medical applications. Notes on IFS **7**(4), 58–64 (2001)
14. E. Szmidt, J. Kacprzyk, Medical diagnostic reasoning using a similarity measure for intuitionistic fuzzy sets. Notes on IFS **10**(4), 61–69 (2004)
15. F. Kutlu, Ö. Atan, T. Bilgin, Comparative Analysis of Performances of Fuzzy and Intuitionistic Fuzzy Similarity Measures on Noise Added Images, in *Conference: FUZZYSS* (2015)
16. E. Szmidt, in *Distances and Similarities in Intuitionistic Fuzzy Sets*. Studies in Fuzziness and Soft Computing (Springer, 2014)
17. F. Tuğrul, B. Yılmaz, M. Çitil, School success ranking in multi criteria decision making. Turk. J. Math. Comput. Sci. **10**, 1–6 (2018). MATDER
18. F. Tuğrul, B. Yılmaz, M. Çitil, Application of ranking with similarity measure in multi criteria decision making. Konuralp J. Math. **7**(2), 438–441 (2019)
19. W. Wang, X. Xin, Distance measure between intuitionistic fuzzy sets. Pattern Recogn. Lett. **26**, 2063–2069 (Elsevier, 2005)
20. Z.S. Xu, J. Chen, An Overview of Distance and Similarity Measures of Intuitionistic Fuzzy Sets. Int. J. Uncertain. Fuzziness Knowl. Based Syst. **16**(4), 529–555 (2008)
21. G. Çuvalcıoğlu, V. Bureva, A. Michalikova, Intercriteria analysis applied to university ranking system of Turkey. Notes on Intuitionistic Fuzzy Sets **25**(4), 90–97 (2019)
22. R. Parvathi, V. Atanassova, L. Doukovska, C. Yuvapriya, K. Indhurekha, Intercriteria analysis of rankings of Indian Universities. Notes on Intuitionistic Fuzzy Sets. **24**(1), 99–109 (2018)
23. F. Tuğrul, M. Çitil, B. Karasolak, M. Dağlı, Interpretation of physical conditions of schools with fuzzy multi criteria decision making. J. Univ. Math. **3**(1), 46–52 (2020)
24. E. Szmidt, J. Kacprzyk, On measuring distances between intuitionistic fuzzy sets. Notes on IFS **3**(4), 1–3 (1997)
25. E. Szmidt, J. Kacprzyk, Distances between intuitionistic fuzzy sets. Fuzzy Sets Syst. **114**(3), 505–518 (2000)
26. S. Chen, J. Tan, Handling multicriteria fuzzy decision-making problems based on vague set theory. Fuzzy Sets Syst. **67**, 163–172 (1994)
27. M. Çitil, Application of the intuitionistic fuzzy in education. Commun. Math. Appl. **10**(1), 131–143 (2019)
28. M. Çitil, F. Tuğrul, B. Yılmaz, An application of multi criteria decision making: ranking of school success. Celal Bayar Univ. J. Sci. **15**(1), 45–50 (2019)
29. E. Szmidt, J. Kacprzyk, An application of intuitionistic fuzzy set similarity measures to a multi-criteria decision making problem (Springer, Heidelberg, 2006), pp. 314-323

On Intuitionistic Fuzzy Abstract Algebras

Gökhan Çuvalcıoğlu and Sinem Tarsuslu

Abstract The concept of abstract algebra on intuitionistic fuzzy sets introduced and some basic theorems prove. Homomorphism between intuitionistic fuzzy abstract algebras defined, intuitionistic fuzzy function examined and then intuitionistic fuzzy congruence relations defined on intuitionistic fuzzy abstract algebra. First and third isomorphism theorems on intuitionistic abstract algebras introduced.

Keywords Intuitionistic fuzzy sets · Intuitionistic fuzzy abstract algebra · Intuitionistic fuzzy function · Intuitionistic fuzzy isomorphism theorems

1 Introduction

Fuzzy set theory was introduced as an extension of crisp sets by Zadeh [1]. As a natural continuation of this study, the generalization of fuzzy set theory called intuitionistic fuzzy set theory was propounded by Atanassov [2]. Both fuzzy sets and intuitionistic fuzzy sets attract the attention of many researchers [3–9]. Intuitionistic fuzzy sets have various application areas; like intuitionistic fuzzy expert systems, intuitionistic fuzzy neural networks, intuitionistic fuzzy generalized nets etc. [10, 11].

Intuitionistic fuzzy group defined by Biswas as an generalized algebraic structure in 1989 [12]. Intuitionistic M-fuzzy groups was introduced by Zhan and Than [13]. The concept of intuitionistic fuzzy rings was propounded by Yan [14]. Intuitionistic L-fuzzy subgroups were studied in 2009 [15]. Intuitionistic fuzzy semigroups were examined by Melliani and his colleagues [16]. Later years, different intuitionistic fuzzy algebraic sturucters had studied by several authors.

Abstract algebra (or algebra) is a set with finitary operations defined on it. By working on universal algbera, the common properties of algebraic structures can be

G. Çuvalcıoğlu (✉)
Department of Mathematics, Faculty of Arts and Sciences, Mersin University, Mersin, Turkey
e-mail: gcuvalcioglu@mersin.edu.tr

S. Tarsuslu
Department of Natural and Mathematical Sciences, Faculty of Engineering, Tarsus University, 33400 Tarsus, Turkey
e-mail: sinemtarsuslu@tarsus.edu.tr

O. Castillo and P. Melin (eds.), *Fuzzy Logic and Neural Networks for Hybrid Intelligent System Design*, Studies in Computational Intelligence 1061,
https://doi.org/10.1007/978-3-031-22042-5_2

examined. The generalization of universal algebra on fuzzy set theory was studied by Murali, in 1987 [17].

In this study, the fundamental concepts on intuitionistic fuzzy abstract algebras were introduced; these are intuitionistic fuzzy homomorphism, intuitionistic fuzzy congruence relation on intuitionistic fuzzy abstract algebra, isomorphisim theorems on intuitionistic fuzzy abstract algebras.

2 Preliminaries

Atanassov introduced the intuitionistic fuzzy set theory in 1983 [2] as an extension of fuzzy sets by enlarging the truth value set to the lattice $[0, 1] \times [0, 1]$ is defined as following.

Definition 1 Let $L = [0, 1]$ then $L^* = \{(a_1, a_2) \in L^2 \ : \ a_1 + a_2 \leq 1\}$ is a lattice with

$$(a_1, a_2) \leq (b_1, b_2) :\Leftrightarrow \ a_1 \leq b_1 \text{ and } a_2 \geq b_2.$$

The operations $\wedge$ and $\vee$ on $(L^*, \leq)$ are defined as following;
For $(a_1, b_1), (a_2, b_2) \in L^*$, $(a_1, b_1) \ \wedge (a_2, b_2) = (\min(a_1, a_2), \max(b_1, b_2))$.

$$(a_1, b_1) \vee (a_2, b_2) = (\max(a_1, a_2), \min(b_1, b_2))$$

For each $J \subseteq L^*$

$$\sup J = (\sup\{a \ : (a, \ b \in L), ((a, \ b) \in J)\}, \inf\{b \ : a, \ b \in L)((a, \ b) \in J)\}) \text{ and}$$
$$\inf J = (\inf\{a \ : (a, \ b \in L)((a, \ b) \in J)\}, \sup\{b \ : (a, \ b \in L)((a, \ b) \in J)\}).$$

Definition 2 [2] Let a crisp set X be fixed. An intuitionistic fuzzy set (shortly IFS) in X is an object of the following form

$A = \{< x, \mu_A(x), \nu_A(x), >: \ x \in X\}$, where functions $\mu_A(x)$, $(\mu_A \ : \ X \to [0, 1])$ and $\nu_A(x)$, $(\nu_A \ : \ X \to [0, 1])$ are called degree of membership and the degree of non- membership of $x \in X$ to the set A, respectively, and $0 \leq \mu_A(x) + \nu_A(x) \leq 1$, for all $x \in X$.

$\pi_A(x) = 1 - \mu_A(x) - \nu_A(x)$ is the definition of hesitation degree of $x \in X$.

Basic definitions are given as following.

Definition 3 [2] Let a set X be fixed. An IFS A is contained in an IFS B (notation $A \sqsubseteq B$) if and only if, for all $x \in X \ : \ \mu_A(x) \leq \mu_B(x)$ and $\nu_A(x) \geq \nu_B(x)$.

Clearly, $A = B$ if and only if $A \sqsubseteq B$ and $B \sqsubseteq A$.

Definition 4 [2] Let a set X be fixed, $A \in IFS(X)$ and $A = \{< x, \mu_A(x), \nu_A(x) >: x \in X\}$ then the complement of A defined as follow:

$$A^c = \{< x, \nu_A(x), \mu_A(x) >: \ x \in X\}$$

Definition 5 [2] Let a crisp set X be fixed and $A, B \in IFS(X)$ is an intuitionistic fuzzy set on X.

i. $A \sqcap B = \{< x, \min(\mu_A(x), \mu_B(x)), \max(\nu_A(x), \nu_B(x)) >: \ x \in X\}$
ii. $A \sqcup B = \{< x, \max(\mu_A(x), \mu_B(x)), \min(\nu_A(x), \nu_B(x)) >: \ x \in X\}$
iii. $\Box A = \{< x, \mu_A(x), 1 - \mu_A(x) >: \ x \in X\}$
iv. $\Diamond A = \{< x, 1 - \nu_A(x), \nu_A(x) >: \ x \in X\}$

Level sets have important role on intuitionistic fuzzy set theory. This concept was defined by Atanassov and main properties were studied.

Definition 6 [2] Let a set X be fixed and $A \in IFS(X)$. The $(t, s)-$ cut and strong $(t, s)-$ cut of A are crisp subsets $A_{(t,s)}$ and $A_{\langle t,s\rangle}$ of the X, respectively are given by

$$A_{(t,s)} = \{x \ : \ x \in X \text{ such that } \mu_A(x) \geq t, \ \nu_A(x) \leq s\}$$
$$A_{\langle t,s\rangle} = \{x \ : \ x \in X \text{ such that } \mu_A(x) > t, \ \nu_A(x) < s\}$$

where $t, s \in [0, 1]$ with $t + s \leq 1$.

Burille and Bustince were introduced the definitions of intuitionistic fuzzy relation and intuitionistic fuzzy equivalence relation

Definition 7 [18] An intuitionistic fuzzy relation (shortly IFR) is an intuitionistic fuzzy subset of $X \times Y$ that is, is an expression R given by

$$R = \{\langle (x, y), \mu_R(x, y), \nu_R(x, y)\rangle \ : \ x \in X, \ y \in Y\}$$

where $\mu_R : X \times Y \to [0, 1]$ $\nu_R : X \times Y \to [0, 1]$ with $0 \leq \mu_R(x, y) + \nu_R(x, y) \leq 1$ for any $(x, y) \in X \times Y$.

Definition 8 [19] Let X be a universal and $R \in IFR(X)$.

(1) For every $x \in X$, $\mu_R(x, x) = 1$ and $\nu_R(x, x) = 0$ then R is called an intuitionistic fuzzy reflexsive.
(2) For every $x, y \in X$, $\mu_R(x, y) \leq \mu_R(y, x)$ and $\nu_R(x, y) \geq \nu_R(y, x)$ then R is called an intuitionistic fuzzy symmetric.
(3) For every $x, y, z \in X$,

$$\mu_R(x, y) \wedge \mu_R(y, z) \leq \mu_R(x, z) \text{ and } \nu_R(x, y) \vee \nu_R(y, z) \geq \nu_R(x, z)$$

then R is called an intuitionistic fuzzy transitive.

If an intuitionistic fuzzy relation satisfies the previous properties then it is called an intuitionistic fuzzy equivalence relation (IFE(X)).

Theorem 1 [20] *Let X be a non-empty set and* $R \in IFR(X)$. *Then* $R \in IFR(X)$ *if and only if* $R_{(r,s)}$ *is an equivalence relation on X for each* $r, s \in [0, 1]$ *with* $r + s \leq 1$.

Definition 9 [21] Let X be a non-empty set, $R \in IFE(X)$ and $a \in X$.

$$[a]_R = \{< x, \mu_{[a]_R}(x), \nu_{[a]_R}(x) >: x \in X\}$$

where $\mu_{[a]_R}(x) = \mu_R(a, x)$, $\nu_{[a]_R}(x) = \nu_R(a, x)$ is called an intuitionistic fuzzy equivalence class of a w.r.t R.

Level sets of intuitionistic fuzzy equivalence relations were studied by different authors. Here, following definition will be used for equivalence classes.

Definition 10 [22] Let X be a non-empty set and $R \in IFE(X)$. Let $a \in X$ and $r, s \in [0, 1]$ and $r + s \leq 1$;

1. $_r[a]_R = \{x \in X : \mu_{[a]_R}(x) = \mu_R(a, x) \geq r\}$
2. $^s[a]_R = \{x \in X : \nu_{[a]_R}(x) = \nu_R(a, x) \leq s\}$
3. $^s_r[a]_R = \{x \in X : \mu_{[a]_R}(x) = \mu_R(a, x) \geq r, \nu_{[a]_R}(x) = \nu_R(a, x) \leq s\}$

For each $a \in X$, $^s_r[a]_R$ denotes the crisp equivalence class containing a w.r.t $R_{(r,s)}$. The extension of functions on intuitionistic fuzzy sets is given as follow;

Definition 11 [20] Let X and Y be two non-empty sets and $f : X \to Y$ be a mapping. Let $A \in IFS(X)$ and $B \in IFS(Y)$. Then f is extended to a mapping from IFS(X) to IFS(Y) as.

$$f(A)(y) = (\mu_{f(A)}(y), \nu_{f(A)}(y))$$

$$\text{where } \mu_{f(A)}(y) = \begin{cases} \vee\{\mu_A(x) : x \in f^{-1}(y)\} \\ 0; \text{ otherwise} \end{cases} \text{ and}$$

$$\nu f(A)(y) = \begin{cases} \wedge\{\nu A(x) : x \in f^{-1}(y)\} \\ 1; \text{ otherwise} \end{cases}$$

f(A) is called the image of A under the map f Also, the pre-image of B under f is denoted by $f^{-1}(B)$ and defined as

$$f^{-1}(B)(x) = (\mu_{f^{-1}(B)}(x), \nu_{f^{-1}(B)}(x)) \text{where } \mu_{f^{-1}(B)}(x) = \mu_B(f(x)) \text{ and}$$

$$\nu_{f^{-1}(B)}(x) = \nu_B(f(x)).$$

Abstract algebras have a comprehensive study area on crisp sets. Murali introduced the concept of fuzzy abstract algebra using Zadeh's extension principle [17, 23]. Fuzzy congruence relation was defined and some properties of fuzzy congruence relations were studied by same author [16]. Here, the generalization of abstract algebra to intuitionistic fuzzy abstract algebra is studied and theorems on intuitionistic fuzzy abstract algebras are proved.

Let remember the definition of crisp abstract algebra.

Definition 12 [24] An abstract algebra (or algebra) A is a pair $[S, F]$ where S is a non-empty set and F is a specified set of operatians f_α, each mapping a power $S^{n(\alpha)}$ of S into S for some appropriate nonnegative finite integer $n(\alpha)$.

Unless otherwise stated, each operation f_α assigns to every $n(\alpha)-$ ple $(x_1, \ldots, x_{n(\alpha)})$ of elements of S, a value $f_\alpha(x_1, \ldots, x_{n(\alpha)})$ in S, the result of performing the operation f_α on the sequence $x_1, \ldots, x_{n(\alpha)}$. If $n(\alpha) = 1$, the operation f_α is called unary; if $n(\alpha) = 2$, it is called binary; if $n(\alpha) = 3$, it is called ternary, etc. When $n(\alpha) = 0$, the operation f_α is called nullary; it selects a fixed element of S.

$A = [S, F]$ and $B = [T, F']$ are called similar algebras if F and F' are same for each α the types of f_α and f'_α.

Definition 13 [24] Let $A = [S, F]$ and $B = [T, F']$ be two similar algebras. A function $\varphi : S \to T$ is called a homomorphism of A into B if and only if for all $f_\alpha \in F$ and $x_i \in S$, $i = 1, 2, \ldots, n(a)$,

$$f'_\alpha\left(\varphi(x_1), \varphi(x_2), \ldots, \varphi\left(x_{n(a)}\right)\right) = \varphi\left(f_\alpha\left(x_1, x_2, \ldots, x_{n(a)}\right)\right).$$

A crisp congruence relation on an algebraic system $A = [S, F]$ is an equivalence relation θ on $A = [S, F]$ which has the substitution property for its operations. It means that, for all $f_\alpha \in F$ and $a_i, b_i \in S$, $i = 1, 2, \ldots, n(a)$,

$$a_i \equiv b_i(\theta) \Rightarrow f_\alpha\left(a_1, a_2, \ldots, a_{n(a)}\right) \equiv f_\alpha\left(b_1, b_2, \ldots, b_{n(a)}\right)(\theta).$$

3 Intuitionistic Fuzzy Abstract Algebras

Algebraic structures extended on intuitionistic fuzzy sets by many author [25–30] and main theorems were studied. The concept of intuitionistic fuzzy abstract algebra defined as follows.

Definition 14 Let $S = [X, F]$ be an algebra where X is a non-empty set and F is a specified set of finite operations f_α, each mapping a power $X^{n(\alpha)}$ of X into X, for some appropriate nonnegative finite integer $n(\alpha)$. For each f_α, a corresponding operation ω_α on IFS(X) as follows;

$$\omega_{f_\alpha} : IFS(X) \times IFS(X) \times \ldots \times IFS(X) \to IFS(X),\ \omega_{f_\alpha}\left(A_1, A_2, \ldots, A_{n(\alpha)}\right) = A$$

such that

$$A(x) = \begin{cases} \sup\left\{A_1(x_1) \wedge A_2(x_2) \wedge \ldots \wedge A_{n(\alpha)}\left(x_{n(\alpha)}\right)\right\}; & f_\alpha\left(x_1, x_2, \ldots, x_{n(\alpha)}\right) = x \\ \theta = (0, 1); & \text{other wise} \end{cases}$$

Shortly, $A = \omega_{f_\alpha}\left(A_1, A_2, \ldots, A_{n(\alpha)}\right)$.

Let $\Omega = \{\ \omega_{f_\alpha}$: corresponding operation for each $f_\alpha \in F\}$ then $\mathscr{L} = \left[(I \times I)^X, \Omega\right]$ is called intuitionistic fuzzy abstract algebra (or intuitionistic fuzzy algebra).

If $n(\alpha) = 0$ then $f_\alpha(x) = e$ that e is a fixed element of X. So, ω_{f_α} is defined as following:

$$\omega_{f_\alpha}(A) = A_e, A_e(x) = \begin{cases} \sup_{x \in X} A(x), & x = e \\ (0, 1), & x \neq e \end{cases}$$

Definition 15 Let X be a non-empty set and $A \in IFS(X)$. A is called an intuitionistic fuzzy subalgebra (IF− subalgebra) of $\mathscr{L} = [IFS(X), \Omega]$ intuitionistic fuzzy algebra if and only if for nonnegative finite integer $n(\alpha)$, $\omega_{f_\alpha}(A, A, \ldots, A) \sqsubseteq A$, for every ω_{f_α}.

Theorem 2 *Let* $S = [X, F]$ *be an algebra,* $f_\alpha \in F$ *and* $A, A_1, A_2, \ldots, A_{n(\alpha)}$ *be* IF− *subalgebras.*

$\omega_{f_\alpha}\left(A_1, A_2, \ldots, A_{n(\alpha)}\right) \sqsubseteq A$ if and only if $A\left(f_\alpha\left(x_1, x_2, \ldots, x_{n(\alpha)}\right)\right) \geq \min_{1 \leq i \leq n(\alpha)} A_i(x_i)$ is true for every $\left(x_1, x_2, \ldots, x_{n(\alpha)}\right) \in X^{n(\alpha)}$.

Proof

(1) Let $n(\alpha) \neq 0$ and $\omega_{f_\alpha}\left(A_1, A_2, \ldots, A_{n(\alpha)}\right) \leq A$.

$$\omega_{f_\alpha}\left(A_1, A_2, \ldots, A_{n(\alpha)}\right)(x) \leq A(x), \quad \text{for all } x \in X.$$

So, $\sup_{f_\alpha(x_1, x_2, \ldots, x_{n(\alpha)}) = x}\left(A_1(x_1) \wedge A_2(x_2) \wedge \ldots \wedge A_{n(\alpha)}\left(x_{n(\alpha)}\right)\right) \leq A(x)$, for all $\left(x_1, \ldots, x_{n(\alpha)}\right) \in X^{n(\alpha)}$.

$$A\left(f_\alpha\left(x_1, x_2, \ldots, x_{n(\alpha)}\right)\right) \geq \omega_{f_\alpha}\left(A_1, A_2, \ldots, A_{n(\alpha)}\right)\left(f_\alpha\left(x_1, x_2, \ldots, x_{n(\alpha)}\right)\right)$$

$$\geq A_1(x_1) \wedge A_2(x_2) \wedge \ldots \wedge A_{n(\alpha)}\big(x_{n(\alpha)}\big)$$
$$= \min_{1 \leq i \leq n(\alpha)} A_i(x_i)$$

Conversely, let $f_\alpha\big(x_1, x_2, \ldots, x_{n(\alpha)}\big) = x$. It is clear that, since $\min_{1 \leq i \leq n(\alpha)}$ for all $f_\alpha\big(x_1, x_2, \ldots, x_{n(\alpha)}\big) = x$ then

$$\sup_x\big(\min_{1 \leq i \leq n(\alpha)} A_i(x_i)\big) \leq A(x).$$

That is, $\omega_{f_\alpha}\big(A_1, A_2, \ldots, A_{n(\alpha)}\big)(x) \leq A(x)$, for all $f_\alpha\big(x_1, x_2, \ldots, x_{n(\alpha)}\big) = x$.
If for some x there exists no such n(a)− tuples then $\omega_{f_\alpha}\big(A_1, A_2, \ldots, A_{n(\alpha)}\big)(x) = (0, 1) \leq A(x)$.

(2) If $n(\alpha) = 0$ then $f_\alpha(x) = e$, e is a fixed element of X.

$$\omega_{f_\alpha}(A_1)(x) \leq A(x) \Leftrightarrow A(e) \geq \omega_{f_\alpha}(A_1) = \sup_x A_1(x)$$
$$\Leftrightarrow A\big(\omega_{f_\alpha}(x)\big) \geq A_1(x) \text{ for all } x \in X.$$

Example 1 A group $S = [G, F]$ is an algebra where $F = \{., e\}$ include one binary operation and one nullary operation respectively. Let $L = [IFS(G), \Omega]$ and $A_1, A_2 \in IFS(G)$, $x, x_1, x_2 \in G$ then with corresponding operations defined as follow;

$$A_1A_2(x) = \big(\mu_{A_1}\mu_{A_2}(x), \nu_{A_1}\nu_{A_2}(x)\big)$$

such that

$$\mu_{A_1}\mu_{A_2}(x) = \sup_{x=x_1x_2}\big(\mu_{A_1}(x_1) \wedge \mu_{A_2}(x_2)\big),$$
$$\nu_{A_1}\nu_{A_2}(x) = \inf_{x=x_1x_2}\big(\nu_{A_1}(x_1) \vee \nu_{A_2}(x_2)\big)$$

L is intuitionistic fuzzy algebra.

Example 2 Let G be a group. $A \in IFS(G)$ intuitionistic fuzzy subgroup defined as follow:

for all $x, y \in G$,

$$A(xy) \geq A(x) \wedge A(y)$$
$$A\big(x^{-1}\big) \geq A(x)$$

that is,

$$\mu_A(xy) \geq \mu_A(x) \wedge \mu_A(y) \text{ and } \nu_A(xy) \leq \nu_A(x) \vee \nu_A(y)$$

$$\mu_A(x^{-1}) \geq \mu_A(x) \text{ and } \nu_A(x^{-1}) \leq \nu_A(x).$$

So, this is an intuitionistic fuzzy subalgebra.

Proposition 1 *Let* $S = [X, F]$ *be an algebra and* A *is an* IF− *subalgebra of* $L = [IFS(X), \Omega]$. *For any* $r, s \in [0, 1]$ *with* $r + s \leq 1$, $A_{(r,s)}$ *is a crisp subalgebra of* S.

Proof Let $f_\alpha \in F$. If $x_1, x_2, \ldots, x_{n(a)} \in A_{(r,s)}$ then $A(x_i) \geq (r, s)$ for each $i = 1, 2, \ldots, n(a)$ Since A is an IF− subalgebra,

$$\begin{aligned} A(f_\alpha(x_1, x_2, \ldots, x_{n(\alpha)})) &\geq \min_{1 \leq i \leq n(\alpha)} A(x_i) \geq (r, s) \\ &\Rightarrow f_\alpha(x_1, x_2, \ldots, x_{n(\alpha)}) \in A_{(r,s)} \end{aligned}$$

hence, $A_{(r,s)}$ is a crisp subalgebra of S.

Theorem 3 *et* $L = [IFS(X), \Omega]$ *be an* IF− *algebra. If* $\{A_i\}$ *is a family of* IF− *subalgebras of* L *then*

$$A = \bigcap_{i \in \Lambda} A_i$$

is a IF− subalgebra of L.

Proof Let $f_\alpha \in F$ and $(x_1, x_2, \ldots, x_{n(\alpha)}) \in S^{n(\alpha)}$ for the corresponding $n(\alpha)$.

$$\begin{aligned} A(f_\alpha(x_1, x_2, \ldots, x_{n(\alpha)})) &= \bigcap_{i \in \Lambda} A_i(f_\alpha(x_1, x_2, \ldots, x_{n(\alpha)})) \\ &\geq \bigcap_{i \in \Lambda} (\min_{1 \leq j \leq n(\alpha)} A_i(x_j)) \\ &= \min_{1 \leq j \leq n(\alpha)} (\inf_{i \in \Lambda} A_i(x_j)) \\ &= \min_{1 \leq j \leq n(\alpha)} A(x_j) \end{aligned}$$

So, A is a IF− subalgebra of S.

Proposition 2 Let $S = [X, F]$ be an algebra. If A is an IF− subalgebra of $L = [IFS(X), \omega]$ then so are $\Box$ A and $\Diamond$ A.

Proof If A is an IF− subalgebra then $\omega_{f_\alpha}(A, A, \ldots, A) \sqsubseteq A$ for all ω_{f_α}. So,

$$sup_{f_\alpha(x_1, x_2, \ldots, x_{n(\alpha)})} = x \left(\begin{array}{c} min \\ 1 \leq i \leq n(\alpha) \end{array} \mu_\alpha(x_i) \right) \leq \mu_A(x)$$

$$inf_{f_\alpha(x_1,x_2,\ldots,x_{n(\alpha)})} = x\left(\underset{1\le i\le n(\alpha)}{max}\right)1-\mu_\alpha(x_i)$$
$$= 1-\left(sup_{f_\alpha(x_1,x_2,\ldots,x_{n(\alpha)})}\right) = x\left(min_{1\le i\le n(\alpha)}\right)\mu_A(x)$$
$$\geq 1-\mu_A(x)$$

and

$$\omega_{f_\alpha} = (\Box A, \Box A, \ldots, \Box A)(x) = \begin{bmatrix} sup_{f_\alpha(x_1,x_2,\ldots,x_{n(\alpha)})} = x\left(\min_{1\le i\le n(\alpha)}\mu_\alpha(x_i)\right) \\ inf_{f_\alpha(x_1,x_2,\ldots,x_{n(\alpha)})} = x\left(\max_{1\le i\le n(\alpha)} 1-\mu_\alpha(x_i)\right)\end{bmatrix} \leq \Box A$$

$\Diamond$ A can be proved similarly.

Proposition 3 *Let* S = [X, F], T = [Y, F′] *be two similar algebras and* ϕ *be a homomorphism of* S *into* T. *The exstention of* ϕ *from* IF− *algebra* L = [IFS(X), Ω] *to* IF− *algebra* K = [IFS(Y), Ω′] *is a homomorphism of intuitionistic fuzzy algebras* L *to* K.

Proof Let $Z = \{\phi(x) : x \in X\}$. It is clear that [Z, F′] is a subalgebra of T.

Now, we can suppose that

$$\omega_{f_\alpha}\left(A_1, A_2, \ldots, A_{n(\alpha)}\right) = A \text{ where } A_1, A_2, \ldots, A_{n(\alpha)}, A \in IFS(X)$$

and

$$\omega_{f_\alpha}\left(\phi(A_1), \phi(A_2), \ldots, \phi\left(A_{n(\alpha)}\right)\right) = B, B \in IFS(Y).$$

We will prove that $\phi(A)(y) = B(y)$ for all $y \in Y$.

(I) If $y \notin Z$ then $\phi(A)(y) = \theta = (0, 1)$. That is, if $y = f_\alpha\left(y_1, y_2, \ldots, y_{n(\alpha)}\right)$ then $\exists y_i$, $i = 1, 2, \ldots, n(a)$ such that $y_i \notin Z$.

Let for some j with $1 \leq j \leq n(a)$, $y_j \notin Z$. So, $\phi(A)\left(y_j\right) = \theta$. Furthermore,

$$\phi(A_1)\left(y_1\right) \wedge \phi(A_2)\left(y_2\right) \wedge \ldots \wedge \phi\left(A_{n(\alpha)}\right)\left(y_{n(\alpha)}\right) = \theta$$

and

$$\begin{aligned} B(y) =&\, \omega_{f_\alpha}\left(\phi(A_1), \phi(A_2), \ldots, \phi\left(A_{n(\alpha)}\right)\right)(y) \\ =&\left(sup_{y=f_\alpha(y_1,y_2,\ldots,y_{n(\alpha)})}\left\{\begin{matrix}\mu_{\phi(A_1)}\left(y_1\right) \wedge \mu_{\phi(A_2)}\left(y_2\right) \\ \wedge \ldots \wedge \mu_{\phi(A_{n(\alpha)})}\left(y_{n(\alpha)}\right)\end{matrix}\right\},\right. \\ &\left. inf_{y=f_\alpha(y_1,y_2,\ldots,y_{n(\alpha)})}\left\{\begin{matrix}\nu_{\phi(A_1)}\left(y_1\right) \wedge \nu_{\phi(A_2)}\left(y_2\right) \\ \wedge \ldots \wedge \nu_{\phi(A_{n(\alpha)})}\left(y_{n(\alpha)}\right)\end{matrix}\right\}\right)\end{aligned}$$

$$=\Theta.$$

So, $\phi(A)(y) = B(y)$.

(II) If $y \in Z$ then $\phi(A)(y) = \sup_{y=\phi(x)} A(x) = (a_1, a_2)$. without losing generality, let $(a_1, a_2) > \theta$.

Given small enough $\varepsilon, \varepsilon' > 0$ such that $(a_1 - \varepsilon, a_2 + \varepsilon') < (a_1, a_2)$, there exsit $x \in X$ with $\phi(x) = y$ and $A(x) > (a_1 - \varepsilon, a_2 + \varepsilon')$.

So,

$$\omega_{f_\alpha}\left(A_1, A_2, \ldots, A_{n(\alpha)}\right)(x) = A(x) > \left(a_1 - \varepsilon, a_2 + \varepsilon'\right).$$

Therefore, $x_1, \ldots, x_{n(\alpha)} \in X$ such that $f_\alpha\left(x_1, x_2, \ldots, x_{n(a)}\right) = x$ and

$$A_1(x_1) \wedge A_2(x_2) \wedge \ldots \wedge A_{n(\alpha)}\left(x_{n(\alpha)}\right) > \left(a_1 - \varepsilon, a_2 + \varepsilon'\right).$$

Because ϕ is a homomorphism,

$$\begin{aligned} y =&\phi(x) = \phi(f_\alpha(x_1, x_2, \ldots, x_{n(a)})) \\ =&f_\alpha\left(\phi(x_1), \phi(x_2), \ldots, \phi(x_{n(a)})\right). \end{aligned}$$

Hence,

$$\begin{aligned} B(y) =&\omega_{f_\alpha}\left(\phi(A_1), \phi(A_2), \ldots, \phi(A_{n(\alpha)})\right)(y) \\ &\geq \phi(A_1)(\phi(x_1)) \wedge \phi(A_2)(\phi(x_2)) \wedge \ldots \wedge \phi(A_{n(\alpha)})\left(\phi(x_{n(a)})\right) \\ &\geq A_1(x_1) \wedge A_2(x_2) \wedge \ldots \wedge A_{n(\alpha)}\left(x_{n(\alpha)}\right) \\ &> \left(a_1 - \varepsilon, a_2 + \varepsilon'\right) \end{aligned}$$

Since $\varepsilon, \varepsilon'$ are small enough, we obtain that $B(y) \geq (a_1, a_2) = \phi(A)(y)$.

Let $y_1, y_2, \ldots, y_{n(\alpha)} \in Y$ with $y = f_\alpha\left(y_1, y_2, \ldots, y_{n(\alpha)}\right)$ and $B(y) = (c_1, c_2)$. Then, given small enough $\delta, \delta' > 0$ such that $\left(c_1 - \delta, c_2 + \delta'\right) < (c_1, c_2)$. Now,

$$\left(c_1 - \delta, c_2 + \delta'\right) < \phi(A_1)\left(y_1\right) \wedge \phi(A_2)\left(y_2\right) \wedge \ldots \wedge \phi(A_{n(\alpha)})\left(y_{n(\alpha)}\right)$$

If $B(y) = 0$ then $\phi(A)(y) \geq B(y)$.

Let $0 < B(y)$.

$$\left(c_1 - \delta, c_2 + \delta'\right) < A_1(x_1) \wedge A_2(x_2) \wedge \ldots \wedge A_{n(\alpha)}\left(x_{n(\alpha)}\right).$$

If we use $\phi(f_\alpha\left(x_1, x_2, \ldots, x_{n(a)}\right)) = f_\alpha\left(\phi(x_1), \phi(x_2), \ldots, \phi(x_{n(a)})\right)$ and $\phi(A)(y) > \left(c_1 - \delta, c_2 + \delta'\right)$ then $\phi(A)(y) \geq (c_1, c_2)$ since δ, δ' are small enough. Therefore, $\phi(A)(y) = B(y)$, for all $y \in Z$.

Theorem 4 *Let* $S = [X, F]$, $T = [Y, F']$ *be two similar algebras and* ϕ *be a homomorphism of* S *into* T. *Then* ϕ *extends to a homomorphism between intuitionistic fuzzy algebras* $L = [IFS(X), \Omega]$ *and* $K = \left[IFS(Y), \Omega'\right]$. *If* A *is an intuitionistic fuzzy subalgebra of* L *then* $\phi(A)$ *is an intuitionistic fuzzy subalgebra of* K. *On the otherhand, if* B *is an intuitionistic fuzzy subalgebra of* K *then* $\phi^{-1}(B)$ *is an intuitionistic fuzzy subalgebra of* L.

Proof Since A is an intuitionistic fuzzy subalgebra of L,

$$\omega_{f_\alpha}(A, A, \ldots, A) \sqsubseteq A, \text{ for all } f_\alpha \in F.$$

Also we know that $\phi(\omega_{f_\alpha}(A, A, \ldots, A)) \leq \phi(A)$. So we obtain that

$$\begin{aligned}\phi(\omega_{f_\alpha}(A, A, \ldots, A)) &= \omega_{f_\alpha}(\phi(A), \phi(A), \ldots, \phi(A)) \\ &= \sup_{f_\alpha(x_1, x_2, \ldots, x_{n(\alpha)})=x}(\phi(A)(x_1) \wedge \phi(A)(x_2) \wedge \ldots \wedge \phi(A)(x_{n(\alpha)})) \\ &\leq \phi(A)\end{aligned}$$

and that is $\phi(A)$ is an intuitionistic fuzzy subalgebra of K.

Here, $f_\alpha \in F$ and for every $n(\alpha)-$ tuples $\left(x_1, x_2, \ldots, x_{n(\alpha)}\right) \in X^{n(\alpha)}$,

$$\begin{aligned}&\phi^{-1}(B)\left(f_\alpha\left(x_1, x_2, \ldots, x_{n(a)}\right)\right) \\ &= B\left(\phi\left(f_\alpha\left(x_1, x_2, \ldots, x_{n(a)}\right)\right)\right) \\ &= B\left(f_\alpha\left(\phi(x_1), \phi(x_2), \ldots, \phi(x_{n(a)})\right)\right) \\ &\geq B(\phi(x_1)) \wedge B(\phi(x_2)) \wedge \ldots \wedge B\left(\phi(x_{n(a)})\right) \\ &= \phi^{-1}(B)(x_1) \wedge \phi^{-1}(B)(x_2) \cdots \wedge \phi^{-1}(B)\left(x_{n(a)}\right) \\ &\Rightarrow \omega_{f_\alpha}\left(\phi^{-1}(B), \phi^{-1}(B), \ldots, \phi^{-1}(B)\right) \\ &\leq \phi^{-1}(B).\end{aligned}$$

$\phi^{-1}(B)$ is an intuitionistic fuzzy subalgebra of L.

Definition 16 Let $S = [X, F]$ and $T = [Y, F]$ be two similar algebras. The intuitionistic fuzzy function ϕ from S to T is called an intuitionistic fuzzy homomorphism if and only if for each $f_\alpha \in F$.

$$\varpi_{f_\alpha}(\phi, \phi, \ldots, \phi) \sqsubseteq \phi$$

That is, ϕ is an intuitionistic fuzzy algebra of $S \times T$. If $S = T$ then ϕ called an intuitionistic fuzzy endomorphism.

3.1 Intuitionistic Fuzzy Congruence Relations on Abstract Algebras

The definition of intuitionistic fuzzy congruence relation on abstract algebras is;

Definition 17 Let $S = [X, F]$ be an algebra and $f_\alpha \in F$. For any $(A_1, A_2, \ldots, A_{n(\alpha)}) \in IFR(X)^{n(\alpha)}$ and for any $x, y \in S$, $\overline{\omega}_{f_\alpha}(A_1, A_2, \ldots, A_{n(\alpha)})$ to be an element of $IFR(X)$ defined by

$$\overline{\omega}_{f_\alpha}(A_1, A_2, \ldots, A_{n(\alpha)})(x, y) = \sup_{x, y}(\min_{1 \le i \le n(\alpha)} A_i(x_i, y_i))$$

such that the supremum is taken over all representations of $f_\alpha(x_1, x_2, \ldots, x_{n(\alpha)}) = x$ and $f_\alpha(y_1, y_2, \ldots, y_{n(\alpha)}) = y$. Therefore, $[IFR(X), \Omega]$ is an intuitionistic fuzzy algebra on intuitionistic fuzzy relations.

Definition 18 Let $S = [X, F]$ be an algebra. $A \in IFE(X)$ is an intuitionistic fuzzy congruence relation on S if and only if, for each $f_\alpha \in F$, $\overline{\omega}_{f_\alpha}(A, A, \ldots, A) \sqsubseteq A$.

Proposition 4 *Let A be an intuitionistic fuzzy congruence relation on* $S = [X, F]$ *algebra then* $A_{((r\,s))}$(shortly $\sim$) *is a crisp congruence relation on S for each* $(r\,s) \in [0, 1]$ *with* $r + s \le 1$.

The properties of intuitionistic fuzzy congruence relation were examined in detailed and the following main results were obtained.

Theorem 5 *Let* $S = [X, F]$ *be an algebra,* $f_\alpha \in F$, *and* $A_1, A_2, \ldots, A_{n(\alpha)}, A$ *be intuitionistic fuzzy relations on S.*

$$\overline{\omega}_{f_\alpha}(A_1, A_2, \ldots, A_{n(\alpha)}) \sqsubseteq A \Leftrightarrow A(f_\alpha(x_1, x_2, \ldots, x_{n(\alpha)}), f_\alpha(y_1, y_2, \ldots, y_{n(\alpha)})) \ge \min_{1 \le i \le n(\alpha)} A_i(x_i, y_i)$$

for all pairs of $n(\alpha)-$ tuples $(x_1, x_2, \ldots, x_{n(\alpha)})$ and $(y_1, y_2, \ldots, y_{n(\alpha)})$.

The proof of this theorem can be seen easily from Theorem 1 and following Corollary is clear.

Corollary 1 *Let* $S = [X, F]$ *be an algebra. An intuitionistic fuzzy equivalence relation A on S is an intuitionistic fuzzy congruence relation on S if and only if*

$$A(f_\alpha(x_1, x_2, \ldots, x_{n(\alpha)}), f_\alpha(y_1, y_2, \ldots, y_{n(\alpha)})) \ge \min_{1 \le i \le n(\alpha)} A_i(x_i, y_i)$$

for all $f_\alpha \in F$, and for all $n(\alpha)-$ tuples $(x_1, x_2, \ldots, x_{n(\alpha)}), (y_1, y_2, \ldots, y_{n(\alpha)}) \in X^{n(\alpha)}$.

Example 3 Let consider the group $G = \{e, a, b\}$ determined by following operation $*$. It is clear that G is an algebra given by a binary operation, a unary operation(inversion) and a constant operation(the neutral element) satisfying well known laws.

*	e	a	b
e	e	a	b
a	a	b	e
b	b	e	a

R	e	a	b
e	(1, 0)	(0.6, 0.2)	(0.6, 0.2)
a	(0.6, 0.2)	(1, 0)	(0.6, 0.2)
b	(0.6, 0.2)	(0.6, 0.2)	(1, 0)

R is an intuitionistic fuzzy congruence relation on G. That is, R is an intuitionistic fuzzy equivalence relation on G and for any $x_1, x_2, y_1, y_2 \in G$,

$$R(x_1 * x_2, y_1 * y_2) \geq R(x_1, y_1) \wedge R(x_2, y_2) \text{ and } R(x_1^{-1}, y_1^{-1}) \geq R(x_1, y_1)$$

Proposition 5 *Let* $S = [X, F]$ *be an algebra with subalgebra* $\{e\}$ *and A be an intuitionistic fuzzy congruence relation on S. Then* $A_{\bar{e}}$ *is an intuitionistic fuzzy subalgebra of S such that*

$$\bar{e} = \{x \ : \ A(x, e) = (1, 0), x \in X\}.$$

Proof Let $f_\alpha \in F$ and $(x_1, x_2, \ldots, x_{n(\alpha)}) \in X^{n(\alpha)}$. Then,

$$\begin{aligned} A_{\bar{e}}(f_\alpha(x_1, x_2, \ldots, x_{n(\alpha)})) &= A(f_\alpha(x_1, x_2, \ldots, x_{n(\alpha)}), e) \\ &= A(f_\alpha(x_1, x_2, \ldots, x_{n(\alpha)}), f_\alpha(e, e, \ldots, e)) \\ &\geq \min_{1 \leq i \leq n(\alpha)} A(x_i, e) \\ &= \min_{1 \leq i \leq n(\alpha)} A_{\bar{e}}(x_i) \end{aligned}$$

So,$f_\alpha(A_{\bar{e}}, A_{\bar{e}}, \ldots, A_{\bar{e}}) \leq A_{\bar{e}}$.

Proposition 6 *Let* $S = [X, F]$ *be an algebra. If A is an intuitionistic fuzzy congruence relation on S then so are* $\Box A$ *and* $\Diamond A$.

Proof Let A be an intuitionistic fuzzy congruence relation.

(i) For all $a \in X$,

$$\mu_{\Box A}(a, a) = \mu_A(a, a) = 1 \text{ and } \nu_{\Box A}(a, a) = 1 - \mu_A(a, a) = 0$$

intuitionistic fuzzy reflexive property has been provided.

(ii) Let a, b ∈ X,

$$\mu_{\Box A}(a, b) = \mu_A(a, b) = \mu_A(b, a) = \mu_{\Box A}(b, a) \text{ and}$$
$$\nu\Box A(a, b) = 1 - \mu_A(a, b) = 1 - \mu_A(b, a) = \nu\Box A(b, a)$$

intuitionistic fuzzy symmetric has been provided.

(iii) Let a, b and A is an intuitionistic fuzzy congruence relation then

$$\sup_{z\in X}\{\mu_A(a, c) \wedge \mu_A(c, b)\} \le \mu_A(a, b) \Rightarrow \inf_{c\in X}\{1 - \mu_A(a, c) \vee 1 - \mu_A(c, b)\} \ge 1 - \mu_A(a, b)$$

is proved.

(iv) For each $\overline{\omega}_{f_\alpha}$,

$$A(a, b) \ge \overline{\omega}_{f_\alpha}(A, A, \ldots, A)(a, b) = \sup_{a, b}\left(\min_{1\le i\le n(\alpha)} A(a_i, b_i)\right)\text{and}$$
$$\sup_{a, b}\left(\min_{1 \le i \le n(\alpha)} \mu_A(a_i, b_i)\right) \le \mu_A(a, b)$$
$$\Rightarrow \inf_{a, b}\left(\max_{1 \le i \le n(\alpha)} 1 - \mu_A(a_i, b_i)\right) \ge 1 - \mu_A(a, b)$$
$$\Rightarrow \overline{\omega}_{f_\alpha}(\Box A, \Box A, \ldots, \Box A) \le \Box A$$

So, proof is completed. The operator ◇ A can be examined similarly.

We obtained some results about intuitionistic fuzzy congruence relations under homomorphsim.

Theorem 6 *Let* S = [X, F], T = [Y, F] *be two similar algebras and* ϕ *be a homomorphism of S into T. If B is an intuitionistic fuzzy congruence relation on T then* ϕ^{-1}(B) *is an intuitionistic fuzzy congruence relation on S.*

Proof

(i) For all a ∈ X,

$$\phi^{-1}(B)(a, a) = B(\phi(a), \phi(a)) = (\mu_B(\phi(a), \phi(a)), \nu_B(\phi(a), \phi(a))) = (1, 0)$$

ϕ^{-1}(B) is intuitionistic fuzzy reflexsive.

(ii) Let a, b ∈ X,

$$\begin{aligned}\phi^{-1}(B)(a,b) &= B(\phi(a),\ \phi(b)) = (\mu_B(\phi(a),\ \phi(b)),\ \nu_B(\phi(a),\ \phi(b)))\\ &= (\mu_B(\phi(b),\ \phi(a)),\ \nu_B(\phi(b),\ \phi(a)))\\ &= B(\phi(b),\ \phi(a)) = \phi^{-1}(B)(b,a)\end{aligned}$$

$\phi^{-1}(B)$ is intuitionistic fuzzy symmetric.

(iii) Let $a, b \in X$,

$$\begin{aligned}\left(\phi^{-1}(B)\circ\phi^{-1}(B)\right)(a,b) &= \sup_{c\in X}\left\{\phi^{-1}(B)(a,c)\wedge\phi^{-1}(B)(c,b)\right\}\\ &= \sup_{c\in X}\{B(\phi(a),\ \phi(c))\wedge B(\phi(c),\ \phi(b))\}\\ &= \begin{pmatrix}\sup_{c\in X}\{\mu_B(\phi(a),\ \phi(c))\wedge\mu_B(\phi(c),\ \phi(b))\},\\ \inf_{c\in X}\{\nu_B(\phi(a),\ \phi(c))\vee\nu_B(\phi(c),\ \phi(b))\}\end{pmatrix}\\ &\le \begin{pmatrix}\sup_{y\in Y}\{\mu_B(\phi(a),\ y)\wedge\mu_B(y,\ \phi(b))\},\\ \inf_{y\in Y}\{\nu_B(\phi(a),\ y)\vee\nu_B(y,\ \phi(b))\}\end{pmatrix}\\ &= (B\circ B)(\phi(a),\ \phi(b)) = \phi^{-1}(B)(a,b)\end{aligned}$$

So, $\phi^{-1}(B)$ is intuitionistic fuzzy equivalence relation.

(iv) Let $f_\alpha \in F$ and $a, b \in X$,

$$\begin{aligned}&\overline{\omega}_{f_\alpha}\left(\phi^{-1}(B),\ \phi^{-1}(B),\ \ldots,\ \phi^{-1}(B)\right)(a,b)\\ &= \sup_{\substack{a=f_\alpha(a_1,\ a_2,\ \ldots,\ a_{n(\alpha)})\\ b=f_\alpha(b_1,\ b_2,\ \ldots,\ b_{n(\alpha)})}}\left\{\min_{1\le i\le n(\alpha)}\left(\phi^{-1}(B)\right)(a_i,b_i)\right\}\\ &= \sup_{\substack{a=f_\alpha(a_1,\ a_2,\ \ldots,\ a_{n(\alpha)})\\ b=f_\alpha(b_1,\ b_2,\ \ldots,\ b_{n(\alpha)})}}\left\{\min_{1\le i\le n(\alpha)} B(\phi(a_i),\ \phi(b_i))\right\}\\ &= \begin{pmatrix}\sup_{\substack{a=f_\alpha(a_1,\ a_2,\ \ldots,\ a_{n(\alpha)})\\ b=f_\alpha(b_1,\ b_2,\ \ldots,\ b_{n(\alpha)})}}\left\{\min_{1\le i\le n(\alpha)}\mu_B(\phi(a_i),\ \phi(b_i))\right\},\\ \inf_{\substack{a=f_\alpha(a_1,\ a_2,\ \ldots,\ a_{n(\alpha)})\\ b=f_\alpha(b_1,\ b_2,\ \ldots,\ b_{n(\alpha)})}}\left\{\max_{1\le i\le n(\alpha)}\nu_B(\phi(a_i),\ \phi(b_i))\right\}\end{pmatrix}\\ &\le \begin{pmatrix}\sup_{\substack{\phi(a)=f_\alpha(\phi(a_1),\ \phi(a_2),\ \ldots,\ \phi(a_{n(\alpha)}))\\ \phi(b)=f_\alpha(\phi(b_1),\ \phi(b_2),\ \ldots,\ \phi(b_{n(\alpha)}))}}\left\{\min_{1\le i\le n(\alpha)}\mu_B(\phi(a_i),\ \phi(b_i))\right\},\\ \inf_{\substack{\phi(a)=f_\alpha(\phi(a_1),\ \phi(a_2),\ \ldots,\ \phi(a_{n(\alpha)}))\\ \phi(b)=f_\alpha(\phi(b_1),\ \phi(b_2),\ \ldots,\ \phi(b_{n(\alpha)}))}}\left\{\max_{1\le i\le n(\alpha)}\nu_B(\phi(a_i),\ \phi(b_i))\right\}\end{pmatrix}\\ &= \overline{\omega}_{f_\alpha}(B,\ B,\ \ldots,\ B)(\phi(a),\ \phi(b))\end{aligned}$$

the substitution property has been provided.

Theorem 7 *Let* $S = [X, F]$, $T = [Y, F]$ *be two similar algebras and* ϕ *be a monomorphism of S into T. If A is an intuitionistic fuzzy congruence relation on S then* $\phi(A)$ *is an intuitionistic fuzzy congruence relation on T.*

Proof

(i) For all $b \in Y$, $\phi(A)(b, b) = \sup_{a, c \in \phi^{-1}(b)} A(a, c) = (1, 0)$

$\phi(A)$ is intuitionistic fuzzy reflexsive and the symmetric relation is clear.

(ii) Let $b_1, b_2 \in Y$,

$$(\phi(A) \circ \phi(A))(b_1, b_2) = \sup_{b_3 \in Y}(\phi(A)(b_1, b_3) \wedge \phi(A)(b_3, b_2))$$

$$= \sup_{b_3 \in Y}\left(\left(\sup_{\substack{a_1 \in \phi^{-1}(b_1) \\ a_3 \in \phi^{-1}(b_3)}} A(a_1, a_3)\right) \wedge \left(\sup_{\substack{a_2 \in \phi^{-1}(b_2) \\ a_3 \in \phi^{-1}(b_3)}} A(a_3, a_2)\right)\right)$$

$$\leq \sup_{b_3 \in Y}\left(\sup_{\substack{a_1 \in \phi^{-1}(b_1),\, a_2 \in \phi^{-1}(b_2) \\ a_3 \in \phi^{-1}(b_3)}} A(a_1, a_3) \wedge A(a_3, a_2)\right)$$

$$\leq \sup_{\substack{a_1 \in \phi^{-1}(b_1) \\ a_2 \in \phi^{-1}(b_2)}} \left(\sup_{a_3 \in X} A(a_1, a_3) \wedge A(a_3, a_2)\right)$$

$$= \sup_{\substack{a_1 \in \phi^{-1}(b_1) \\ a_2 \in \phi^{-1}(b_2)}} A(a_1, a_2) = \phi(A)(b_1, b_2)$$

So, $\phi(A)$ is intuitionistic fuzzy transitive.

(iii) Let $f_\alpha \in F$ and $a', b' \in Y$,

$$\varpi_{f_\alpha}(\phi(A), \phi(A), \ldots, \phi(A))(a', b')$$
$$= \sup_{\substack{a' = f_\alpha\left(a'_1, a'_2, \ldots, a'_{n(\alpha)}\right) \\ b' = f_\alpha\left(b'_1, b'_2, \ldots, b'_{n(\alpha)}\right)}} \left(\min_{1 \leq i \leq n(\alpha)} \phi(A)\left(a'_i, b'_i\right)\right)$$

$$= \sup_{\substack{a' = f_\alpha\left(a'_1, a'_2, \ldots, a'_{n(\alpha)}\right) \\ b' = f_\alpha\left(b'_1, b'_2, \ldots, b'_{n(\alpha)}\right)}} \left(\min_{1 \leq i \leq n(\alpha)}\left(\sup_{\substack{a_i \in \phi^{-1}\left(a'_i\right) \\ b_i \in \phi^{-1}\left(b'_i\right)}} A(a_i, b_i)\right)\right)$$

$$\leq \sup_{\substack{a' = f_\alpha(\phi(a_1),\ \phi(a_2),\ \ldots,\ \phi(a_{n(\alpha)}))\\ b' = f_\alpha(\phi(b_1),\ \phi(b_2),\ \ldots,\ \phi(b_{n(\alpha)}))}} \left(\min_{1\leq i\leq n(\alpha)} A(a_i, b_i)\right)$$

$$= \sup_{\substack{a' = \phi(f_\alpha(a_1,\ a_2,\ \ldots,\ a_{n(\alpha)}))\\ b' = \phi(f_\alpha(b_1,\ b_2,\ \ldots,\ b_{n(\alpha)}))}} \left(\min_{1\leq i\leq n(\alpha)} A(a_i, b_i)\right)$$

$$= \varpi_{f_\alpha}(A,\ A,\ \ldots,\ A)\left(\phi^{-1}(a'),\ \phi^{-1}(b')\right).$$

$$\leq A\left(\phi^{-1}(a'),\ \phi^{-1}(b')\right)$$

Therefore $\phi(A)$ is an intuitionistic fuzzy congruence relation on T.

Theorem 8 *Let* $\{A_j\ :\ j \in J\}$ *be a non-empty family of intuitionistic fuzzy congruence relation on algebra* $S = [X,\ F]$. *Then*

$$A = \inf_{j\in J} A_j = \wedge_{j\in J} A_j$$

is an intuitionistic fuzzy congruence relation on S.

Proof

(i) For all $a \in X$,

$$A(a,a) = \inf_{j\in J} A_j(a,a) = \left(\inf_{j\in J} \mu_{A_j}(a,a),\ \sup_{j\in J} \nu_{A_j}(a,a)\right) = (1,\ 0)$$

(ii) Let $a,\ b \in X$,

$$\begin{aligned} A(a,b) &= \inf_{j\in J} A_j(a,\ b) = \left(\inf_{j\in J} \mu_{A_j}(a,b),\ \sup_{j\in J} \nu_{A_j}(a,b)\right) \\ &= \left(\inf_{j\in J} \mu_{A_j}(b,a),\ \sup_{j\in J} \nu_{A_j}(b,a)\right) = \inf_{j\in J} A_j(b,a) = A(b,a) \end{aligned}$$

(iii) Let $a,\ b \in X$,

$$(A \circ A)(a,\ b) = \sup_{c\in X}(A(a,\ c) \wedge A(c,\ b)) = \sup_{c\in X}\left(\inf_{j\in J} A_j(a,\ c) \wedge \inf_{j\in J} A_j(c,\ b)\right)$$

$$= \sup_{c\in X}\left(\left(\inf_{j\in J} \mu_{A_j}(a,\ c),\ \sup_{j\in J} \nu_{A_j}(a,\ c)\right) \wedge \left(\inf_{j\in J} \mu_{A_j}(c,\ b),\ \sup_{j\in J} \nu_{A_j}(c,\ b)\right)\right)$$

$$= \sup_{c\in X}\left(\inf_{j\in J} \mu_{A_j}(a,\ c) \wedge \inf_{j\in J} \mu_{A_j}(c,\ b),\ \sup_{j\in J} \nu_{A_j}(a,\ c) \vee \sup_{j\in J} \nu_{A_j}(c,\ b)\right)$$

$$= \left(\sup_{c\in X}\left(\inf_{j\in J} \mu_{A_j}(a,\ c) \wedge \inf_{j\in J} \mu_{A_j}(c,\ b)\right),\ \inf_{z\in X}\left(\sup_{j\in J} \nu_{A_j}(a,\ c) \vee \sup_{j\in J} \nu_{A_j}(c,\ b)\right)\right)$$

$$= \left(\sup_{c\in X}\left(\inf_{j\in J}\left(\inf_{k\in J} \mu_{A_j}(a,\ c) \wedge \mu_{A_k}(c,\ b)\right)\right),\ \inf_{c\in X}\left(\sup_{j\in J}\left(\sup_{k\in J} \nu_{A_j}(a,\ c) \vee \nu_{A_k}(c,\ b)\right)\right)\right)$$

$$\leq \left(\sup_{c\in X}\left(\inf_{j\in J}\left(\mu_{A_j}(a,\ c) \wedge \mu_{A_j}(c,\ b)\right)\right),\ \inf_{c\in X}\left(\sup_{j\in J}\left(\nu_{A_j}(a,\ c) \vee \nu_{A_j}(c,\ b)\right)\right)\right)$$
$$= \left(\inf_{j\in J} \mu_{A_j}(a,\ b),\ \sup_{j\in J} \nu_{A_j}(a,\ b)\right) = A(a,\ b)$$

(iv) Let $f_\alpha \in F$ and $\left(a_1,\ a_2, \ldots,\ a_{n(\alpha)}\right),\ \left(b_1,\ b_2, \ldots,\ b_{n(\alpha)}\right) \in X^{n(\alpha)}$.

By corollary,

$$\begin{aligned}
&A\left(f_\alpha\left(a_1,\ a_2, \ldots,\ a_{n(\alpha)}\right),\ f_\alpha\left(b_1,\ b_2, \ldots,\ b_{n(\alpha)}\right)\right)\\
&= \wedge_{j\in J} A_j\left(f_\alpha\left(a_1,\ a_2, \ldots,\ a_{n(\alpha)}\right),\ f_\alpha\left(b_1,\ b_2, \ldots,\ b_{n(\alpha)}\right)\right)\\
&= \inf_{j\in J}\left(A_j\left(f_\alpha\left(a_1,\ a_2, \ldots,\ a_{n(\alpha)}\right),\ f_\alpha\left(b_1,\ b_2, \ldots,\ b_{n(\alpha)}\right)\right)\right)\\
&\geq \inf_{j\in J}\left(\min_{1\leq i\leq n(\alpha)} A_j(a_i, b_i)\right)\\
&= \min_{1\leq i\leq n(\alpha)}\left(\inf_{j\in J} A_j(a_i, b_i)\right) = \min_{1\leq i\leq n(\alpha)} A(a_i, b_i)
\end{aligned}$$

3.2 Isomorphism Theorems on Intuitionistic Fuzzy Universal Algebras

3.2.1 Intuitionistic Fuzzy Functions

Before introduce the definition of intuitionistic fuzzy function we need to prove following propositions. We will use ${}^0_1[a]_A = \bar{a}$ presentation for $A \in IFE(X)$ and define an intuitionistic fuzzy subset $A_{\bar{a}} : X \to I \times I$ as follows:

$$A_{\bar{a}}(z) = A(a,\ z) \text{ for all } z \in X.$$

Proposition 7 $A_{\bar{a}} : X \to I \times I$ *defined as above is well-defined.*

Proof For all $x \in X$, $0 \leq \mu_{A_{\bar{a}}}(x) + \nu_{A_{\bar{a}}}(x) \leq 1$. If $y \in \bar{a}$ then $A(a,\ y) = (1,\ 0)$.

$$A_{\bar{a}}(x) = A(y,\ x) \leq (1,\ 0) \text{ and } A_{\bar{a}}(x) = A(a,\ x) \leq (1,\ 0).$$

So, $A(a,\ x) = A(y,\ x)$. That is $A_{\bar{a}}$ is well defined.

Proposition 8 *For each fixed* $((r\,s))$ *such that* $(r\,s) \in [$ *with* $r + s \leq 1$, $\left\{(A_{\bar{a}})_{((r\,s))} : a \in X\right\}$ *the set of* $(r\,s)-$ *cuts of* A *is a crisp partition of* X.

Proof

(i) For all $a \in X$ there is at least one $(r\,s) \in [$ with $r+s \leq 1$ such that $a \in (A_{\bar{a}})_{((r\,s))}$. So, $\bigcup_{a\in X} (A_{\bar{a}})_{((r\,s))} = X$.

(ii) Let a, b $\in$ X and a $\in \bar{a}$. Suppose that $\bar{a} \neq \bar{b}$. If $z \in (A_{\bar{a}})_{((r\,s))} \cap (A_{\bar{b}})_{((r\,s))}$ then

$$A_{\bar{a}}(z) \geq ((r\,s)) \text{ and } A_{\bar{b}}(z) \geq ((r\,s)) \Rightarrow A(a, z) \geq ((r\,s)) \text{ and } A(b, z) \geq ((r\,s))$$
$$\Rightarrow A(a, b) \geq ((r\,s))$$

If $x \in (A_{\bar{a}})_{((r\,s))}$ then $A(a, x) \geq ((r\,s))$. It is clear by symmetry that $A(x, a) \geq ((r\,s))$. So, $A(b, x) \geq ((r\,s))$ and $x \in (A_{\bar{b}})_{((r\,s))}$. We obtain that $(A_{\bar{a}})_{(r\,s)} \subseteq (A_{\bar{b}})_{(r\,s)}$. Similarly we can show that $(A_{\bar{b}})_{(r,\,s)} \subseteq (A_{\bar{a}})_{(r,\,s)}$.

Therefore $(A_{\bar{a}})_{(r\,s)} \cap (A_{\bar{b}})_{(r\,s)} = \emptyset$ or $(A_{\bar{a}})_{(r\,s)} = (A_{\bar{b}})_{(r\,s)}$.

Definition 19 Let X and Y be non-empty sets and f be an intuitionistic fuzzy relation from X to Y. Each y $\in$ Y determines an intuitionistic fuzzy subset of X as follows:

$$A_y \;:\; X \to I \times I,\; A_y(x) = f(x, y) \text{ for all } x \in X.$$

The intuitionistic fuzzy relation is called an intuitionistic fuzzy function from X to Y if.

1. For each x $\in$ X, $\exists!$ y $\in$ Y such that f(x, y) = (1, 0),
2. For each (r s) such that r s $\in$ [with $r + s \leq 1$ the set $\left\{(A_y)_{(r\,s)} \;:\; y \in Y\right\}$ is a crisp partition of X.

If for each y $\in$ Y, $\exists$x $\in$ X such that f(x, y) = (1, 0) then f called onto function.

If each pair $x_1, x_2 \in X$ such that $f(x_1, y) = f(x_2, y) = (1, 0) \Rightarrow x_1 = x_2$ then f called one-to-one function.

After this definition we can give the following concepts.

Let $f \;:\; X \times Y \to I \times I$ be an intuitionistic fuzzy function then converse of f defined as $\breve{f} \;:\; Y \times X \to I \times I$, $\breve{f}(y, x) = f(x, y)$ for x $\in$ X, y $\in$ Y.

Let $A \;:\; X \to I \times I$ and $B \;:\; Y \to I \times I$ be two intuitionistic fuzzy sets then image and preimage under f as follows:

$$f(A)\,(y) = \sup_{x \in X}(A(x) \wedge f(x, y)),\; y \in Y \text{ and}$$
$$f^{-1}(B)\,(x) = \sup_{x \in X}(B(y) \wedge (y, x)),\; x \in X.$$

Example 4 Let X = {a, b, c} and Y = {d, e} be universal sets then the intuitionistic fuzzy relation f defined as follows;

$$f = \left\{ \begin{array}{c} ((a, d), 1, 0), ((a, e), 0.6, 0.3), ((b, d), 0.6, 0.3), \\ ((b, e), 1, 0), ((c, d), 1, 0), ((c, e), 0.6, 0.3) \end{array} \right\}$$

So, we can examine (A_d) and (A_e) for all $(r\,s)$ such that $r\,s \in [$ with $r + s \leq 1$. If $(r\,s) \leq (0.6, 0.3)$ then $(A_d)_{(r,s)} = (A_e)_{(r,s)} = X$. For other situations of $(r\,s)$, it is clear that $(A_d)_{(r,s)} = \{a, c\}$ and $(A_e)_{(r,s)} = \{b\}$. As a result, for all $((r\,s))$ such that $(r\,s) \in [0, 1]$ with $r+s \leq 1$, $\left\{(A_y)_{((r\,s))} : y \in Y\right\}$ is a crisp partition of X. Hence, f is an intuitionistic fuzzy function from X to Y.

Proposition 9 *Let* X *and* Y *be non-empty sets and* $f : X \times Y \to I \times I$ *be an intuitionistic fuzzy function The composition of* $f : X \to Y$ *and* $\widehat{f} : Y \to X$ *such that*

$$\widehat{f} \circ f(x_1, x_2) = \sup_y (f(x_1, y) \wedge (y, x_2)), \quad \text{for all } x_1, x_2 \in X$$

is an intuitionistic fuzzy equivalence relation on X.

Proof Let A be the relation $\widehat{f} \circ f$. For all $x \in X$, $A(x, x) = (1, 0)$ so, A is intuitionistic fuzzy reflexive. From the definition A is intuitionistic fuzzy symmetric. Now, let for $x_1, x_2 \in X$,

$$(\alpha_1, \beta_1) = (A \circ A)(x_1, x_2) > \theta$$

and let given small enough $\varepsilon, \varepsilon' > 0$ such that $(\alpha_1 - \varepsilon, \beta_1 + \varepsilon') < (\alpha_1, \beta_1)$.

There exist $z \in X$ such that $A(x_1, z) > (\alpha_1 - \varepsilon, \beta_1 + \varepsilon')$ and $A(z, x_2) > (\alpha_1 - \varepsilon, \beta_1 + \varepsilon')$.

Hence, there are $y, w \in Y$ with

$f(x_1, y) \wedge f(z, y) > (\alpha_1 - \varepsilon, \beta_1 + \varepsilon')$ and $f(z, w) \wedge f(x_2, w) > (\alpha_1 - \varepsilon, \beta_1 + \varepsilon')$. So,

$$z \in (A_y)_{\langle \alpha_1 - \varepsilon, \beta_1 + \varepsilon' \rangle} \cap (A_w)_{\langle \alpha_1 - \varepsilon, \beta_1 + \varepsilon' \rangle} \Rightarrow (A_y)_{\langle \alpha_1 - \varepsilon, \beta_1 + \varepsilon' \rangle} = (A_w)_{\langle \alpha_1 - \varepsilon, \beta_1 + \varepsilon' \rangle}$$
$$\Rightarrow x_2 \in (A_w)_{\langle \alpha_1 - \varepsilon, \beta_1 + \varepsilon' \rangle} \text{and } f(x_2, y) > (\alpha_1 - \varepsilon, \beta_1 + \varepsilon').$$

So, $A(x_1, x_2) \geq f(x_1, y) \wedge f(x_2, y) > (\alpha_1 - \varepsilon, \beta_1 + \varepsilon')$. Since $\varepsilon, \varepsilon'$ are small enough, $A(x_1, x_2) \geq (\alpha_1, \beta_1)$.

The intuitionistic fuzzy equivalence relation A, as defined above, is called **kernel** of f.

Theorem 9 *Let* X *and* Y *be non-empty sets,* $f : X \times Y \to I \times I$ *be an intuitionistic fuzzy function and* A *be the kernel of* f. *Then there is a decomposition of* f *given by following diagram where* X/A *is the class of intuitionistic fuzzy subsets* $(A_{\overline{x}})_{x \in X}$, $f(X) = \{y \in Y : \exists x \in X \text{ with } f(x, y) = (1, 0)\}$ *and* ε *is a natural mapping given by* $\varepsilon(x, A_{\overline{x}}) = A_{\overline{x}}(x)$. *Also,* f' *and i are crisp mappings defined by* $f'(A_{\overline{x}}) = y$ *where* $\overline{x}$ *is the class given by x such that* $f(x, y) = (1, 0)$ *and i is given by inclusion,*

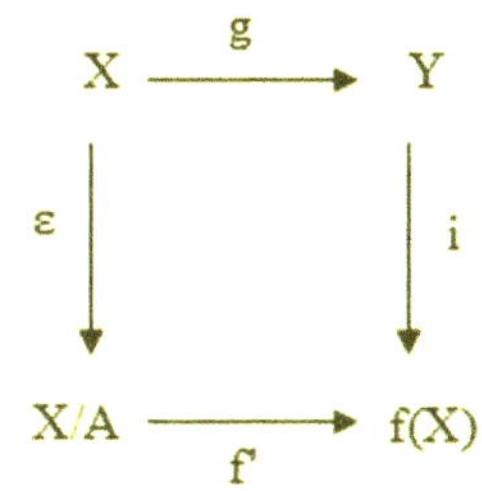

Fig. 1 Decomposition Diagram of Functions with Natural Mapping

$i\ :\ \hookrightarrow Y,\ i(y) = y$ and $g(x) = y$ is given by $f(x,\ y) = (1,\ 0)$ (Fig. 1).

Proof It is enough to show that f' is well-defined. Let $A(x,\ x') = (1,\ 0)$ and $y,\ y' \in T$ such that $f(x,\ y) = f(x',\ y') = (1,\ 0)$.

$$\begin{aligned} A(x,\ x') =& (1,\ 0) \Rightarrow \overline{x} = \overline{x}' \\ \Rightarrow& f(x,\ y) = f(x,\ y') = (1,\ 0) \\ \Rightarrow& y = y' \end{aligned}$$

Murali examined *Isomorphism Theorems* on fuzzy algebra[4] and our aim is extend these theorems to intuitionistic fuzzy algebra.

Let $S = [X,\ F]$ be an algebra and A be an intuitionistic fuzzy congruence relation on S. X/A is the class of intuitionistic fuzzy subsets $(A_{\overline{x}})_{x \in X}$ such that ${}_1^0[x]_A = \overline{x}$, $x \in X$. Now, for $f_\alpha \in F$ let $A_{\overline{x_1}},\ A_{\overline{x_2}},\ \ldots,\ A_{\overline{x_{n(\alpha)}}} \in X/A$ then f'_α on X/A is defined as follow;

$$f'_\alpha\left(A_{\overline{x}_1},\ A_{\overline{x}_2},\ \ldots,\ A_{\overline{x}_{n(\alpha)}}\right) = A_{\overline{z}},\ A_{\overline{z}}(t) = A(z,\ t)$$

where $f_\alpha\left(x_1,\ x_2,\ \ldots,\ x_{n(\alpha)}\right) = z,\ x_i \in \overline{x_i}$ for each $i = 1,\ 2,\ \ldots,\ n(\alpha)$. Suppose that $y_i \in \overline{x_i}$ for each $i = 1,\ 2,\ \ldots,\ n(\alpha)$ Then $x_i \sim_{A_{(1,0)}} y_i$. It is clear that,

$$f_\alpha\left(x_1,\ x_2,\ \ldots,\ x_{n(\alpha)}\right) \sim_{A_{(1,0)}} f_\alpha\left(y_1,\ y_2,\ \ldots,\ y_{n(\alpha)}\right) \Rightarrow z \sim_{A_{(1,0)}} z'$$

where $f_\alpha\left(y_1,\ y_2,\ \ldots,\ y_{n(\alpha)}\right) = z'$. So,

$$z' \in_1^0 [z]_A \text{ and } A_{\overline{z'}}(t) = A(z',\ t) = A(z,\ t) = A_{\overline{z}}(t).$$

We obtain that, $S/A = [X/A,\ F']$ is an algebra similar to $S = [X,\ F]$.

Theorem 10 *Let* ε *be a mapping from* S *to* S/A, $\varepsilon(x) = A_{\overline{x}}$ *then* ε *is a homomorphism.*

Proof If $\overline{z} =_1^0 \left[f_\alpha\left(x_1,\ x_2,\ \ldots,\ x_{n(\alpha)}\right)\right]_A$ then

$$\begin{aligned}\varepsilon(f_\alpha(x_1, x_2, \ldots, x_{n(\alpha)})) &= A_{\overline{z}} = f_\alpha(A_{\overline{x_1}}, A_{\overline{x_2}}, ldots, A_{\overline{x_{n(\alpha)}}}) \\ &= f_\alpha(\varepsilon(x_1), \varepsilon(x_2), \ldots, \varepsilon(x_{n(\alpha)})).\end{aligned}$$

So, ε is a homomorphism.

Proposition 10 *Let* $S = [X, F]$, $T = [Y, F]$ *be two similar algebras and* f *is an intuitionistic fuzzy homomorphism from* S *to* T. *The kernel of* f *is an intuitionistic fuzzy congruence relation on* S.

Proof For any $f_\alpha \in F$ and $(x_1, \ldots, x_{n(\alpha)}), (y_1, \ldots, y_{n(\alpha)}) \in X^{n(\alpha)}$, we have to prove the substitution property. From definition,

$$\begin{aligned}&A(f_\alpha(x_1, x_2, \ldots, x_{n(\alpha)}), f_\alpha(y_1, y_2, \ldots, y_{n(\alpha)})) \\ &= \widehat{f} \circ f(f_\alpha(x_1, x_2, \ldots, x_{n(\alpha)}), f_\alpha(y_1, y_2, \ldots, y_{n(\alpha)})) \\ &= \sup_{z \in Y}(f(f_\alpha(x_1, x_2, \ldots, x_{n(\alpha)}), z) \wedge (z, f_\alpha(y_1, y_2, \ldots, y_{n(\alpha)}))) \\ &= \sup_{z \in Y}(f(f_\alpha(x_1, x_2, \ldots, x_{n(\alpha)}), z) \wedge f(f_\alpha(y_1, y_2, \ldots, y_{n(\alpha)}), z))\end{aligned}$$

and $A_i(x_i, y_i) = \widehat{f} \circ f(x_i, y_i) = \sup_{z \in Y}(f(x_i, z) \wedge f(y_i, z))$. Let presume that $A(f_\alpha(x_1, x_2, \ldots, x_{n(\alpha)}), f_\alpha(y_1, y_2, \ldots, y_{n(\alpha)})) \not\geq \min_1 \leq i \leq n(\alpha) A(x_i, y_i)$.. There exist (β_1, β_2) such that $\beta_1, \beta_2 \in [0, 1]$ with $\beta_1 + \beta_2 \leq 1$,

$$\begin{aligned}A(f_\alpha(x_1, x_2, \ldots, x_{n(\alpha)}), f_\alpha(y_1, y_2, \ldots, y_{n(\alpha)})) &< (\beta_1, \beta_2) \\ &< A(x_i, y_i)\end{aligned}$$

that is

$$\sup_z \in Y(f(f_\alpha(x_1, x_2, \ldots, x_n(\alpha)), z) \wedge f(f_\alpha(y_1, y_2, \ldots, y_n(\alpha)), z)) < (\beta_1, \beta_2)$$

and for each $i = 1, 2$

$$\sup_{z \in Y}(f(x_i, z) \wedge f(y_i, z)) > (\beta_1, \beta_2).$$

It is clear that, there exist a z_i for each $i = 1, 2, \ldots, n$ such that $f(x_i, z_i) > (\beta_1, \beta_2)$ and $f(y_i, z_i) > (\beta_1, \beta_2)$.

Let $z = f_\alpha(z_1, z_2, \ldots, z_{n(\alpha)})$.

$$\begin{aligned}f(f_\alpha(x_1, x_2, \ldots, x_{n(\alpha)}), f_\alpha(z_1, z_2, \ldots, z_{n(\alpha)})) &\geq \min_{1 \leq i \leq n(\alpha)} f(x_i, z_i) \\ &> (\beta_1, \beta_2)\end{aligned}$$

and

$$f(f_\alpha(y_1, y_2, \ldots, y_{n(\alpha)}), f_\alpha(z_1, z_2, \ldots, z_{n(\alpha)})) \geq \min_{1 \leq i \leq n(\alpha)} f(y_i, z_i).$$

Hence,

$$(f(f_\alpha(x_1, x_2, \ldots, x_{n(\alpha)}), z) \wedge f(f_\alpha(y_1, y_2, \ldots, y_{n(\alpha)}), z)) > (\beta_1, \beta_2)\text{and}$$
$$\sup_{z \in Y}(f(f_\alpha(x_1, x_2, \ldots, x_{n(\alpha)}), z) \wedge f(f_\alpha(y_1, y_2, \ldots, y_{n(\alpha)}), z)) > (\beta_1, \beta_2).$$

Theorem 11 (First Isomorphism Theorem) *Let* S = [X, F], T = [Y, F] *be two similar algebras and* f *is an intuitionistic fuzzy homomorphism from* S *to* T. *If* A *is the kernel of* f *then there exist a decomposition of* f *defined by following diagram.*

Proof Thanks to previous proposition and Theorem 1 it is enough to prove f' is an isomorphism. $f'(A_{\overline{x}}) = y$ where $\overline{x}$ is the class given by x such that $f(x, y) = (1, 0)$. From definition f' is bijective. Also (Fig. 2),

$$f'(f_\alpha(A_{\overline{x_1}}, A_{\overline{x_2}}, \ldots, A_{\overline{x_{n(\alpha)}}}))$$
$$= f_\alpha(f'(A_{\overline{x_1}}), f'(A_{\overline{x_2}}), \ldots, f'(A_{\overline{x_{n(\alpha)}}}))$$

that is f' is homomorphism.

Murali [17] proved that second isomorphism theorem could not extend to fuzzy algebras. Likewise we need a restriction of an intuitionistic fuzzy congruence relation to intuitionistic fuzzy subalgebra to introduced second isomorphism theorem on intuitionistic fuzzy algebra. But, we could not define a natural way for this restriction. So, we will deal on third isomorphism theorem.

Let S = [X, F] be an algebras and A, B be two intuitionistic fuzzy congruence relations on S such that B ≤ A. Let $C(A)_{(1,0)}$ and $C(B)_{(1,0)}$ denotes the crisp (1, 0)−equivalence class of A and B, respectively. It is clear that $C(B)_{(1,0)}$ is a refinement of $C(A)_{(1,0)}$. We define a congruence relation on following set;

$$S/B = \{B_{\overline{x}} \ : \ \overline{x} \in C(B)_{(1,0)}, \ x \in X\}$$

Definition 20 Let S = [X, F] be an algebras and A, B be two congruence relations on S such that B ≤ A.

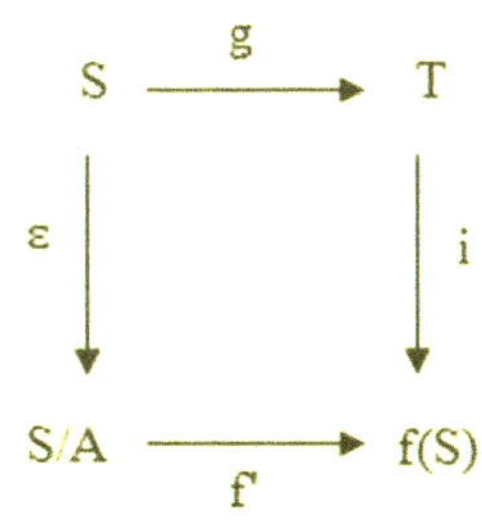

Fig. 2 Decomposition Diagram for Similar Algebras

$$A/B : S/B \times S/B \rightarrow I \times I$$
$$A/B\left(B_{\overline{x}}, B_{\overline{y}}\right) = A(x, y) \text{ for } x \in \overline{x}, y \in \overline{y}$$

is called the quotient intuitionistic fuzzy congruence relation.

Theorem 12 *Let* $S = [X, F]$ *be an algebras and* A, B *be two intuitionistic fuzzy congruence relations on* S *such that* $B \leq A$. *The intuitionistic fuzzy quotient congruence relation* A/B *is an intuitionistic fuzzy congruence relation on* S/B.

Proof

(i) If $x, x' \in \overline{x}$ and $y, y' \in \overline{y}$ then x, x' belong to the same $(1, 0)-$ equivalence class of A and also y, y' belong to the same class. Since $A(x, y) = A(x', y')$ then

$$A/B\left(B_{\overline{x}}, B_{\overline{y}}\right) = A(x, y) = A(x', y').$$

We obtain that A/B is well-defined.

(ii) It is clear that A/B is reflexive and symmetric.
(iii) Shortly $A/B = \rho$. Let $x \in \overline{x}$, $y \in \overline{y}$ and $z \in \overline{z}$,

$$\begin{aligned}
\rho \circ \rho\left(B_{\overline{x}}, B_{\overline{y}}\right) &= \sup_{\overline{z} \in C(B)_{(1,0)}}\left(\rho(B_{\overline{x}}, B_{\overline{z}}) \wedge \rho\left(B_{\overline{z}}, B_{\overline{y}}\right)\right) \\
&= \sup_{\overline{z} \in C(B)_{(1,0)}}\{A(x, z) \wedge A(z, y) \ : \ x \in \overline{x}, y \in \overline{y}, z \in \overline{z}\} \\
&\leq \sup_{z \in X}(A(x, z) \wedge A(z, y)) \\
&= A \circ A(x, y) \leq A(x, y) = \rho\left(B_{\overline{x}}, B_{\overline{y}}\right)
\end{aligned}$$

so, A/B transitive.

(iv) Let $f_\alpha \in F$ and $\left(B_{\overline{x_1}}, B_{\overline{x_2}}, \ldots, B_{\overline{x_{n(\alpha)}}}\right), \left(B_{\overline{y_1}}, B_{\overline{y_2}}, \ldots, B_{\overline{y_{n(\alpha)}}}\right) \in (S/B)^{n(\alpha)}$. Suppose that there exists a $(\beta_1, \beta_2) > \theta$ with $\beta_1, \beta_2 \in [0, 1]$ with β_1+ such that

$$\rho\left(f_\alpha\left(B_{\overline{x_1}}, B_{\overline{x_2}}, \ldots, B_{\overline{x_{n(\alpha)}}}\right), f_\alpha\left(B_{\overline{y_1}}, B_{\overline{y_2}}, \ldots, B_{\overline{y_{n(\alpha)}}}\right)\right) < (\beta_1, \beta_2)$$

and

$$(\beta_1, \beta_2) < \rho\left(B_{\overline{x_i}}, B_{\overline{y_i}}\right) \text{ for each } i = 1, 2, \ldots, n(\alpha).$$

From Theorem 2, $f_\alpha(B_{\overline{x_1}}, B_{\overline{x_2}}, \ldots, B_{\overline{x_{n(\alpha)}}}) = B_{\overline{x}}$ and $f_\alpha(B_{\overline{y_1}}, B_{\overline{y_2}}, \ldots, B_{\overline{y_{n(\alpha)}}}) = B_{\overline{y}}$ such that $\overline{x} =^0_1 [x]_B$ and $\overline{y} =^0_1 [y]_B$ with $x = f_\alpha(x_1, x_2, \ldots, x_{n(\alpha)})$, $x_i \in \overline{x_i}$ and $y = f_\alpha(y_1, y_2, \ldots, y_{n(\alpha)})$, $y_i \in \overline{y_i}$ for each $i = 1, 2$ Now,

$$\begin{aligned} A(x, y) =& A(f_\alpha(x_1, x_2, \ldots, x_{n(\alpha)}), f_\alpha(y_1, y_2, \ldots, y_{n(\alpha)})) \\ &\geq \min_{1 \leq i \leq n(\alpha)} A(x_i, y_i) \\ \Rightarrow& A(x, y) > (\beta_1, \beta_2) \end{aligned}$$

Hence, $\rho(f_\alpha(B_{\overline{x_1}}, B_{\overline{x_2}}, \ldots, B_{\overline{x_{n(\alpha)}}}), f_\alpha(B_{\overline{y_1}}, B_{\overline{y_2}}, \ldots, B_{\overline{y_{n(\alpha)}}})) > (\beta_1, \beta_2)$
This contradicted our acceptance.

Corollary 2 (Third Isomorphism Theorem) *Let* $S = [X, F]$ *be an algebras and* A, B *be two intuitionistic fuzzy congruence relations on* S *such that* $B \leq A$. *Then*

$$(S/B)/(A/B) \cong S/A.$$

Proof Let take into account that $^0_1[B_{\overline{x}}]_{A?B} = \{B_{\overline{y}} \in S/B \ : \ A/B(B_{\overline{x}}, B_{\overline{y}}) = (1, 0)\}$. By the definition of A/B, $^0_1[B_{\overline{x}}]_{A?B} \cong \overline{x}$ as crisp set. Therefore,

$$(A/B)_{^0_1[B_{\overline{x}}]_{A?B}} \cong A_{\overline{x}}$$

as an intuitionistic fuzzy subsets. We obtain that $(S/B)/(A/B) \cong S/A$.

4 Conclusion

In this study, we studied intuitionistic fuzzy algebraic properties on intuitionistic fuzzy algebra. Thus, the comman properties of intuitionistic fuzzy algebraic structures like groups, rings were extended.

References

1. L.A. Zadeh, Fuzzy sets. Inf. Control **8**, 338–353 (1965)
2. K.T. Atanassov, Intuitionistic fuzzy sets, VII ITKR session, Sofia, 20–23 June (1983), (Deposed in Centr. Sci.-Techn. Library of the Bulg. Acad. of Sci., 1697/84) (in Bulgarian). Reprinted: Int. J. Bioautomation **20**(S1), S1–S6 (2016)
3. S. Melliani, O. Castillo, in *Recent Advances in Intuitionistic Fuzzy Logic Systems: Theoretical Aspects and Applications*. Studies in Fuzziness and Soft Computing (Springer, 2019)

4. S. Melliani, O. Castillo, in *Recent Advances in Intuitionistic Fuzzy Logic Systems and Mathematics*. Studies in Fuzziness and Soft Computing (Springer, 2021)
5. F. Tuğrul, B. Yılmaz, M. Çitil, Application of ranking with similarity measure in multi criteria decision making. Konuralp J. Math. **7**(2), 438–441 (2019)
6. F. Tuğrul, M. Çitil, S. Balcı, On the determination students' aptitude using intuitionistic fuzzy logic. J. Univ. Math. **2**(1), 36–41 (2019)
7. F. Tuğrul, M. Çitil, B. Karasolak, M. Dağlı, Interpretation of physical conditions of schools with fuzzy multi criteria decision making. J. Univ. Math. **3**(1), 46–52 (2020)
8. F. Tuğrul, M. Çitil, A new perspective on evaluation system in education with intuitionistic fuzzy PROMETHEE algorithm. J. Univ. Math. **4**(1), 13–24 (2021)
9. F. Tuğrul, Connectedness in temporal intuitionistic fuzzy topology in chang's sense. J. Univ. Math. **4**(2), 157–165 (2021)
10. D. Sánchez, P. Melin, O. Castillo, Optimization of modular granular neural networks using a firefly algorithm for human recognition. Eng. Appl. Artif. Intell. **64**, 172–186 (2017)
11. S. Sotirov, E. Sotirova, P. Melin, O. Castilo, K. Atanassov, Modular neural network preprocessing procedure with intuitionistic fuzzy intercriteria analysis method, in *Flexible Query Answering Systems*, ed. by T. Andreasen et al. Advances in Intelligent Systems and Computing, vol. 400 (Springer, Cham, 2015)
12. R. Biswas, Intuitionistic fuzzy subgroups. Math. Forum **X**, 37–46 (1989)
13. J. Zhan, Z. Tan, Intuitionistic M-fuzzy groups. Soochow J. Math. **30**, 85–90 (2004)
14. L. Yan, Intuitionistic fuzzy ring and its homomorphism image, in *2008 International Seminar on Future BioMedical Information Engineering*, pp. 75–77 (2008)
15. N. Palaniappan, S. Naganathan, K. Arjunan, A study on intuitionistic L-fuzzy subgroups. Appl. Math. Sci. **3**(53), 2619–2624 (2009)
16. S. Melliani, M. Elomari, R. Ettoussi, L.S. Chadli, Intuitionistic fuzzy semigroup. Notes on IFSs **21**(2), 43–50 (2015)
17. V. Murali, A study of universal algebra in fuzzy set theory. Rhodes University, Department of Mathematics Ph.D. Thesis, p. 104 (1987)
18. P. Burillo, H. Bustince, Intuitionistic fuzzy relations (part I). Mathware Soft Comput. **2**, 5–38 (1995)
19. H. Bustince, P. Burillo, Structures on intuitionistic fuzzy relations. Fuzzy Sets Syst. **78**, 293–303 (1996)
20. G. Deschrijver, E.E. Kerre, On the composition of intuitionistic fuzzy relations. Fuzzy Sets Syst. **136**, 333–361 (2003)
21. K. Hur, S.Y. Jang, Y.S. Ahn, Intuitionistic fuzzy equivalence relations. Honam Math. J. **27**(2), 163–181 (2005)
22. G. Çuvalcıoğlu, S. Yılmaz, Some properties of intuitionistic fuzzy equivalence relations and class trees w.r.t. intuitionistic fuzzy equivalence relations. Adv. Stud. Contemp. Math. **24**(1), 77–86 (2014)
23. V. Murali, Fuzzy congruence relations. Fuzzy Sets Syst. **41**(3), 359–369 (1991)
24. G. Birkhoff, in *Lattice Theory*. American Mathematical Society, United States of America, p. 418 (1940)
25. P.M. Cohn, in *Universal Algebra* (D. Reidel Publishing Company, Dordrecht, 1981), p. 412
26. G. Çuvalcıoğlu, S. Tarsuslu (Ylmaz), Universal algebra in intuitionistic fuzzy set theory. Notes Intuitionistic Fuzzy Sets **23**(1), 1–5 (2017)
27. K. Hur, Y.J. Su, W.K. Hee, The lattice of intuitionistic fuzzy congruences. Int. Math. Forum **5**(1), 211–236 (2006)
28. P.K. Sharma, (α, β)—Cut of intuitionistic fuzzy groups. Int. Math. Forum **53**(6), 2605–2614 (2011)

29. S. Tarsuslu (Yılmaz), G. Çuvalcıoğlu, (T, S)—Intuitionistic fuzzy algebras. J. Intell. Fuzzy Syst. **36**(1), 139–147 (2019)
30. S. Tarsuslu (Yılmaz), M. Çitil, Intuitionistic fuzzy congruence relations on intuitionistic fuzzy abstract algebras. Gazi Univ. J. Sci. **32**(2), 649–658 (2019)

Generalization of Intuitionistic Fuzzy Submodules of a Module by Using Triangular Norms and Conorms and (T,S)-L Subrings

Ümit Deniz

Abstract This study consists of two parts. In first part; It is built on the definition of Intuitionistic Fuzzy Submodules of a Module. Many researchers have used the definition of Atanassov's (Fuzzy Sets Syst 20:87–96, [1]) Intuitionistic fuzzy sets definition to move the definitions in classical algebra to intuitionistic fuzzy algebra. Davvaz et al. (Inf Sci 176:1447–1454, [2]) defined the Intuitionistic fuzzy submodules of a module. They used minimum and maximum operations to give that definition. In this study we replace minimum operation with triangular norms and maximum operation with triangular conorms for giving the definition of Intuitionistic (T, S)-fuzzy submodule of a module. By using this definition, we move some definition and theorems in classical algebra to Intuitionistic fuzzy algebra. In the second part It is built on the definition of intuitionistic L-fuzzy rings and ideals. Many researchers have used the definition of Atanassov's (Fuzzy Sets Syst 20:87–96, [1]) intuitionistic fuzzy sets to move the definitions in classical algebra to intuitionistic fuzzy algebra (Davvaz et al. in Inf Sci 176:1447–1454, [2]; Çuvalcıoğlu et al. in Notes Intuitionistic Fuzzy Sets 20:9–16, [3]; Çuvalcıoğlu and Aykut in NIFS 20:57–61, [4]; Isaac and Pearly in Int J Math Sci Appl 1:1447–1454, [5]). When K. Atanassov gave the definition of intuitionistic fuzzy sets he used the closed interval [0, 1]. Then Meena and Thomas (Int Math Forum 6:2561–2572, [6]) replaced the closed interval [0, 1] with L-lattice. In that study they used $\wedge$ $\wedge$-infimum and $\vee$-supremum operations to give the intuitionistic L-fuzzy rings and intuitionistic L-fuzzy ideals. In this study we replace $\wedge$-infimum with triangular norms and we replace $\vee$-supremum with triangular conorms and give the definition of intuitionistic (T,S)-L fuzzy rings and ideals. By using this definitions, we move some definition and theorems in classical algebra to intuitionistic fuzzy algebra.

Keywords Triangular norms · Triangular conforms · TL-submodules · TL · Subrings · TL-ideals

---

Ü. Deniz (✉)
Recep Tayyip Erdogan University, Rize, Turkey
e-mail: umit.deniz@erdogan.edu.tr
URL: https://avesis.erdogan.edu.tr/umit.deniz.

O. Castillo and P. Melin (eds.), *Fuzzy Logic and Neural Networks for Hybrid Intelligent System Design*, Studies in Computational Intelligence 1061,
https://doi.org/10.1007/978-3-031-22042-5_3

1 Introduction

In 1965 Zadeh [7] introduced the notion of a fuzzy set as a method of representing uncertainly in real physical world. Since then this definition has been used to many mathematical branches. In 1971 Rosenfeld [8] applied the notion of fuzzy sets to algebra and introduced fuzzy subgroups of a group. Negoita and Ralesque [9] introduced and examined the notion of a fuzzy submodule of a module. Wang [10] gave the definition of TL-submodules and TL-subspaces. As a generalization of a fuzzy set, the concept of an intuitionistic fuzzy set introduced by Atanassov [1, 11]. By using this definition Davvas [2] established the intuitionistic fuzzification of the concept of submodules of an R-module. Then Isaac and John [5] studied the notion of an intuitionistic fuzzy submodules and some related properties.

The definition of intuitionistic fuzzy sets is using in set theory, group theory, topological space, knowledge engineering, neural network etc. Wang and Yu [12] gave the TL-subring and TL-ideal definitions. Meena and Thomas [6] gave the definition of intuitionistic L-fuzzy subrings. In that study they used $\wedge$-infimum and $\vee$-supremum operations to give that definition. In this study we improve the definition to intuitionistic (T,S)-L fuzzy subrings. We use t-norm instead of $\wedge$-infimum and we use conorm instead of $\vee$-supremum as operations. By using this definition, we give and prove some theorems similar to classic rings.

Throughout this paper we denote by R a commutative ring with unity 1 and by M a unitary R-module. L always represents any given complete lattice with maximal element 1 and minimal element 0.

2 Preliminaries

Definition 2.1 [13] A triangular norm (t-norm, for short) is a function.

$T : [0, 1] \times [0, 1] \to [0, 1]$ satisfying the following properties: for all $x, y, z \in [0, 1]$,

(i) existence of neutral element 1: $T(x, 1) = x$,
(ii) monotonicity: $x \leq y \Rightarrow T(x, z) \leq T(y, z)$,
(iii) commutativity; $T(x, y) = T(y, x)$,
(iv) associativity: $T(x, T(y, z)) = T(T(x, y), z)$.

Conditions (i) and (ii) imply that for any t-norm $T(x, y) \leq x$ and $T(x, y) \leq y$ hold,

thus $T(x, y) \leq x \wedge y$, for all $x, y \in [0, 1]$.

Example 2.2 [14]

(1) Standard intersection T-norm $T_m(x, y) = \min\{x, y\}$.
(2) Bounded sum T-norm $T_b(x, y) = \max\{0, x + y - 1\}$.
(3) Algebraic product T-norm $T_p(x, y) = xy$.

(4) Drastic T-norm

$$T_D(x, y) = \begin{cases} y & if x = 1, \\ x & if y = 1, \\ 0 & otherwise. \end{cases}$$

(5) Nilpotent minimum T-norm

$$T_{nM}(x, y)\begin{cases} \min\{x, y\} & if\ x + y > 1 \\ 0 & otherwise. \end{cases}$$

(6) Hamacher product T-norm

$$T_{Ho}(x, y) = \begin{cases} 0 & if\ x = y = 0, \\ \frac{xy}{x+y-xy} & otherwise. \end{cases}$$

The drastic t-norm is the pointwise smallest t-norm and the minimum is the pointwise largest t-norm: $T_D(x, y) \leq T(x, y) \leq T_{\min}(x, y)$ for all $x, y \in [0, 1]$.

Theorem 2.3 [13] *Let T be a t-norm. Then.*

$$\mathrm{T}(\mathrm{T}(x, y), \mathrm{T}(w, z)) = \mathrm{T}(\mathrm{T}(x, w), \mathrm{T}(y, z)),$$

$$\text{for all } x, y, w, z \in [0, 1].$$

Definition 2.4 [13]

(i) A t-norm T on a lattice L is called ∨-distributive if

$$T(x, y \vee z) = T(x, y) \vee T(x, z).$$

(ii) A t-norm T on a complete lattice L is called infinitely ∨-distributive if

$T\left(a, \bigvee_Q b_\tau\right) = \bigvee_Q T(a, b_\tau)$ for any subset $a, b_\tau \in L, \tau \in Q$ of L.

Theorem 2.5 [14] *Let L be a triangular norm on lattice L then.*

(i) $\mathrm{T}(x, 0) = \mathrm{T}(0, x) = 0$ for all $x \in L$,
(ii) if $a \leq b$ and $x \leq y$ then $\mathrm{T}(a, x) \leq \mathrm{T}(b, y)$,
(iii) if$\{a_i | i \in I\}, \{b_i | i \in I\} \subset L$ then $T\left(\bigwedge_{i\in I} a_i, \bigwedge_{i\in I} b_i\right) \leq \bigwedge_{i\in I} T(a_i, b_i)$.

Theorem 2.6 [15] *Let L be a complete lattice. If T is an infinitely ∨-distributive t-norm then.*

$$\bigvee_{i\in I} \bigvee_{j\in J} T(a_i, b_j) = T\left(\bigvee_{i\in I} a_i, \bigvee_{j\in J} b_j\right).$$

Definition 2.7 [14] A t-conorm is a mapping $S : [0, 1] \times [0, 1] \to [0, 1]$ satisfying. for all $x, y, z \in [0, 1]$,

(i) $S(x, 0) = x$,
(ii) $S(x, y) = S(y, x)$,
(iii) $S(x, S(y, z)) = S(S(x, y), z)$,
(iv) $S(x, y) \leq S(x, z)$whenever $y \leq z$.

Conditions (i) and (iv) imply that for any t-conorm $x \leq S(x, y)$and $y \leq S(x, y)$. hold, thus $x \vee y \leq S(x, y)$, for all $x, y \in [0, 1]$.

Example 2.8 [14] The following are the four basic t-conorms, namely, the maximum S_M, the probabilistic sum S_p, the Lukasiewicz t-conorm or (bounded sum) S_L, and the drastic sum S_D which are given by, respectively:

(i) $S_M(x, y) = \max(x, y)$,
(ii) $S_P(x, y) = x + y - x.y$,
(iii) $S_L(x, y) = \min(x + y, 1)$,
(iv) $S_D(x, y) = \begin{cases} 1 & if(x, y) \in [0, 1]^2, \\ \max(x, y) & otherwise. \end{cases}$

Definition 2.9 [13]

(i) A t-conorm S on a lattice L is called $\wedge$-distributive if

$$S(x, y \wedge z) = S(x, y) \wedge S(x, z).$$

(ii) A t-conorm S on a complete lattice L is called infinitely $\wedge$-distributive if

$S(a, \bigwedge_Q b_\tau) = \wedge_Q S(a, b_\tau)$ for any subset $a, b_\tau \in L, \tau \in Q$ of L.

Theorem 2.10 [14] *Let S be a triangular conorm on lattice L then.* if $\{a_i | i \in I\}, \{b_i | i \in I\} \subset L$ then $\bigvee_{i \in I} S(a_i, b_i) \leq S(\bigvee_{i \in I} a_i, \bigvee_{i \in I} b_i)$.

Theorem 2.11 [15] *Let L be a complete lattice. If S is an infinitely $\wedge$-distributive t-conorm then.*

$$\bigwedge_{i \in I} \bigwedge_{j \in J} S(a_i, b_j) = S\left(\bigwedge_{i \in I} a_i, \bigwedge_{j \in J} b_j\right).$$

Definition 2.12 [7] Let X be a non-empty set. A fuzzy set μ of X is a function. $\mu : X \to [0, 1]$.

Definition 2.13 [10] Let $\mu, \vartheta \in F(M, L)$ and M is a R-module. Define $\mu +_T \vartheta, -\mu \in F(M, L)$ as follows:

$$(\mu +_T \vartheta)(x) = \bigvee \{T(\mu(y), \vartheta(z)) | y, z \in M, y + z = x\}$$

$$(-\mu)(x) = \mu(-x),$$

where x is an arbitrary element in M. Then $\mu +_T \vartheta$ is called the T-sum of μ and ϑ and $-\mu$ the negative of μ. When $T = \wedge$, $\mu +_T \vartheta$ is referred to as the sum of μ *and* ϑ and can be written as $\mu + \vartheta$.

Clearly the operation $+_T$ of L-subsets of M satisfies the associative and commutative laws. Hence, we can consider $\mu_1 +_T \mu_2 +_T \cdots +_T \mu_n$ where $\mu_i \in F(M, L)$, $1 \le i \le n$ and $n \in N$ and abbreviate it as $\sum_{i=1}^{n(T)} \mu_i$.

Definition 2.14 [10] Let $r \in R$ *and* $\mu \in F(M, L)$ and M is a R-module. Define $r\mu \in F(M, L)$ as follows:

$$(r\mu)(x) = \bigvee\{\mu(y) | y \in M, ry = x\} \quad \forall x \in M.$$

Then $r\mu$ is called the product of the scalar r and the L-subset μ, or simply the scalar product of r and μ.

Definition 2.15 [10] Let M be a R-module. By a TL-submodule of M, we mean an L-subset μ of M which satisfies the following conditions.

$$(M1)\mu(\theta) = 1,$$

$$(M2)\mu(rx) \ge \mu(x) \forall r \in R \text{ and } x \in M,$$

$$(M3)_T \mu(x + y) \ge T(\mu(x), \mu(y)) \forall x, y \in M.$$

When $T = \wedge$, a TL-submodule of M is called an L-submodule of M.

As usual when $L = [0, 1]$, aTL-submodule and an L-submodule are referred to as a T-fuzzy module and a fuzzy submodule, respectively. The set of all TL-submodules of M and the set of all L-submodules of M are denoted by $TL(M)$ and $L(M)$, respectively.

Theorem 2.16 [10] *Let* $\mu \in F(M, L)$. *Then* $\mu \in TL(M)$ *if and only if* μ *satisfies condition (M1) and the following condition*:
$(M4)_T \mu(rx + sy) \ge T(\mu(x), \mu(y)) \forall r, s \in R$ and $x, y \in M$.

Definition 2.17 [6] An intuitionistic fuzzy set (in short IFS) A of a non-empty set X is an object having the form.

$$\mathrm{A} = \{\langle \mathrm{x}, \mu_{\mathrm{A}}(\mathrm{x}), \nu_{\mathrm{A}}(\mathrm{x})\rangle | \mathrm{x} \in \mathrm{X}\}$$

where the functions $\mu_A: X \to [0,1]$ and $\nu_A: X \to [0,1]$ denote respectively the degree of membership (namely $\mu_A(x)$) and the degree of non-membership (namely $\nu_A(x)$) each of element $x \in X$ to the set A, and $0 \le \mu_A(x) + \nu_A(x) \le 1$ for all $x \in X$.

For the sake of simplicity, we will denote the set of all IFS's of X as IFS(X).

Definition 2.18 [6] Let X be a non-empty set and $A = (\mu_A, \nu_A)$, $B = (\mu_B, \nu_B)$ be IFS's of X. Then for A and B some relations and operations can be defined as follows;

(1) $A \subseteq B$ iff for all $x \in X$, $\mu_A(x) \leq \mu_B(x)$and $\nu_A(x) \geq \nu_B(x)$,
(2) $A = B$ iff for all $x \in X$, $\mu_A(x) = \mu_B(x)$and $\nu_A(x) = \nu_B(x)$,
$A \cup B = \{\langle x, (\mu_A \cup \mu_B)(x), (\nu_A \cap \nu_B)(x)\rangle | x \in X\}$
(3) where $(\mu_A \cup \mu_B)(x) = \mu_A(x) \vee \mu_B(x)$,
$(\nu_A \cap \nu_B)(x) = \nu_A(x) \wedge \nu_B(x)$,
$A \cap B = \{\langle x, (\mu_A \cap \mu_B)(x), (\nu_A \cup \nu_B)(x)\rangle | x \in X\}$
(4) where $(\mu_A \cap \mu_B)(x) = \mu_A(x) \wedge \mu_B(x)$,
$(\nu_A \cup \nu_B)(x) = \nu_A(x) \vee \nu_B(x)$,
$A + B = \left\{\langle x, \mu_{A_{+B}}(x), \nu_{A_{+B}}(x)\rangle | x \in R\right\}$
(5) where $\mu_{A+B}(x) = \vee\{\mu_A(y) \wedge \mu_B(z) | y, z \in R, y + z = x\}$
and $\nu_{A+B}(x) = \wedge\{\nu_A(y) \vee \nu_B(z) | y, z \in R, y + z = x\}$,
(6) $A^C = (\nu_A, \mu_A)$.

Definition 2.19 [16] Let $A_i, i \in J$ be an arbitrary family of IFS's in X, where. $A_i = (\mu_{Ai}, \nu_{Ai})$ for each $i \in J$. Then

(i)

$$\bigcap_{i \in J} A_i = \left(\mu\left(\bigcap_{i \in J} A_i\right), \nu\left(\bigcap_{i \in J} A_i\right)\right) = \left\{\left(x, \bigwedge_{i \in J} \mu_{A_i}(x), \bigvee_{i \in J} \nu_{A_i}(x)\right) : x \in X\right\}$$

(ii)

$$\bigcup_{i \in J} A_i = \left(\mu\left(\bigcup_{i \in J} A_i\right), \nu\left(\bigcup_{i \in J} A_i\right)\right) = \left\{\left(x, \bigvee_{i \in J} \mu_{A_i}(x), \bigwedge_{i \in J} \nu_{A_i}(x)\right) : x \in X\right\}$$

Definition 2.20 [12] Let R be a ring and $\mu : R \to L, \vartheta : R \to L$. Define $\mu +_T \vartheta, -\mu, \mu -_T \vartheta, \mu ._T \vartheta \in F(R, L)$ as follows.

$$(\mu +_T \vartheta)(x) = \bigvee\{T(\mu(y), \vartheta(z)) | y, z \in R, y + z = x\}$$

$$(-\mu)(x) = \mu(-x)$$

$$(\mu -_T \vartheta)(x) = \bigvee\{T(\mu(y), \vartheta(z)) | y, z \in R, y - z = x\}$$

$$(\mu ._T \vartheta)(x) = \bigvee\{T(\mu(y), \vartheta(z)) | y, z \in R, y.z = x\}$$

where x is any element of R. $\mu +_T \vartheta$, $\mu -_T \vartheta$, $\mu ._T \vartheta$ are called the T-sum, T-difference, and T-product of μ *and* ϑ, respectively, and $-\mu$ is called the negative of μ.

Definition 2.21 [12] Let R be a ring. We call $\mu : R \to L$ L-subset of R TL-subring if it satisfies the following conditions.

$$(\text{R1}) \quad \mu(0) = 1$$

$$(\text{R2}) \quad \mu(-x) \geq \mu(x) \quad \forall x \in R$$

$$(\text{R3}) \quad \mu(x + y) \geq T(\mu(x), \mu(y)) \quad \forall x, y \in R$$

$$(\text{R4}) \quad \mu(xy) \geq T(\mu(x), \mu(y)) \quad \forall x, y \in R$$

In particular, a TL-subring is simply called an L-subring when$T = \wedge$. The set of all TL-subrings of R and the set of all L-subrings of R are denoted by the symbols TL(R) and L(R), respectively.

Definition 2.22 [12] Let $\mu : R \to L$ and μ satisfy conditions (R1), (R2) and (R3). Then μ is called a TL-left ideal of R if it also satisfies the condition.

$$(R5)_l \quad \mu(xy) \geq \mu(y) \quad \forall x, y \in R;$$

a TL-right ideal of R if it is also satisfying the condition

$$(R5)_r \quad \mu(xy) \geq \mu(x) \quad \forall x, y \in R;$$

and a TL-two sided ideal or TL-ideal of R if it is also satisfying the condition

$$(R5)\mu(xy) \geq \mu(x) \vee \mu(y) \quad \forall x, y \in R.$$

In particular, when$T = \wedge$, a TL-left ideal, TL-right ideal and TL-ideal of R are referred to as an L-left ideal, L-right ideal and L-ideal of R, respectively we denote by $\text{TLI}_l(\text{R})$, $\text{TLI}_r(\text{R})$ and TLI(R), respectively, the set of all TL-left ideals of R, the set of all TL-right ideals of R and the set of all TL-ideals of R.

3 Intuitionistic (T,S)-Fuzzy Submodules

In this section we denote by R a commutative ring with unit 1 and by M a unitary R-module.

Definition 3.1 Let M be a module over a ring R and let T and S be respectively a t-norm and a t-conorm on [0,1] and an intuitionistic fuzzy subset $A = \{\langle x, \mu_A(x), \nu_A(x)\rangle | x \in X\}$ of M is said to be an intuitionistic (T,S)-fuzzy submodule of M (ITSFSM) if for all x, y ∈ M and r ∈ R,

(i) $\mu_A(0) = 1$ and $\nu_A(0) = 0$,
(ii) $\mu_A(x + y) \geq T(\mu_A(x), \mu_A(y))$,
(iii) $\mu_A(rx) \geq \mu_A(x)$,
(iv) $\nu_A(x + y) \leq S(\nu_A(x), \nu_A(y))$,
(v) $\nu_A(rx) \leq \nu_A(x)$.

Definition 3.2 [5] Let $A = (\mu_A, \nu_A)$, $B = (\mu_B, \nu_B)$ be IFS's of M. Then the *(T, S)*—intersection of A and B and the *(T, S)*—sum of A and B can be defined as follows respectively.

(1) $A \sqcap B = \{\langle x, (\mu_A \sqcap \mu_B)(x), (\nu_A \sqcup \nu_B)(x)\rangle \mid x \in X\}$ where $(\mu_A \sqcap \mu_B)(x) = T(\mu_A(x), \mu_B(x))$, $(\nu_A \sqcup \nu_B)(x) = S(\nu_A(x), \nu_B(x))$
(2) $A \oplus B = \{\langle x, (\mu_{A\oplus B})(x), (\nu_{A\oplus B})(x)\rangle \mid x \in M\}$ where $(\mu_{A\oplus B})(x) = \vee\{T(\mu_A(y), \mu_B(z)) \mid y, z \in M, y + z = x\}$ and $(\nu_{A\oplus B})(x) = \wedge\{S(\nu_A(y), \nu_B(z)) \mid y, z \in M, y + z = x\}$.

Definition 3.3 [5] Let $A = (\mu_A, \nu_A)$ be IFS's of M. Then.
$-A = (\mu_{-A}, \nu_{-A})$, an IFS in M is defined as.
$-A = \{(x, \mu_{-A}(x), \nu_{-A}(x)) : x \in M\}$ where,

$$\mu_{-A}(x) = \mu_A(-x) \text{and } \nu_{-A}(x) = \nu_A(-x), \forall x \in M.$$

Definition 3.4 [5] For an IFS $A = (\mu_A, \nu_A)$ in M and for any $r \in R$, we define the IFS $rA = (\mu_{rA}, \nu_{rA})$ in M as $rA = \{(x, \mu_{rA}(x), \nu_{rA}(x)) : x \in M\}$ where for each $x \in M$, $\mu_{rA}(x) = \vee\{\mu_A(y) : y \in M, x = ry\}$ and $\nu_{rA}(x) = \wedge\{\nu_A(y) : y \in M, x = ry\}$.

Proposition 3.5 [5] *If* $A = (\mu_A, \nu_A)$ *is an IFS in a R-module M, then* $1.A = A$ *and* $(-1).A = -A$.

Proposition 3.6 [5] *If* $A = (\mu_A, \nu_A)$, $B = (\mu_B, \nu_B)$ *be IFS's of M with* $A \subseteq B$, *then* $rA \subseteq rB$ *for any* $r \in R$.

Proposition 3.7 [5] *If* $A = (\mu_A, \nu_A)$ *is an IFS in M, then* $r(sA) = (rs)A$ *for any* $r, s \in R$.

Theorem 3.8 *Let* $A = (\mu_A, \nu_A)$, $B = (\mu_B, \nu_B)$ *be IFS's of M and T be an infinitely* $\vee$*-distributive t-norm and S be an infinitely* $\wedge$*-distributive t-conorm then,*
$r(A \oplus B) = rA \oplus rB$ for any $r \in R$.

Proof We have $r(A \oplus B) = (\mu_{r(A \oplus B)}, \nu_{r(A \oplus B)})$ and $rA \oplus rB = (\mu_{rA \oplus rB}, \nu_{rA \oplus rB})$.
Now,

$$\begin{aligned}\mu_{r(A\oplus B)}(x) &= \vee\{\mu_{(A\oplus B)}(y) : y \in M, x = ry\} \\ &= \vee\{\vee\{T(\mu_A(z_1), \mu_B(z_2)) : z_1, z_2 \in M, y = z_1 + z_2\} : y \in M,\ x = ry\} \\ &= \vee\{T(\mu_A(z_1), \mu_B(z_2)) : z_1, z_2 \in M, x = rz_1 + rz_2\}\end{aligned}$$

$$= \vee\{T(\vee\{\mu_A(z_1) : z_1 \in M, x_1 = rz_1\}, \vee\{\mu_B(z_2) : z_2 \in M, x_2 = rz_2\}) : x_1 + x_2 = x\}$$
$$= \vee\{T(\mu_{rA}(x_1), \mu_{rB}(x_2)) : x_1, x_2 \in M, x_1 + x_2 = x\}$$
$$= \mu_{rA+rB}(x), \forall x \in M.$$

Similarly we can show that $\nu_{r(A+B)}(x) = \nu_{rA+rB}(x), \forall x \in M$. Hence $r(A \oplus B) = rA \oplus rB$.

Proposition 3.9 [5] *If $A = (\mu_A, \nu_A)$ is an IFS in M, then.*
$\mu_{rA}(rx) \geq \mu_A(x)$ and $\nu_{rA}(rx) \leq \nu_A(x), \forall x \in M$.

Proposition 3.10 [5] *Let $A = (\mu_A, \nu_A), B = (\mu_B, \nu_B)$ be IFS's of M then,*

(1) $\mu_B(rx) \geq \mu_A(x), \forall x \in M \Leftrightarrow \mu_{rA} \subseteq \mu_B$,
(2) $\nu_B(rx) \leq \nu_A(x), \forall x \in M \Leftrightarrow \nu_{rA} \supseteq \nu_B$.

Theorem 3.11 *Let $A = (\mu_A, \nu_A)$, $B = (\mu_B, \nu_B)$ be IFS's of M and T be an infinitely $\vee$-distributive t-norm and S be an infinitely $\wedge$-distributive t-conorm then,*

(1) $\mu_{rA\oplus rB}(rx + sy) \geq T(\mu_A(x), \mu_B(y))$,
(2) $\nu_{rA\oplus rB}(rx + sy) \leq S(\nu_A(x), \nu_B(y)), \forall x, y \in M, r, s \in R$.

Proof

(1) $\mu_{rA\oplus sB}(rx + sy) = \vee\{T(\mu_{rA}(z_1), \mu_{sB}(z_2)): z_1, z_2 \in M, z_1 + z_2 = rx + sy\} \geq T(\mu_{rA}(rx), \mu_{sB}(sy)) \geq T(\mu_A(x),\mu_B(y)) \forall x, y \in M, r, s \in R$.
(2) Similarly we have,

$$\nu_{rA\oplus sB}(rx + sy) = \wedge\{ S(\nu_{rA}(z_1), \nu_{sB}(z_2)): z_1, z_2 \in M, z_1 + z_2 = rx + sy\}.$$
$$\leq S(\nu_{rA}(rx), \nu_{sB}(sy)) \leq S(\nu_A(x), \nu_B(y)) \forall x, y \in M, r, s \in R$$

Theorem 3.12 *Let $A = (\mu_A, \nu_A), B = (\mu_B, \nu_B)$ and $C = (\mu_C, \nu_C)$ be IFS's of M and T be an infinitely $\vee$-distributive t-norm and S be an infinitely $\wedge$-distributive t-conorm then, $\forall r,s \in R$,*

(1) $\mu_C(rx + sy) \geq T(\mu_A(x), \mu_B(y)) \forall x, y \in M \Leftrightarrow \mu_{rA\oplus sB} \subseteq \mu_C$,
(2) $\nu_C(rx + sy) \leq S(\nu_A(x), \nu_B(y)) \forall x, y \in M \Leftrightarrow \nu_{rA\oplus sB} \supseteq \mu_C$

Proof

(1) Suppose we have $\mu_C(rx + sy) \geq T(\mu_A(x),\mu_B(y)) \forall x, y \in M$. Then,

$$\mu_{rA\oplus sB}(z) = \vee\{T(\mu_{rA}(z_1), \mu_{sB}(z_2)) : z_1, z_2 \in M, z_1 + z_2 = z\}$$
$$= \vee\{T(\vee\{\mu_A(x) : x \in M, z_1 = rx\}, \vee\{\mu_B(y) : y \in M, z_2 = sy\}) \; z_1, z_2 \in M, z_1 + z_2 = z\}$$
$$= \vee\{T(\mu_A(x), \mu_B(y)) : x, y \in M, rx + sy = z\}$$
$$\leq \vee\{\mu_C(rx + sy) : x, y \in M, rx + sy = z\}$$
$$\leq \mu_C(z), \forall z \in M.$$

Thus $\mu_{rA\oplus sB} \subseteq \mu_C$. Now conversely suppose $\mu_{rA\oplus sB} \subseteq \mu_C$. Then $\mu_C(rx + sy) \geq \mu_{rA\oplus sB}(rx + sy) \geq T(\mu_A(x),\mu_B(y))$ by Theorem 3.11.

(2) Similarly suppose

$\nu_C(rx + sy) \leq S(\nu_A(x), \nu_B(y)), \forall x, y \in M$. Then,

$$\begin{aligned}\nu_{rA\oplus sB}(z) &= \wedge\{S(\nu_{rA}(z_1), \nu_{sB}(z_2)) : z_1, z_2 \in M, z_1 + z_2 = z\}\\ &= \wedge\{S(\wedge\{\nu_A(x), x \in M, z_1 = rx\}, \wedge\{\nu_B(y) : y \in M, z_2 = sy\})\\ &\quad : z_1, z_2 \in M, z_1 + z_2 = z\}\\ &= \wedge\{S(\nu_A(x), \nu_B(y)) : x, y \in M, rx + sy = z\}\\ &\geq \wedge\{\nu_C(rx + sy) : x, y \in M, rx + sy = z\}\\ &\geq \nu_C(z), \forall z \in M. \text{Thus } \nu_{rA\oplus sB} \supseteq \nu_C.\end{aligned}$$

Now conversely suppose $\nu_{rA\oplus sB} \supseteq \nu_C$. Then $\nu_C(rx + sy) \leq \nu_{rA\oplus sB}(rx + sy) \leq S(\nu_A(x), \nu_B(y))\ \forall x, y \in M$ by Theorem 3.11.

Proposition 3.13 [5] *If $A_i = (\mu_{Ai}, \nu_{Ai}), i \in J$, is a collection of IFS's on R-module M, then,*

$$r\left(\bigcup_{i\in J} Ai\right) = \bigcup_{i\in J} rA_i \quad \forall r \in R.$$

Proposition 3.14 [5] *Let $r \in R$ and $A = (\mu_A, \nu_A)$ be an IFS in M. Then,*

(1) $\mu_{rA} \subseteq \mu_A \Leftrightarrow \mu_A(rx) \geq \mu_A(x)$,
(2) $\nu_{rA} \supseteq \nu_A \Leftrightarrow \nu_A(rx) \leq \nu_A(x) \forall x \in M$.

Proposition 3.15 *Let $r, s \in R$ and $A = (\mu_A, \nu_A)$ be an IFS in M. Then,*

(1) $\mu_{rA\oplus sA} \subseteq \mu_A \Leftrightarrow \mu_A(rx + sy) \geq T(\mu_A(x), \mu_A(y)$,
(2) $\nu_{rA\oplus sA} \supseteq \nu_A \Leftrightarrow \nu_A(rx + sy) \leq S(\nu_A(x), \nu_A(y), \forall x, y \in M$.

Proof Proof follows from Theorem 3.11 and Proposition 3.10.

Theorem 3.16 *Let $A = (\mu_A, \nu_A)$ and $B = (\mu_B, \nu_B)$ be IFS's of M and T be an infinitely $\vee$-distributive t-norm and S be an infinitely $\wedge$-distributive t-conorm. Then A is an ITSFSM of M if and only if A satisfies the following conditions.*

(1) $\mu_A(0) = 1, \nu_A(0) = 0$
(2) $\mu_A(rx + sy) \geq T(\mu_A(x), \mu_A(y))$and
$\nu_A(rx + sy) \leq S(\nu_A(x), \nu_A(y)) \forall x, y \in M, r, s \in R$.

Proof Let $A = (\mu_A, \nu_A)$ be an IFS in M. Then obviously $\mu_A(0) = 1$ and $\nu_A(0) = 0$. Now,

$$\mu_A(rx + sy) \geq T(\mu_A(rx), \mu_A(sy)) \geq T(\mu_A(x), \mu_A(y)) \text{ and}$$

$$\mu_A(0) = 1 \text{ and } \nu_A(0) = 0.$$

Now, $\nu_A(rx + sy) \leq S(\nu_A(rx), \nu_A(sy)) \leq S(\nu_A(x), \nu_A(y))\ \forall x, y \in M, r, s \in R$.

Conversely suppose A satisfies the conditions (1) and (2). Then we have, $\mu_A(0) = 1$ and $\nu_A(0) = 0$, $\mu_A(x + y) = \mu_A(1.x + 1.y) \geq T(\mu_A(x), \mu_A(y))$,

$$\nu_A(x + y) = \nu_A(1.x + 1.y) \leq S(\nu_A(x), \nu_A(y)),$$

$$\mu_A(rx) = \mu_A(rx + r0) \geq T(\mu_A(x), \mu_A(0)) = \mu_A(x) \text{and}$$

$$\nu_A(rx) = \nu_A(rx + r0) \leq S(\nu_A(x), \nu_A(0)) = \nu_A(x) \forall x, y \in M, r \in R.$$

Hence $A = (\mu_A, \nu_A)$ is an ITSFSM of M.

Theorem 3.17 *If $A = (\mu_A, \nu_A)$ and $B = (\mu_B, \nu_B)$ be ITSFSM of M, then $A \sqcap B$ is ITSFSM of M.*

Proof $(\mu_A \sqcap \mu_B)(0) = T(\mu_A(0), \mu_B(0)) = 1$ and $(\nu_A \sqcup \nu_B)(0) = S(\nu_A(0), \nu_B(0)) = 0$.

Now $(\mu_A \sqcap \mu_B)(x + y) = T(\mu_A(x + y), \mu_B(x + y)) \geq T(T(\mu_A(x), \mu_A(y)), T(\mu_B(x), \mu_B(y)))$.

$= T(T(\mu_A(x), \mu_B(x)), T(\mu_A(y), \mu_B(y))) = T((\mu_A \sqcap \mu_B)(x), (\mu_A \sqcap \mu_B)(y))$.

Similarly, we can show that.

$(\nu_A \sqcup \nu_B)(x + y) \leq S((\nu_A \sqcup \nu_B)(x), (\nu_A \sqcup \nu_B)(y)) \forall x, y \in M$.

Further $(\mu_A \sqcap \mu_B)(rx) = T(\mu_A(rx), \mu_B(rx)) \geq T(\mu_A(x), \mu_B(x)) = (\mu_A \sqcap \mu_B)(x)$ $\forall x \in M, r \in R$.

Similarly we get $(\nu_A \sqcup \nu_B)(rx) \leq (\nu_A \sqcup \nu_B)(x)\ \forall x \in M, r \in R$.

Hence $A \sqcap B$ is an ITSFSM of M.

4 Intuitionistic (T, S)-L Fuzzy Subrings

Definition 4.1 Let T and S be respectively a t-norm and a t-conorm on L and an intuitionistic L fuzzy subset $A = \{x, \mu_A(x), \nu_A(x) | x \in X\}$ of R is said to be an intuitionistic (T,S)-L fuzzy subring of R (ITSLFSR) if for all $x, y \in R$,

(i) $\mu_A(0) = 1 and \nu_A(0) = 0$,
(ii) $\mu_A(x - y) \geq T(\mu_A(x), \mu_A(y))$,
(iii) $\mu_A(xy) \geq T(\mu_A(x), \mu_A(y))$,
(iv) $\nu_A(x - y) \leq S(\nu_A(x), \nu_A(y))$,
(v) $\nu_A(xy) \leq S(\nu_A(x), \nu_A(y))$.

Lemma 4.2 *If* $A = \{x, \mu_A(x), \nu_A(x) | x \in X\}$ *is an ITSLFSR then.*

$$\mu_A(0) \geq \mu_A(x) \ and \ \nu_A(0) \leq \nu_A(x) for \ all \ x \in R.$$

Definition 4.3 Let T and S be respectively a t-norm and a t-conorm on L and an intuitionistic L fuzzy subset $A = \{x, \mu_A(x), \nu_A(x) | x \in X\}$ of R is said to be an intuitionistic (T,S)-L fuzzy ideal of R (ITSLFI) if for all x, y ∈ R,

(i) $\mu_A(0) = 1 \ and \ \nu_A(0) = 0$,
(ii) $\mu_A(x - y) \geq T(\mu_A(x), \mu_A(y))$,
(iii) $\mu_A(xy) \geq \mu_A(x)$,
(iv) $\nu_A(x - y) \leq S(\nu_A(x), \nu_A(y))$,
(v) $\nu_A(xy) \leq \nu_A(x)$.

Theorem 4.4 *Every ITSLFI is an ITSLFSR.*

Proof Let $A = \{x, \mu_A(x), \nu_A(x) | x \in X\}$ be an ITSLFI then we proof the conditions of ITSLFSR.

The conditions (i), (ii) and (iv) of ITSLFSR are equivalent the conditions (i), (ii) and (iv) of ITSLFI. So we only proof the conditions (iii) and (v).

(iii) $\mu_A(xy) \geq \mu_A(x)$, and because of R is a commutative ring,
$\mu_A(xy) = \mu_A(yx) \geq \mu_A(x)$. Hence $\mu_A(xy) \geq T(\mu_A(x), \mu_A(y))$.
(v) $\nu_A(xy) \leq \nu_A(x)$ and because of R is a commutative ring,
$\nu_A(xy) = \nu_A(yx) \leq \nu_A(y)$. Hence $\nu_A(xy) \leq S(\nu_A(x), \nu_A(y))$.

Theorem 4.5 *Let* $A = \{x, \mu_A(x), \nu_A(x) | x \in X\}$ *and* $B = \{x, \mu_B(x), \nu_B(x) | x \in X\}$ *be two ITSLFI. Then* $A \cap B$ *is an ITSLFI of R.*

Proof Let's proof the conditions of ITSLFI.

(i)
$$(\mu_A \cap \mu_B)(0) = \mu_A(0) \wedge \mu_B(0) = 1 \wedge 1 = 1,$$
$$(\nu_A \cup \nu_B)(0) = \nu_A(0) \vee \nu_B(0) = 0 \vee 0 = 0.$$

(ii) Let $x, y \in R$

$$\begin{aligned}(\mu_A \cap \mu_B)(x - y) &= \mu_A(x - y) \wedge \mu_B(x - y) \\ &\geq T(\mu_A(x), \mu_A(y)) \wedge T(\mu_B(x), \mu_B(y)) \\ &\geq T(\mu_A(x) \wedge \mu_B(x), \mu_A(y) \wedge \mu_B(y)) \\ &= T((\mu_A \cap \mu_B)(x), (\mu_A \cap \mu_B)(y)).\end{aligned}$$

(iii) Let $x, y \in R$

$$(\mu_A \cap \mu_B)(xy) = \mu_A(xy) \wedge \mu_B(xy) \geq \mu_A(x) \wedge \mu_B(x) = (\mu_A \cap \mu_B)(x)$$

(iv) Let $x, y \in R$

$$
\begin{aligned}
(\nu_A \cup \nu_B)(x-y) &= \nu_A(x-y) \vee \nu_B(x-y) \\
&\leq S(\nu_A(x), \nu_A(y)) \vee S(\nu_B(x), \nu_B(y))
\end{aligned}
$$

$$
\leq S(\nu_A(x) \vee \nu_B(x), \nu_A(y) \vee \nu_B(y)) = S((\nu_A \cup \nu_B)(x), (\nu_A \cup \nu_B)(y)).
$$

(v) Let $x, y \in R$

$$
(\nu_A \cup \nu_B)(xy) = \nu_A(xy) \vee \nu_B(xy) \leq \nu_A(x) \vee \nu_B(x) = (\nu_A \cup \nu_B)(x).
$$

Theorem 4.6 *Let* $A = \{\langle x, \mu_A(x), \nu_A(x) | x \in X \rangle\}$ *and.* $B = \{x, \mu_B(x), \nu_B(x) | x \in X\}$ *be two ITSLFI. If T is an infinitely* $\vee$*-distributive t-norm and S is an infinitely* $\wedge$*- distributive conorm then* $A + B$ *is an ITSLFI of* R.

Proof Lets proof the conditions of ITSLFI.

(i)

(for $y = 0$ and $z = 0$ $\mu_A(0) \wedge \mu_B(0) = 1 \wedge 1 = 1$ and $\nu_A(0) \vee \nu_B(0) = 0 \vee 0 = 0$ then)

$$(\mu_A + \mu_B)(0) = \vee\{\mu_A(y) \wedge \mu_B(z) | y + z = 0\} = 1$$

$$(n_A + n_B)(0) = \wedge\{n_A(y) \wedge n_B(z) | y + z = 0\} = 0$$

(ii)

$$
\begin{aligned}
(\mu_A + \mu_B)(x - x') &= \vee\{\mu_A(y) \wedge \mu_B(z) | y + z = x - x'\} \\
&\geq \vee\{\mu_A(y' - y'') \wedge \mu_B(z' - z'') | y' - y'' + z' - z'' = x - x'\} \\
&\geq \vee\{T(\mu_A(y'), \mu_A(y'')) \wedge T(\mu_B(z'), \mu_B(z''))\} \\
&\geq \vee\{T(\mu_A(y')) \wedge \mu_B(z'), T(\mu_A(y'') \wedge \mu_B(z''))\} \\
&= T(\vee\{\mu_A(y') \wedge \mu_B(z') | y' + z' = x\}, \\
&\quad \vee\{\mu_A(y'') \wedge \mu_B(z'') | y'' + z'' = x'\}) \\
&= T((\mu_A + \mu_B)(x), (\mu_A + \mu_B)(x')).
\end{aligned}
$$

(iii)

$$
\begin{aligned}
(\mu_A + \mu_B)(xx') &= \vee\{\mu_A(y) \wedge \mu_B(z) | y + z = xx'\} \\
&\geq \vee\mu_A(y'x') \wedge \mu_B(z') | y' + z' = x \\
&= (\mu_A + \mu_B)(x)
\end{aligned}
$$

(iv)

$$
\begin{aligned}
(v_A + v_B)(x - x') &= \wedge\{\mu_A(y) \vee \mu_B(z) | y + z = x - x'\} \\
&\le \wedge\{v_A(y' - y'') \vee v_B(z' - z'') | y' - y'' + z' - z'' = x - x'\} \\
&\le \wedge\{S(v_A(y'), v_A(y'')) \vee S(v_B(z'), v_B(z'')) | (y' + z') \\
&\quad -(y'' + z'') = x - x'\} \\
&\le S\{\wedge\{v_A(y') \vee v_B(z')) | y' + z' = x\}), \\
&\quad \wedge \{v_A(y'') \vee v_B(z'') | y'' + z'' = x'\}) \\
&= S((v_A + v_B)(x), (v_A + v_B)(x'))
\end{aligned}
$$

(v)

$$
\begin{aligned}
(v_A + v_B)(xx') &= \wedge\{\mu_A(y) \vee \mu_B(z) | y + z = xx'\} \\
&\le' \wedge\{v_A(y'x') \vee v_B(z'x') | y'x' + z'x' = x - x'\} \\
&\le \wedge\{(v_A(y') \vee v_B(z') | y' + z' = x\} = (v_A + v_B)(x)
\end{aligned}
$$

Definition 4.7 [6] Let f: R $\rightarrow$ S be a ring homomorphism. Let $A = \{x, \mu_A(x), \nu_A(x) | x \in X\}$ and $B = \{x, \mu_B(x), \nu_B(x) | x \in X\}$ be two ILFS. Then $C = \{y, f(\mu_A)(y), f(\nu_A)(y) | y \in S\}$ is called intuitionistic image of A, where.

$$
f(\mu_A)(y) = \begin{cases} \bigvee\{\mu_A(x) | x \in R, f(x) = y\} & iff^{-1}(y) \neq \emptyset, \\ 0 & otherwise. \end{cases}
$$

$$
f(\nu_A)(y) = \begin{cases} \bigwedge\{\nu_A(x) | x \in R, f(x) = y\} & iff^{-1}(y) \neq \emptyset, \\ 0 & otherwise. \end{cases}
$$

for all y $\in$ S;

and $D = \{x, f^{-1}(\mu_B)(x), f^{-1}(\nu_B)(x) | x \in R\}$ is called intuitionistic inverse image of B, where $f^{-1}(\mu_B)(x) = \mu_B(f(x)) f^{-1}(\nu_B)(x) = \nu_B(f(x))$ for all x $\in$ R.

Here $f(\mu_A)$ and $f(\nu_A)$ are called the image of μ_A and ν_A under f. Also $f^{-1}(\mu_B)$ and $f^{-1}(\nu_B)$ are called the inverse of μ_B and ν_B under f.

Theorem 4.8 *Let* $A = \{x, \mu_A(x), \nu_A(x) | x \in X\}$ *be ban ITSLFI of R and* $B = \{x, \mu_B(x), \nu_B(x) | x \in X\}$ *be an ITSLFI of S. If T is an infinitely* $\vee$*-distributive t-norm and S is an infinitely* $\wedge$*-distributive conorm then* $C = \{y, f(\mu_A)(y), f(\nu_A)(y) | y \in S\}$ *is and ITSLFI of S and* $D = \{x, f^{-1}(\mu_B)(x), f^{-1}(\nu_B)(x) | x \in R\}$ *is an ITSLFI of R.*

Proof First let proof for C the ITSLFI conditions.

(i) Since f is a ring homomorphism f(0) = 0 then $f^{-1}(0) \neq \emptyset$;
since $\mu_A(0) = 1$ and $\nu_A(0) = 0$ then;

$$f(\mu_A)(0) = \bigvee\{\mu_A(x)|x \in R, f(x) = 0\} = 1,$$

$$f(\nu_A)(0) = \bigwedge\{\nu_A(x)|x \in R, f(x) = 0\} = 0.$$

(ii) for $f^{-1}(y - y') = \emptyset f^{-1}(y) = \emptyset and f^{-1}(y') = \emptyset$ because if both them don't be an empty set then $f(x) = y, f(x') = y' \Rightarrow f(x) - f(x') = y - y' \Rightarrow f(x - x') = y - y'$ and there we obtain $f^{-1}(y - y') \neq \emptyset$.

if one of them is empty and the other isn't then we obtain.

$f(\mu_A)(y - y') \geq T(f(\mu_A)(y), f(\mu_A)(y')) \geq 0 \geq T(0, a) = 0$ the inequality is satisfied. Now we consider that $f^{-1}(y - y') \neq \emptyset$.

$$\begin{aligned} f(\mu_A)(y - y') &= \bigvee\{\mu_A(x)|f(x) = y - y'\} \\ &\geq \bigvee\{\mu_A(x' - x'')|f(x' - x'') = y - y'\} \\ &= \bigvee\{\mu_A(x' - x'')|f(x') - f(x'') = y - y'\} \\ &\geq T\left(\bigvee\{\mu_A(x')|f(x') = y\}, \bigvee\{\mu_A(x'')|f(x'') = y'\}\right) \\ &= T(f(\mu_A)(y), f(\mu_A)(y')). \end{aligned}$$

(iii) for $f^{-1}(yy') = \emptyset f^{-1}(y) = \emptyset and f^{-1}(y') = \emptyset$ because if both them don't be an empty set then $f(x) = y, f(x') = y' \Rightarrow f(x)f(x') = yy' \Rightarrow f(xx') = yy'$ and there we obtain $f^{-1}(yy') \neq \emptyset$. Now we consider that $f^{-1}(yy') \neq \emptyset$.

$$f(\mu_A)(yy') = \bigvee\{\mu_A(x)|f(x) = yy'\} \geq \bigvee\{\mu_A(x'x'')|f(x'x'') = yy'\}$$

$$= \bigvee\{\mu_A(x'x'')|f(x')f(x'') = yy'\} \geq \bigvee\{\mu_A(x')|f(x') = y\} \geq f(\mu_A)(y).$$

(iv) Likewise when $f^{-1}(y - y') = \emptyset$ the inequality is satisfied. Now we consider that $f^{-1}(yy') \neq \emptyset$.

$$\begin{aligned} f(\mu_A)(y - y') &= \bigvee\{\mu_A(x)|\ f(x) = y - y'\} \\ &\geq \bigvee\{\mu_A(x' - x'')|\ f(x' - x'') = y - y'\} \\ &= \bigvee\{\mu_A(x' - x'')|\ f(x') - f(x'') = y - y'\} \\ &\geq T\left(\bigvee\{\mu_A(x')|\ f(x') = y\}, \bigvee\{\mu_A(x'')|\ f(x'') = y'\}\right) \\ &= T(f(\mu_A)(y), f(\mu_A)(y')). \end{aligned}$$

(v) Likewise when $f^{-1}(yy') = \emptyset$ the inequality is satisfied. Now we consider that $f^{-1}(yy') \neq \emptyset$.

$$\begin{aligned} f(\nu_A)(yy') &= \bigwedge\{\nu_A(x) | f(x) = yy'\} \le \bigwedge\{\nu_A(x'x'') | f(x'x'') = yy'\} \\ &= \bigwedge\{\nu_A(x'x'') | f(x')f(x'') = yy'\} \le \bigwedge\{\nu_A(x') | f(x') = y\} \\ &= f(\nu_A)(y) \end{aligned}$$

We show that intuitionistic fuzzy set C is an ITSLFI. Now let proof for D then ITSLFI conditions.

(i)

$$\begin{aligned} f^{-1}(\mu_B)(0) &= \mu_B(f(0)) = \mu_B(0) = 1 \\ f^{-1}(\nu_B)(0) &= \nu_B(f(0)) = \nu_B(0) = 0, \end{aligned}$$

(ii)

$$\begin{aligned} f^{-1}(\mu_B)(x - x') &= \mu_B(f(x - x')) = \mu_B(f(x) - f(x')) \\ &\ge T(\mu_B(f(x)), \mu_B(f(x'))) \\ &= T(f^{-1}(\mu_B)(x), f^{-1}(\mu_B)(x')). \end{aligned}$$

(iii)

$$f^{-1}(\mu_B)(xx') = \mu_B(f(xx')) = \mu_B(f(x)f(x')) \ge \mu_B(f(x)) = f^{-1}(\mu_B)(x)$$

(iv)

$$\begin{aligned} f^{-1}(\nu_B)(x - x') &= \nu_B(f(x - x')) = \nu_B(f(x) - f(x')) \\ &\le S(\nu_B(f(x)), \nu_B(f(x'))) \\ &= S(f^{-1}(\nu_B)(x), f^{-1}(\nu_B)(x')). \end{aligned}$$

(v)

$$f^{-1}(\nu_B)(xx') = \nu_B(f(xx')) = \nu_B(f(x)f(x')) \le \nu_B(f(x)) = f^{-1}(\nu_B)(x).$$

References

1. K. Atanassov, Intuitionistic fuzzy sets. Fuzzy Sets Syst. **20**, 87–96 (1986). https://doi.org/10.1016/S0165-0114(86)80034-3
2. B. Davvaz, W.A. Dudek, Y.B. Jun, Inf. Sci. **176**, 285–300, 1447–1454 (2006)
3. G. Çuvalcıoğlu, S. Yılmaz, K.T. Atanassov, Matrix representation of the second type of intuitionistic fuzzy modal operators. Notes Intuitionistic Fuzzy Sets **20**(5), 9–16 (2014)

4. G. Çuvalcıoğlu, E. Aykut, An application of the intuitionistic fuzzy modal operator $E_{\alpha,\beta}$. NIFS **20**(5), 57–61 (2014)
5. P. Isaac, P.J. Pearly, On intuitionistic fuzzy submodules of a module. Int. J. Math. Sci. Appl. **1**(3), 1447–1454 (2011)
6. K. Meena, K.V. Thomas, Intuitionistic L-fuzzy subrings. Int. Math. Forum **6**, 2561–2572 (2011)
7. L.A. Zadeh, Fuzzy sets. Inf. Control **8**, 338–353 (1965)
8. A. Rosenfeld, Fuzzy groups. J. Math. Anal. Appl. **35**, 512–517 (1971)
9. C.V. Negoita, D.A. Ralesque, *Applications of Fuzzy Sets and Systems Analysis* (Birkhauser, Basel, 1975)
10. Z. Wang, TL-submodules and TL-linear spaces. Fuzzy Sets Syst. **68**, 211–225 (1994)
11. K. Atanassov, *On Intuitionistic Fuzzy Sets Theory* (Springer, Berlin, 2012)
12. Z.D. Wang, Y.D. Yu, TL-subrings and TL-ideals, part 2: generated TL-ideals. Fuzzy Sets Syst. **87**, 209–217 (1997)
13. B. De Baets, R. Mesiar, Triangular norms on product lattices. Fuzzy Sets Syst. **104**, 61–75 (1999)
14. E.P. Klement, R. Mesiar, E. Pap, *Triangular Norms* (Kluwer Academic Publishers, Dordrecht, 2000)
15. F. Karaal, D. Khadjiev, $\vee$-Distributive and infinitely $\vee$-distributive t-norms on complete lattice. Fuzzy Sets Syst. **151**, 341–352 (2005)
16. D. Coker, An introduction to intuitionistic fuzzy topological spaces. Fuzzy Sets Syst. **88**, 81–89 (1997)
17. P.K. Sharma, Homomorphism of intuitionistic fuzzy groups. Int. Math. Forum **6**(64), 3169–3178 (2011)

Fuzzy Dynamic Parameter Adaptation in the Mayfly Algorithm: Implementation of Fuzzy Adaptation and Tests on Benchmark Functions and Neural Networks

Enrique Lizarraga, Fevrier Valdez, Oscar Castillo, and Patricia Melin

Abstract Inspired by the flight behavior and mating process of mayflies, the Mayfly algorithm combines the main advantages of swarm intelligence and evolutionary algorithms, resulting in better performance than the particle swarm algorithm. So, we proposed a modification of Mayfly algorithm by applying a fuzzy parameter adapter to be able to apply this to neural network problems. We were able to observe that the fuzzy adapter improves the speed of convergence of the mayfly algorithm and when applied to a neural network for the Mackey glass series, it manages to detect the optimal number of neurons of the hidden layer for the network architecture. However, when using the Mayfly algorithm to optimize the architecture of neural networks, the results do not improve much, so we can deduce that this metaheuristic method is not recommended (for the moment) for this type of optimization, due to the fact that the root mean square error did not get below 0.001 even using the modified Mayfly algorithm with the fuzzy adapter.

Keywords Mayfly algorithm · Fuzzy logic · Dynamic parameter adaptation

1 Introduction

Optimization is a process of finding the best solution of a function (either its minimum or maximum value). A lot of real-world problems are represented as optimization problems, for the minimization of a single objective.

Inspired by the flight behavior and mating process of mayflies, the proposed algorithm combines the main advantages of swarm intelligence and evolutionary algorithms. In this work to evaluate the performance of the algorithm, 10 benchmark functions were used [1] and also neural network optimization is considered.

Ephemera is a species of insects whose young would live in the water for several years and when they mature, they will bark and fly through the air. However, these mature insects only have the purpose of reproducing since their life span is short [2].

E. Lizarraga · F. Valdez (✉) · O. Castillo · P. Melin
Tijuana Institute of Technology, Tomas Aquino, Tijuana, B.C, Mexico
e-mail: fevrier@tectijuana.mx

O. Castillo and P. Melin (eds.), *Fuzzy Logic and Neural Networks for Hybrid Intelligent System Design*, Studies in Computational Intelligence 1061,
https://doi.org/10.1007/978-3-031-22042-5_4

In fact, this offers a powerful hybrid algorithmic structure, based on the behavior of mayflies, for researchers trying to improve the performance of the particle swarm optimization (PSO) algorithm using techniques, such as crossover, since it has been shown that PSO needs some modifications to be improved [1].

Fuzzy logic is based on the fuzzy set theory proposed by Lotfi Zadeh, which helps us model human knowledge through the use of fuzzy if–then rules. Fuzzy logic provides a systematic computation for dealing with linguistic information and improves numerical computation through the use of linguistic labels stipulated by membership functions [3].

Solution search algorithms such as PSO, differential evolution (DE), genetic algorithms (GAs), and firefly algorithm (FA) have proven to be efficient in terms of speed of convergence for optimization problems, so we expect that the Mayfly algorithm, being an improvement of the PSO algorithm, to work in an efficient manner. We also expect that it will be efficient when applied to neural network optimization problems.

The structure of this work is as follows. In Sect. 2 we have the theoretical framework, where we describe the background of the Mayfly algorithm (MA), as well as the variants that have emerged from it so far. Then we have Sect. 3, in this section the problem statement is shown as well as the main hypothesis. In Sect. 4 the development of this research will be discussed, starting with the study of the impact of the parameters on the performance of the Mayfly algorithm. In Sect. 5 the design of the fuzzy adapter will be shown where we start with a fuzzy adapter with a single input and a single output. In Sect. 6 we find the creation of the neural network for the Mackey Glass time series. In Sect. 7 we have the statistical tests and the results obtained from the optimization of the neural network for the Mackey Glass series using Mayfly algorithm enhanced with fuzzy logic. Finally, we have the Conclusions Section, where we discuss if the Mayfly algorithm with fuzzy adaptation turned out to improve the performance in comparison with its original version and how efficient its application to the optimization of network architectures was.

2 Theoretical Framework

First of all, we start by mentioning that genetic algorithms are mainly used to solve problems in which classical algorithms would entail a large computational cost, and these algorithms are based on the genetic evolution of Darwin's theory. Therefore, we start with a set of chromosomes, where each one represents a possible solution to the given problem, this chromosome is randomly generated with the dimensions that we consider appropriate for the specific problem that we seek to solve, as well as the population number that we will have.

The mayfly is a species of insects in which the young would live in the water for several years and when they mature, they will bark and fly through the air. However, these mature insects only have the purpose of reproducing since their life span is short [1].

By observing male and female mayflies moving in search of a mate, the MA algorithm is based on this behavior. Built similarly to the PSO algorithm [4, 5], mayflies will also update their positions. However, they converge faster to the optimum, so the Mayfly algorithm, in some way, is an improvement of the PSO algorithm.

$$x_i^{t+1} = x_i^t + v_i^{t+1} \tag{1}$$

$$v_{ij}^{t+1} = v_{ij}^t + d\mathrm{r} \tag{2}$$

$$v_{ij}^{t+1} = v_{ij}^t + a_1 e^{-\beta r_p^2}\left(pbest_{ij} - x_{ij}^t\right) + a_2 e^{-\beta r_g^2}\left(gbest_j - x_{ij}^t\right) \tag{3}$$

Parameters a_1 and a_2 will be adjusted with fuzzy logic, and these parameters represent the constants of positive attraction that are used to scale the contribution of the cognitive and social components, respectively.

Based on the mayfly literature by Konstantinos Zervoudakis, the recommended value for the parameter a_1 is in the range of 1 to 2 and for the parameter a_2 it is in the range of 1.5–2.

Research has shown that not only would the best candidates or best historical trajectories work well in guiding people in swarms to find the best solution, but that the worst and worst historical trajectories would also work well in doing so. Such situations could be treated directly as pairs of oppositions and satisfy the ancient Chinese Yin-Yang philosophy which shows us that there would be a strong relationship between the Yin and Yang sides, or good and evil, negative and positive, fire and water, dry and wet, then, the opposition-based learning (OBL) rule was proposed [6].

This algorithm is a variant of the Mayfly algorithm in which, like its predecessor, population is divided into two types, the male mayfly and the female mayfly, both of which will update their positions with their velocities $v_i(t)$ [6].

$$p_i(t+1) = p_i(t) + v_i(t) \tag{4}$$

$$op_i(t+1) = a + b - p_i(t) \tag{5}$$

The improved mayfly optimization algorithm.

This algorithm has been designed through the explained conception and modeling of three algorithms, the particle swarm optimization algorithm (PSO) [7–9], the firefly optimization algorithm (FOA) and the genetic algorithm (GA) that made it so efficient due to to the use of the three popular advantages of the methods. With the hybridization [10–14] of the algorithms, the exploitation of the PSO [15] algorithm is combined with the exploration features of the other algorithms to achieve greater efficiency in the modeling of mayflies, especially the reproduction model. This model helps the fittest mayflies [9, 16, 17] survive after hatching from the eggs regardless of their lifespan. We first look for the short-lived male with the lowest cost value in the population, then we select the number of seeds, followed by this the short-lived females attract the elite males to spawn new offspring using this as their growth ratio, updating their speed is given by the following form [18].

$$\mathrm{v}_{\mathrm{id}}^{\mathrm{t+1}} = \mathrm{v}_{\mathrm{id}}^{\mathrm{t}} + \mathrm{c}_1\, \mathrm{e}^{-\xi D_{\mathrm{p}}^2}\, \mathrm{gbest}_{\mathrm{d}} - \mathrm{p}_{\mathrm{id}}^{\mathrm{t}} + \mathrm{c}_2 \mathrm{e}^{-\xi D_{\mathrm{g}}^2}\, \mathrm{gbest}_{\mathrm{d}} - \mathrm{p}_{\mathrm{id}}^{\mathrm{t}} \tag{6}$$

Parameters c_1 and c_2 are the constants of positive attraction that represent the participation of the cognitive and social elements, respectively. In addition, p represents the position, v is the velocity, t represents the iteration.

Algorithm NMO

This is a negative ephemeral optimization algorithm. In which the male mayflies would update their speeds according to the worst candidates along with their worst trajectories. Contrary to the normal positive interpretation, in the NMO algorithm, male mayflies would flee their worst trajectories and global worst candidates, female mayflies have a duty to reproduce, thus they are all in a rush to mate. Their speeds would update according to their peers. That is, the speed of the i-th female mayfly would be guided to update according to her fitness value f [yi (t)] and her partner f [xi (t)], and the speed will be updated by the following equation well optimizing the unimodal or multimodal reference functions, even for the non-symmetric one. However, for unimodal reference functions, MO would perform better than NMO [19].

$$gv_i^t + a_1 e^{-\beta r_{mf}^2}\left[x_i^t - y_i^t\right] \quad f\left[\left(x_i^t - y_i^t\right)\right] > f\left[\left(x_i^t\right)\right] \tag{7}$$

$$v_i^{t+1} = \begin{cases} gv_i^t + dr_2 & f\left[x_i^t\right] \leq f\left[x_{hi}^t\right] \\ gv_i^t + a_2 e^{-\beta r_{p}^2}\left[x_{hi} - x_i^t\right] + a_3 e^{-\beta r_g^2}\left[x_g - x_i^t\right] & f\left[x_i^t\right] > f\left[x_{hi}^t\right] \end{cases} \tag{8}$$

The parameters $\boldsymbol{a}_2$ and $\boldsymbol{a}_3$ are constant, $\boldsymbol{r}_{\mathrm{p}}$ and $\boldsymbol{r}_{\boldsymbol{g}}$ represent the Cartesian distance between the current individual and the best historical trajectory, d represents the proportion of the wedding dance around the current position, $\boldsymbol{r}_2$ it is another random number in uniform distribution with interval of -1 and 1 and $\boldsymbol{x}_{\boldsymbol{hi}}$ represents the best historical trajectory.

Balanced Mayfly algorithm (BMA)

The algorithm has been designed by the explained conception and the modeling of three algorithms, Particle Swarm Optimization (PSO) algorithm, Firefly Algorithm (FA), and Genetic Algorithm (GA) that made it so efficient due to the use of the advantages all three popular methods. With hybridizing of the algorithms, the exploitation of the PSO algorithm is combined with exploration features of the other algorithms for achieving higher efficiency in modeling the mayflies, especially the model of breeding. This modeling helps the fittest mayflies to survive after egg hatching without considering their lifespan.

First we look for the ephemeral male with the lowest cost value in the population, then we select the number of seeds, followed by this the ephemeral females attract the elite males to generate a new offspring using this as the growth radius [18].

$$v_{id}^{t+1} = v_{id}^{t} + c_1 e^{-\xi D_p^2} gbest_d - p_{id}^{t} + c_2 e^{-\xi D_g^2} gbest_d - p_{id}^{t} \tag{9}$$

The parameters x_{ij}^t and v_{ij}^t represent the location and velocity of the i-th mayfly in dimension j, β is a coefficient to make the ephemeral visible to others, r_p and r_g represent the restricted Cartesian distance in the range, x_l^i and x_g^j describe the mayfly with the optimal position for the mayfly, b_1 and b_2 represent the constants of positive attraction and social factors, respectively.

This chaotic time series is widely used to evaluate the performance of a neural network, it was invented by mathematicians Michael Mackey and Leon Glass. It is a differential equation and with lags, however it has a chaotic behavior at $\tau = 17$ which will be used to evaluate our neural network and will be calculated with the following equation.

$$\frac{dx(t)}{dt} = \frac{0.2x(-t-\tau)}{1 + X^{10}(t-\tau)} - 0.1_x(t) \tag{10}$$

3 Proposed Method

In this Section, we can see the flow diagram of the proposed method to improve the convergence of the mayfly algorithm in Fig. 1.

In the original method first we start/initialize the female population,/find the best global/evaluate solutions/we check the stop criteria, if its satisfied we find the solution, if not then update velocities and solutions of males and females, evaluate solutions, Rank the mayflies/Mate the mayflies/evaluate offspring/separate offspring to mate and female randomly/replace worst solutions with the best new ones, update pbest and gbest and check the stop criteria again. In the proposed method after checking the stopping criteria, we will add fuzzy adaptation of the parameters that affect the algorithm performance the most.

4 Study of the Impact of the Parameters

We show in Fig. 2 a study of how the parameter values impact the behavior of the algorithm, in this case for the sphere benchmark function.

In Fig. 2 we can see how the benchmark function of the sphere behaves, where the a2 parameter stands out in the impact of the performance of the Mayfly algorithm.

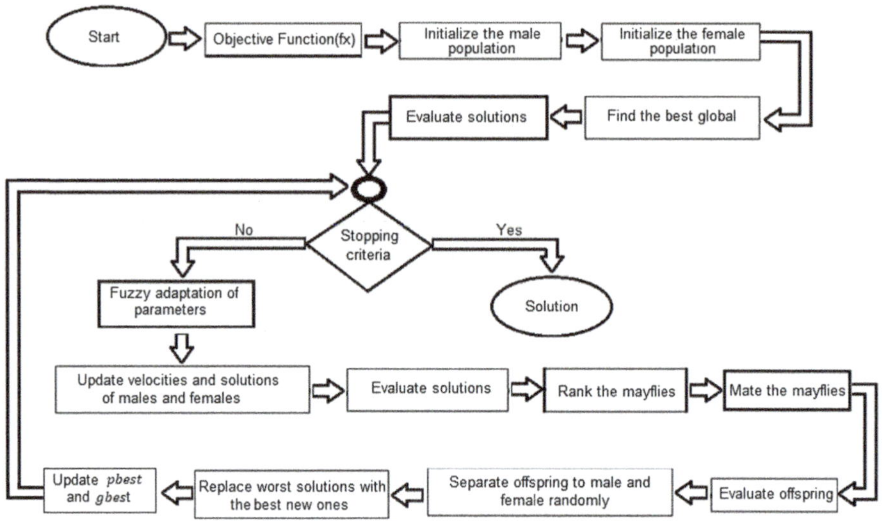

Fig. 1 Diagram of the proposed method

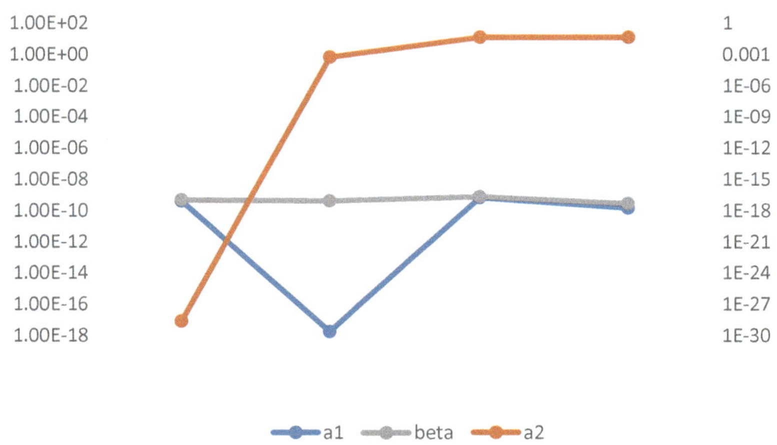

Fig. 2 Impact of sphere benchmark function parameters

In Fig. 3 we analyze the impact of the Mayfly algorithm on the Rastrigin benchmark function, here we observe that a1 represents a greater impact when varying it.

In Fig. 4 we analyze the impact of the Mayfly algorithm on the Griewank benchmark function, where a1 and a2 represent a greater impact when varying them.

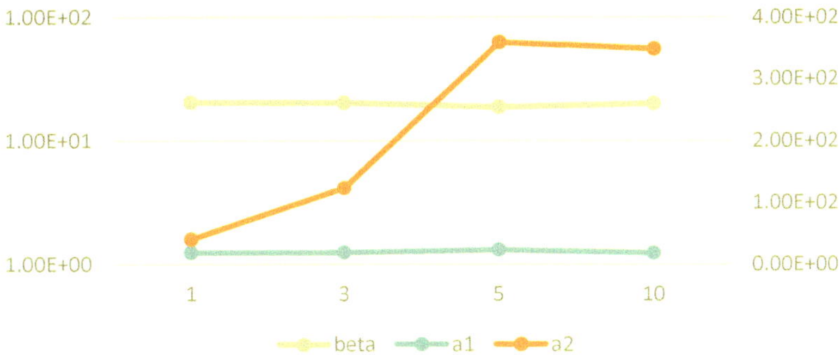

Fig. 3 Impact of Rastrigin benchmark function parameters

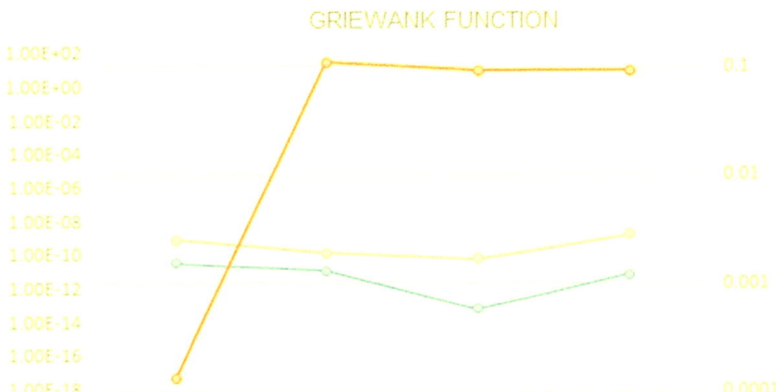

Fig. 4 Impact of Griewank benchmark function parameters

5 Fuzzy Adapter Design

First, a fuzzy adapter was designed with parameter a_2 because it was the parameter that was most relevant in the performance of the Mayfly algorithm, although parameter a1 was very close.

5.1 Fuzzy Adapter with 1 Input and 1 Output

The fuzzy rules for parameter adaptation are presented in Fig. 5.

In Fig. 5 we have the initial fuzzy rules in which we start with a_2 with high values in the first iterations and as they progress a_2 will decrease, in the first fuzzy rule the algorithm is just beginning, while in the last rule it is almost has finished, that makes

Fig. 5 Fuzzy rules

if iteration is very small then a_2 is very large
if iteration is small then a_2 is large
if iteration is median then a_2 is median
if iteration is large then a_2 is small
if iteration is very large then a_2 is very small

the fuzzy rules adjust the value of the parameter a_2 as the iterations go by, and thus get closer to its optimal values for said parameter.

$$\mathbf{f(x; a, b, c)} = \begin{cases} 0, & \mathbf{x \le a} \\ \frac{\mathbf{x-a}}{\mathbf{b-a}}, & \mathbf{a \le x \le b} \\ \frac{\mathbf{c-x}}{\mathbf{c-b}}, & \mathbf{b \le x \le c} \\ 0, & \mathbf{c \le x} \end{cases} \tag{11}$$

In Eq. 11 we have the triangular membership functions in which the values a, b and c will be given by the tables below. Table 1 shows the parameter values for the input membership functions.

Table 2 shows the parameter values for the output membership functions.

In Fig. 6 we have the initial fuzzy adapter which consisted of an input variable and an output variable, in the input variable we only take into account the iterations.

As we can note in Table 3 we have the abbreviation of the benchmark functions.

Table 4 shows a summary of the results for the 10 benchmark functions.

Table 1 Input membership functions in the initial fuzzy adapter

Membership function	a	b	c
IMP (very small iteration)	0	0	0.25
IP (small iteration)	0	0.25	0.5
IM (medium iteration)	0.25	0.5	0.75
IG (big iteration)	0.5	0.75	1
IMG (very big iteration)	0.75	1	1

Table 2 Output membership functions in the initial fuzzy adapter

Membership function	A	B	C
A2MP (a_2 very small)	0	0	0.25
A2P (a_2 small)	0	0.25	0.5
A2M (a_2 medium)	0.25	0.5	0.75
A2G (a_2 big)	0.5	0.75	1
A2MG (a_2 very big)	0.75	1	1

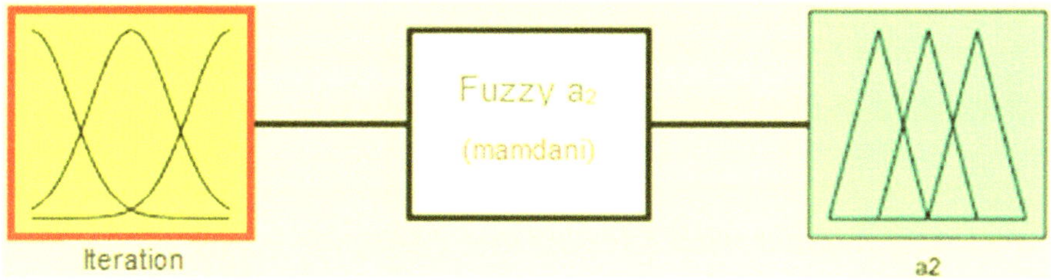

Fig. 6 Initial structure of the fuzzy adapter

Table 3 Representation of the benchmark functions

Abbreviation	Name of function
F1	Sphere
F2	Rastrigin
F3	Ackley
F4	Dejong1
F5	Griewank
F6	Schwefel
F7	Styblinskitank
F8	Wood
F9	Hyper-ellipsoid
F10	Powell

Table 4 Fuzzy adapter applied to a_2 and a_1 individually

Benchmark function	Beta parameter/50 dimensions			
	Fuzzy adapter for a_2		Fuzzy adapter for a_1	
	$\bar{x}$	S	$\bar{x}$	S
F1	2.1415×10^{-16}	7.7240×10^{-17}	8.3797×10^{-17}	5.2048×10^{-17}
F2	2.2920×10^{1}	8.9131×10^{0}	2.2685×10^{1}	8.5655×10^{0}
F3	3.0944×10^{0}	5.3400×10^{-2}	3.1116×10^{0}	3.4500×10^{-2}
F4	1.9221×10^{-18}	2.9649×10^{-18}	8.2611×10^{-17}	4.5886×10^{-17}
F5	1.600×10^{-3}	3.6000×10^{-3}	1.700×10^{-3}	3.600×10^{-3}
F6	2.0755×10^{4}	4.0579×10^{0}	2.0754×10^{4}	2.8595×10^{0}
F7	-1.7482×10^{3}	4.1381×10^{1}	-1.7465×10^{3}	4.5256×10^{1}
F8	1.074×10^{-1}	5.8840×10^{-1}	4.6900×10^{-2}	1.7930×10^{-1}
F9	6.5445×10^{-17}	1.0540×10^{-16}	2.5736×10^{-15}	2.1093×10^{-15}
F10	5.1923×10^{-16}	1.2900×10^{-15}	9.2123×10^{-17}	5.1465×10^{-17}

As we can note in Table 4, the results of both initial fuzzy adapters were compared, where it can be seen that there was not a considerable difference, however, it can be seen that using the parameter a2 better results are achieved.

5.2 *Fuzzy Adapter with 2 Inputs and 2 Outputs*

Noting that the results of the initial fuzzy system did not significantly improve the performance of the Mayfly algorithm, even when applying trapezoidal fuzzy rules, it was decided to make a fuzzy adapter with 2 inputs and 2 outputs. In this case, the parameters a_1 and a_2 were used together to improve their effectiveness, since individually no difference was achieved between one parameter and another.

Figure 7 shows, in a general way, the design of the final fuzzy adapter, where we have 2 inputs and 2 outputs respectively to control the parameters a_1 and a_2

$$\mathbf{f}(\mathbf{x};\ \mathbf{a},\ \mathbf{b},\ \mathbf{c},\ \mathbf{d}) = \begin{cases} 0, & si(x < a)o(x > d) \\ \frac{x-a}{b-a}, & sia \le x \le b \\ 1, & sib \le x \le c \\ \frac{d-x}{d-c}, & sic \le x \le d \end{cases} \tag{12}$$

Trapezoidal membership functions are shown in Eq. 12, in which the values a, b and c d will be given by the tables below.

In Table 5 we can find the input membership functions for our final fuzzy adapter.

In Table 6 we can find the output membership functions for our final fuzzy adapter.

We can see that the Eq. 13 represents the diversity calculation.

$$d = \frac{1}{n_s}\sum_{i=1}^{n_s}\sqrt{\sum\nolimits_{j=1}^{n_s}\left(x_{ij}(t) - \overline{X}_j(t)\right)^2} \tag{13}$$

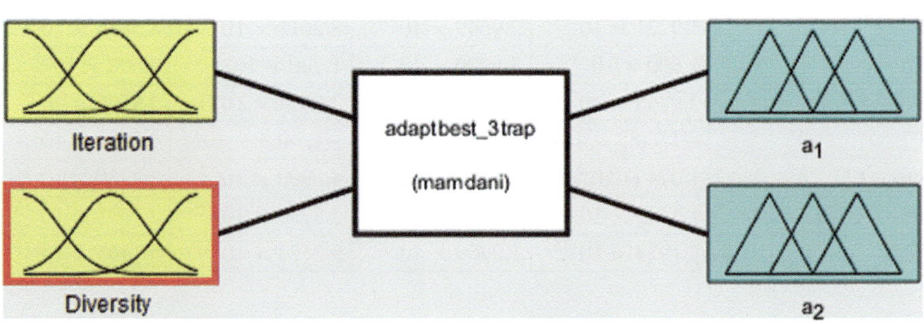

Fig. 7 Two inputs and two outputs fuzzy adapter

Table 5 Input membership functions in the 2-input, 2-output fuzzy adapter

Membership functions	a	b	c	d
IB (small iteration)	0	0	0.1	0.33
IM (medium iteration)	0.25	0.42	0.58	0.75
IA (big iteration)	0.67	0.9	1	1
DB (small diversity)	0	0	0.1	0.33
DM (medium diversity)	0.25	0.42	0.58	0.75
DA (big diversity)	0.67	0.9	1	1

Table 6 Output membership functions in the 2-in, 2-out fuzzy adapter

Membership functions	a	b	c	d
A1B (a_1 small)	1.094	1.241	1.78	1.427
A1MB (a_1 medium small)	1.278	1.399	1.485	1.611
A1M (a_1 medium)	1.475	1.578	1.659	1.779
A1MA (a_1 medium big)	1.648	1.763	1.844	1.982
A1A (a_1 big)	1.833	1.982	2.018	2.167
A2B (a_2 small)	0.42	0.58	0.62	0.7803
A2MB (a_2 medium small)	0.62	0.7803	0.8203	0.9797
A2M (a_2 medium)	0.85	0.9797	1.02	1.15
A2MA (a_2 medium big)	1.02	1.18	1.22	1.38
A2A (a_2 big)	1.22	1.38	1.42	1.58

6 Mackey Glass Neural Network Design

To design the neural network we went through several steps, firstly, we obtained the training data for the Mackey Glass chaotic time series, this with the help of a data generator previously programmed in Matlab, as input parameters we used $\tau = 17$, $x_0 = 1.2$ and generate a sample from 0 to 1200.

As we can see in Fig. 8 we have the Mackey Glass chaotic series with a sample of 1200 data from which we take 800 to train our neural network, without any delay, therefore our network will only be able to predict a few steps ahead. of the unknown data, however the fundamental thing about this neural network is to extract the mean square error in order to determine how effective it is and determine how many neurons the hidden layer requires, said information will be sent to the original Mayfly algorithm and to which we have modified with the fuzzy adapter to verify how efficient the method is and in turn maximize the performance of the previously designed neural network.

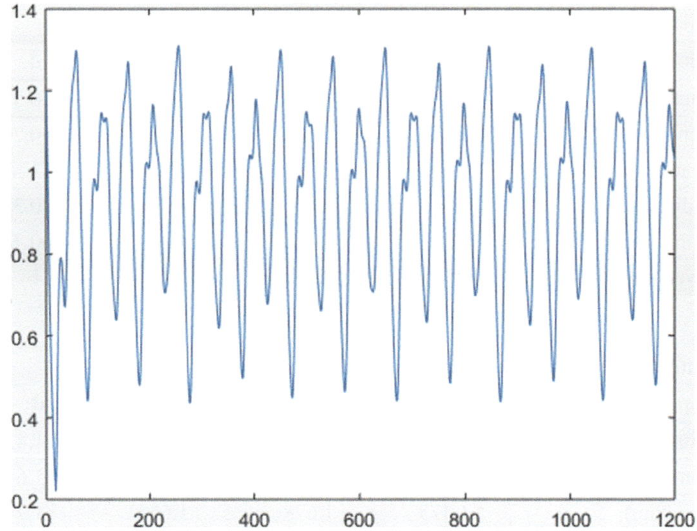

Fig. 8 Mackey Glass series

7 Results

In this section we will present the different results obtained in this investigation, first we will show the evaluations of the fuzzy adapters, after that we will have the statistical tests and neural network to predict the Mackey Glass time series.

As we can see in Table 7 the results for the different fuzzy adapters and the original mayfly algorithm evaluated with benchmark functions are shown.

As we can see in Table 8 we have the results of the statistical tests applied in comparison with the author of Mayfly for 50 dimensions and maximum iterations of 2000, as results we obtained 5 tests with significant evidence that the inclusion of the fuzzy adapter improves the performance of the Mayfly algorithm in the benchmark functions commonly used for this type of tests, of these tests the most notable was the one carried out with the Powell function since it is the only one that passes the statistical test with 95% confidence.

In Fig. 9 we can appreciate a comparison between the chaotic series of Mackey Glass and the results of our previously optimized neural network, in said optimization result we obtained that the adequate number of neurons in the hidden layer is 73, since with it the quadratic error mean is 0.0365.

Table 7 Results of the fuzzy adapters

Benchmark function	50 dimensions			
		Fuzzy adapter initial	Fuzzy adapter final	Mayfly original
F1	$\bar{x}$	2.1415×10^{-16}	8.3455×10^{-17}	1.1777×10^{-7}
	S	7.7240×10^{-17}	5.1033×10^{-17}	4.5203×10^{-7}
F2	$\bar{x}$	1.7254×10^{1}	1.1270×10^{1}	1.1900×10^{1}
	S	5.5900×10^{0}	3.7928×10^{0}	3.8200×10^{0}
F3	$\bar{x}$	3.1267×10^{0}	3.1224×10^{0}	0.0000×10^{0}
	S	3.0100×10^{-2}	2.9100×10^{-2}	0.0000×10^{0}
F4	$\bar{x}$	1.1000×10^{-3}	9.0247×10^{-4}	4.1400×10^{-3}
	S	4.0000×10^{-3}	3.8000×10^{-3}	1.2900×10^{-2}
F5	$\bar{x}$	2.0754×10^{4}	2.0755×10^{4}	3.8800×10^{0}
	S	3.3438×10^{0}	2.9589×10^{0}	8.8300×10^{-1}
F6	$\bar{x}$	1.8052×10^{-15}	3.9987×10^{-127}	5.2800×10^{-49}
	S	5.7657×10^{-15}	2.1902×10^{-122}	1.6500×10^{-48}
F7	$\bar{x}$	5.6907×10^{1}	5.52105×10^{1}	6.7703×10^{1}
	S	2.5495×10^{1}	2.3000×10^{1}	3.9877×10^{1}
F8	$\bar{x}$	3.6000×10^{-3}	7.9186×10^{-4}	3.3465×10^{-8}
	S	3.0000×10^{-3}	2.1000×10^{-3}	1.6242×10^{-7}
F9	$\bar{x}$	5.9987×10^{1}	6.8769×10^{1}	1.7130×10^{-1}
	S	1.0816×10^{2}	1.3059×10^{2}	8.3733×10^{-2}
F10	$\bar{x}$	2.7503×10^{-10}	3.6940×10^{-15}	7.3923×10^{-6}
	S	3.8725×10^{-10}	5.6091×10^{-15}	2.5797×10^{-5}

8 Conclusions

The a_2 parameter mainly affects the performance of the mayfly algorithm, it was observed that its performance increases in values close to 1, so is one of the parameters to which the fuzzy logic was applied. The parameter a_1 increases the performance of the algorithm in values between 3 and 5, so it will be the other parameter to take into account when applying fuzzy logic. As for the parameter β, it produced the best results at a value of 1, however its variation affects the performance of the algorithm very little, so it is not taken into account for the application of fuzzy logic. The results improved when applying a fuzzy adapter with two input and two output parameters. Applying fuzzy adaptation to parameters a_1 and a_2 together significantly improves performance on mathematical functions compared to the original author of the Mayfly method. Better results were obtained with the trapezoidal functions when compared to the triangular ones. Using the Mayfly algorithm to optimize the neural network architecture works well, however, the results do not improve much, so we can conclude that this heuristic is not recommended for this type of optimization at the moment, because the mean square error does not was able to drop below 0.001

Table 8 Hypothesis test results

Benchmark function	50 dimensions				Results statistical test	
	Fuzzy adapter		Mayfly original			
	$\overline{x}$	S	$\overline{x}$	S	Z	S
F1	8.3455×10^{-17}	5.1033×10^{-17}	1.1777×10^{-7}	4.5203×10^{-7}	−1.4270	Pass test
F2	1.1270×10^{1}	3.7928×10^{0}	1.1900×10^{1}	3.8200×10^{0}	−0.6410	Does not pass
F3	3.1224×10^{0}	2.9100×10^{-2}	0.0000×10^{0}	0.0000×10^{0}	0	Does not pass
F4	9.0247×10^{-4}	3.8000×10^{-3}	4.1400×10^{-3}	1.2900×10^{-2}	−1.3186	Pass test
F5	2.0755×10^{4}	2.9589×10^{0}	3.8800×10^{0}	8.8300×10^{-1}	0	Does not pass
F6	3.9987×10^{-127}	2.1902×10^{-122}	5.2800×10^{-49}	1.6500×10^{-48}	−1.7527	Pass test
F7	5.52105×10^{1}	2.3000×10^{1}	6.7703×10^{1}	3.9877×10^{1}	−1.4764	Pass test
F8	7.9186×10^{-4}	2.1000×10^{-3}	3.3465×10^{-8}	1.6242×10^{-7}	2.0652	Does not pass
F9	6.8769×10^{1}	1.3059×10^{2}	1.7130×10^{-1}	8.3733×10^{-2}	2.8771	Pass test
F10	3.6940×10^{-15}	5.6091×10^{-15}	7.3923×10^{-6}	2.5797×10^{-5}	−1.5696	Pass test

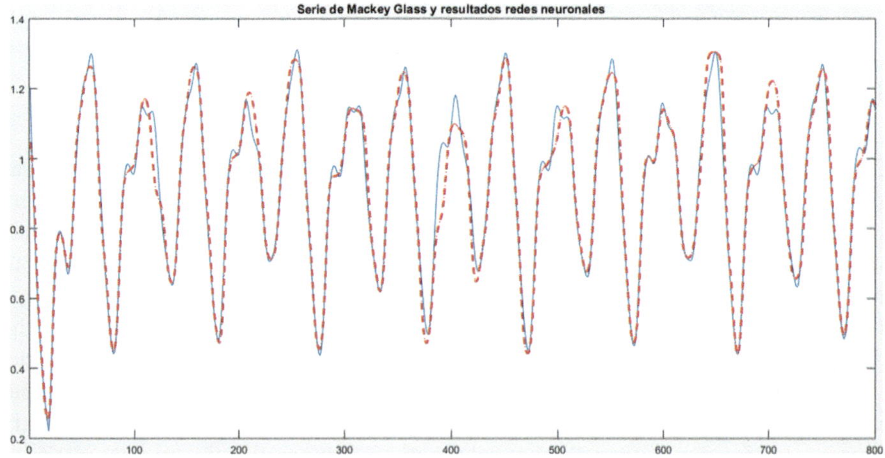

Fig. 9 Optimized network forecast

even using the modified Mayfly algorithm with the fuzzy adapter. Optimizing the architecture of a neural network has a high computational cost, however the cost is lower than if we did it with a classic search method, it is difficult for us to obtain an adequate number of neurons in the hidden layer since its complexity is different from the optimization of mathematical functions, which is what this method has mainly excelled at. As future work, the fuzzy adapter can be optimized by means of a GA to maximize the performance of the fuzzy Mayfly algorithm by finding the optimal values of the parameters in the membership functions.

References

1. J. Zhao, Z. Gao, The negative mayfly optimization algorithm. J. Phys. Conf. Ser. **1693**(1) (2020). https://doi.org/10.1088/1742-6596/1693/1/012098
2. X. Guo, X. Yan, K. Jermsittiparsert, Using the modified mayfly algorithm for optimizing the component size and operation strategy of a high temperature PEMFC-powered CCHP. Energy Rep. **7**, 1234–1245 (2021). https://doi.org/10.1016/j.egyr.2021.02.042
3. K. Rajeswari, P. Lakshmi, PSO optimized fuzzy logic controller for active suspension system, in *Proceedings—2nd International Conference on Advances in Recent Technologies in Communication and Computing, ARTCom* (2010), pp. 278–283. https://doi.org/10.1109/ARTCom.2010.22
4. P. Melin, F. Olivas, O. Castillo, F. Valdez, J. Soria, M. Valdez, Optimal design of fuzzy classification systems using PSO with dynamic parameter adaptation through fuzzy logic. Expert Syst. Appl. **40**(8), 3196–3206 (2013). https://doi.org/10.1016/j.eswa.2012.12.033
5. F. Valdez, O. Castillo, P. Melin, Bio-inspired algorithms and its applications for optimization in fuzzy clustering. Algorithms **14**(4) (2021). https://doi.org/10.3390/a14040122
6. Z. Gao, J. Zhao, S. R. Li, Y. Hu, The improved mayfly optimization algorithm with opposition based learning rules. J. Phys. Conf. Ser. **1693**(1), 1–6 (2020). https://doi.org/10.1088/1742-6596/1693/1/012117
7. M. Kaminski, Neural network training using particle swarm optimization—a case study, in *2019 24th International Conference on Methods and Models in Automation and Robotics (MMAR)* (2019), pp. 115–120. https://doi.org/10.1109/MMAR.2019.8864679
8. F. Olivas, F. Valdez, O. Castillo, P. Melin, Dynamic parameter adaptation in particle swarm optimization using interval type-2 fuzzy logic. Soft. Comput. **20**(3), 1057–1070 (2016). https://doi.org/10.1007/s00500-014-1567-3
9. T. Bhattacharyya, B. Chatterjee, P. Singh, J. Yoon, Z. Geem, R. Sarkar, Mayfly in harmony: a new hybrid meta-heuristic feature selection algorithm. IEEE Access **8**, 195929–195945 (2020). https://doi.org/10.1109/ACCESS.2020.3031718
10. F. Olivas, F. Valdez, O. Castillo, I. Gonzalez, G. Martinez, P. Melin, Ant colony optimization with dynamic parameter adaptation based on interval type-2 fuzzy logic systems. Appl. Soft Comput. J. **53**, 74–87 (2017). https://doi.org/10.1016/j.asoc.2016.12.015
11. J. Zhou, Z. Duan, Y. Li, J. Deng, D. Yu, PSO-based neural network optimization and its utilization in a boring machine. J. Mater. Process. Technol. **178**(1–3), 19–23 (2006). https://doi.org/10.1016/j.jmatprotec.2005.07.002
12. F. Valdez, P. Melin, O. Castillo, A survey on nature-inspired optimization algorithms with fuzzy logic for dynamic parameter adaptation. Expert Syst. Appl. **41**(14), 6459–6466 (Elsevier Ltd) (2014). https://doi.org/10.1016/j.eswa.2014.04.015
13. C. Maa, M. Shanblatt, S. Member, A two-phase optimization neural network. IEEE Trans. Neural Netw. **3**(6) (1992). https://doi.org/10.1109/72.165602

14. F. Valdez, P. Melin, O. Castillo, An improved evolutionary method with fuzzy logic for combining particle swarm optimization and genetic algorithms. Appl. Soft Comput. J. **11**(2), 2625–2632 (2011). https://doi.org/10.1016/j.asoc.2010.10.010
15. D. Cook, C. Ragsdale, R. Major, Combining a neural network with a genetic algorithm for process parameter optimization. Eng. Appl. Artif. Intell. **13**(4), 391–396 (2000). https://doi.org/10.1016/s0952-1976(00)00021-x
16. R. Martinez, O. Castillo. L. Aguilar, P. Melin, Fuzzy logic controllers optimization using genetic algorithms and particle swarm optimization, in ed. by G. Sidorov, A. Hernández Aguirre, C.A. Reyes García *Advances in Soft Computing*. MICAI 2010. *Lecture Notes in Computer Science*, vol. 6438 (Springer, Berlin, Heidelberg, 2010). https://doi.org/10.1007/978-3-642-16773-7_41
17. Z. Konstantinos, T. Stelios, A mayfly optimization algorithm. Comput. Ind. Eng. **145**, 106559. ISSN: 0360-8352. https://doi.org/10.1016/j.cie.2020.106559
18. A. Dey, S. Chattopadhyay, P. Singh, A. Ahmadian, M. Ferrara, R. Sarkar, A hybrid meta-heuristic feature selection method using golden ratio and equilibrium optimization algorithms for speech emotion recognition. IEEE Access **8**, 200953–200970 (2020). https://doi.org/10.1109/ACCESS.2020.3035531
19. L. Chen, C. Xu, H. Song, K. Jermsittiparsert, Optimal sizing and sitting of EVCS in the distribution system using metaheuristics: a case study. Energy Rep. **7**, 208–217 (2021). https://doi.org/10.1016/j.egyr.2020.12.032
20. Z. Gao, J. Zhao, S. R. Li, Y. Hu, The improved mayfly optimization algorithm. J. Phys. Conf. Ser. **1684**(1) (2020). https://doi.org/10.1088/1742-6596/1684/1/012077

Fuzzy Classifier Using the Particle Swarm Optimization Algorithm for the Diagnosis of Arterial Hypertension

Martha Pulido and Patricia Melin

Abstract The main objective of this article is the creation of a new fuzzy classifier, using the optimization algorithm by means of particles, to optimize the structures of the type-1 and type-2 fuzzy systems, (such as parameters and type of membership functions, type of system, and number of rules).Tests were carried out with 40 patients and the blood pressure readings of the patients were taken at a time interval for 24 h, and these were taken through an ambulatory blood pressure monitor (ABPM). In this work, good results for Classification and Diagnosis of Arterial Hypertension with the proposed model are shown.

Keywords Particle swarm optimization · Hypertension · Classification · Diastolic · Systolic · Blood pressure

1 Introduction

Computing is being incorporated into all branches of knowledge, and medicine has been highly favored. One of the first uses was the processing of medical records, epidemiological data, statistical analysis; Subsequently, medical diagnosis and treatment instruments have been added, be it laboratory, imaging, [7, 8, 14, 15, 35], There are also expert systems for medical sciences, we can talk about medical diagnosis packages. Several expert systems have been developed to diagnose diseases and recommend treatments [10, 12, 21, 24, 26].

Blood pressure increases with age. 66% of people > 65 years of age have high blood pressure, and 128/5.000. 90% of 55-year-olds with normal blood pressure are at risk of developing hypertension at some point in their lives, [27], Because high blood pressure becomes so common over the years, age-related increases in blood pressure may seem innocuous, but they actually increase the risk of associated diseases and death [23, 28–30]. Hypertension is probably the most prevalent disease in the world and affects approximately one-third of the population. It is the main cardiovascular

M. Pulido · P. Melin (✉)
Tijuana Institute of Technology/TecNM, Tijuana, Mexico
e-mail: pmelin@tectijuana.mx

O. Castillo and P. Melin (eds.), *Fuzzy Logic and Neural Networks for Hybrid Intelligent System Design*, Studies in Computational Intelligence 1061,
https://doi.org/10.1007/978-3-031-22042-5_5

risk factor. It is known as the "silent killer", because in most cases it does not present symptoms, so it is possible to develop heart, brain, or kidney problems without being aware of having it [1, 2, 9, 32]. The main reason for the paper is the creation of ta fuzzy classifier for experts to know and verify the behavior of a patient's arterial hypertension and from this give a diagnosis if the patient requires it. The blood pressure readings of the patients were taken at a time interval for 24 h, and these were taken through an ambulatory blood pressure monitor (ABPM) [4].

The fuzzy system consists of 64 rules, with 2 input variables each with 8 membership functions, and these are called Systolic and Diastolic, also an output called classification which consists of 11 membership functions, and optimization of the fuzzy system with the particle swarm optimization algorithm as soon as parameters and type of membership functions, type of system, and a number of rules [13, 25, 26].

The Particle Swarm Optimization algorithm is an optimization/search technique. PSO is normally used in search spaces with many dimensions. The Particle Swarm optimization algorithm is inspired by the evolution in collective behavior, mainly tries to imitate the social behavior of various groups of animals such as flocks herds, etc.

Let $x_i(t)$ denoted by the position of the particle i the search space at time step t; unless otherwise indicated, t denotes discrete time steps. The position of the particle is changed by the addition of a velocity, $v_i(t)$, for the current position, i.e.

$$x_i(t+1) = x_i(t) + v_i(t+1) \tag{1}$$

Con $x_i(0) \sim U(X_{\min}, X_{\max})$.

For gbest PSO, the velocity of particle i is calculated as:

$$v_{ij}(t+1) = v_{ij}(t) + c_1 r_1\left[y_{ij}(t) - x_{ij}(t)\right], + c_2 r_2(t)\left[\hat{y}_j(t) - x_{ij}(t)\right] \tag{2}$$

where $v_{ij}(t)$ Velocity of the particle i dimension j In passing time $t, c_1 y c_2$ are positive acceleration constants that serve to dimension the contribution of cognitive and social capacities, capacitates cognitive y socials, y $r_{1j}(t)$, y $r_{2j}(t) \sim U(0,1)$ are random values in the range [0, 1], the sample of a uniform distribution. These values randomly introduce a stochastic element for the algorithm. [11, 22].

The best personal position y_{ij} is associated with particles best position the particle has visited since the first step. Taking into account the minimization problems, the best personal position in the next step of time t + 1 is calculated as the best position the particle has visited since the first step. Taking minimization problems, the best personal position in the next step of time $t + 1$ is calculated as:

$$y_i(t+1 = \begin{cases} y_i(t) & iff(x_i(x_i(t+1)) \geq fy_i(t)) \\ x_i(t+1) & iff(x_i(x_i(t+1)) > fy_i(t)) \end{cases} \tag{3}$$

where $f : \mathbb{R}^{nx} \rightarrow \mathbb{R}$ is the objective function:

The position of the global best, $y_i(t)$, in time step t, is defined as:

$$\hat{y}(t)\epsilon\{y_o(t),\ldots,y_s(t)\}f(y(t)) = \min\{f(y_o(t)),\ldots f(y_s(t)),\} \tag{4}$$

The position of the global best can be selected from the particles of the current cloud, in which case:

$$\hat{y}(t) = \min\{f(x_o(t)),\ldots f(x_{ns}(t)),\} \tag{5}$$

The best overall PSO is summarized in the algorithm. In this algorithm, the notation $S.x_i$ is used to denote the position of particle i in the cloud S.

The position of the glbest:

The velocity is calculated as:

$$v_{ij}(t+1) = v_{ij}(t) + c_1r_1\left[y_{ij}(t) - x_{ij}(t)\right], +c_2r_2(t)\left[\hat{y}_j(t) - x_{ij}(t)\right] \tag{6}$$

where $\hat{y}_j$ is the best position, which is in the neighborhood of the particle i dimension j. The position of the best local particle $\hat{y}_i$, Example the best position found in the neighborhood $N_{i,}$ is define as:

$$\hat{y}(t+1) \in \left\{N_1 | f\left(\hat{y}_i(t+1)\right) = \min\{f(x)\}, \quad \forall x \in N_i\right\} \tag{7}$$

With the neighborhood defined as:

$$N_i = \{y_i - nN_i(t), y_i - nN_i + 1(t), \ldots y_i - 1(t), y_i + 1(t), \ldots, y_i + nN_i(t)\} \tag{8}$$

For neighborhoods of nN_i. The position of the best local is also known as the best-positioned neighborhood.

Fuzzy Sets and Membership Functions

If X is a collection of objects denoted by x, then a fuzzy set A in X is defined as a set of ordered pairs [34].

$$\mathrm{A} = \{(\mathrm{x}, \mu\ \mathrm{A(x)}) \in \mathrm{x} \in \mathrm{X}\} \tag{9}$$

where μA is the membership function for the fuzzy set A. The membership function maps each element of X to a membership degree of between 0 and 1 [31–33].

Type-2 Fuzzy Systems

Type-2 fuzzy systems are an extension of type-1 fuzzy systems, the fuzzy rules are the same, the difference is that the membership functions are used to model uncertainty and inaccuracy in a better way, since they also have the ability to model complex nonlinear systems, achieving better performance [3, 16].

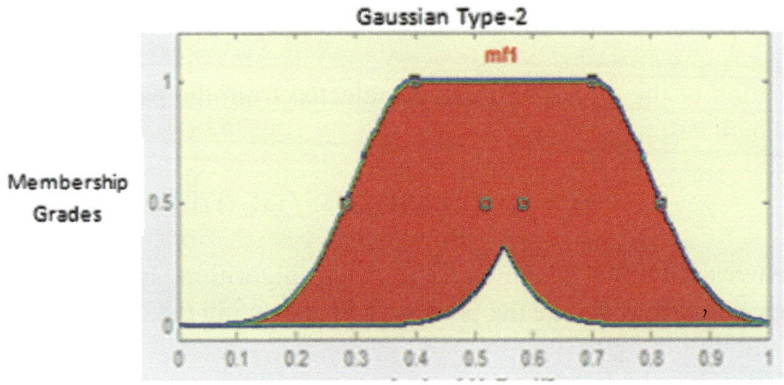

Fig. 1 A type-2 fuzzy set represents a type-1 fuzzy set with an uncertain standard deviation

As an example: Consider the case of a fuzzy set characterized by a Gaussian membership function with mean m and a standard deviation that can take values in $[\sigma_1, \sigma_2]$, i.e.,

$$\mu(x) = \exp\left\{-{}^1/_2\left[\frac{x-m}{\sigma}\right]^2\right\}; \quad \sigma\epsilon[\sigma_1, \sigma_2] \tag{10}$$

Corresponding to each value of σ we will get a different membership curve (as shown in Fig. 1). So, the membership grade of any particular $x(except x = m)$ can take any of a number of possible values depending upon the value of σ i.e., the membership grade is not a crisp number, it is a fuzzy set. Figure 1 shows the domain of the fuzzy set associated with $x = 0.7$; however, the membership function associated with this fuzzy set is not shown in the Fig. 1.

Gaussian type-2 fuzzy set

A Gaussian type-2 fuzzy set is one in which the membership grade of every domain point is a Gaussian type-1 set contained in [0, 1].

Interval type-2 fuzzy set

An interval type-2 fuzzy set is one in which the membership grade of every domain point is a crisp set whose domain is some interval contained in [0, 1].

Footprint of uncertainty

Uncertainty in the primary memberships of a type-2 fuzzy set, Ã, consists of a bounded region that we call the "footprint of uncertainty" (FOU). Mathematically, it is the union of all primary membership functions [18–20].

Upper and lower membership functions

An "upper membership function" and a "lower membership functions" are two type-1 membership functions that are bounds for the FOU of a type-2 fuzzy set Ã. The upper membership function is associated with the upper bound of the FOU(Ã). The lower membership function is associated with the lower bound of the FOU(Ã).

Operations of Type-2 Fuzzy Sets

Union of type-2 fuzzy sets

The union of $\widetilde{A}_1$ and $\widetilde{A}_2$ is another type-2 fuzzy set, just as the union of type-1 fuzzy sets A_1 and A_2 is another type-1 fuzzy set. More formally, we have the following expression:

$$\widetilde{A}_1 \cup \widetilde{A}_2 = \int_{x \in X} \mu_{\tilde{A}_1 \cup \tilde{A}_2}(x)/x \tag{11}$$

Intersection of type-2 fuzzy sets

The intersection of $\widetilde{A}_1$ and $\widetilde{A}_2$ is another type-2 fuzzy set, just as the intersection of type-1 fuzzy sets A_1 and $\widetilde{A}_2$ is another type-1 fuzzy set. More formally, we have the following expression:

$$\widetilde{A}_1 \cap \widetilde{A}_2 = \int_{x \in X} \mu_{\tilde{A}_1 \cap \tilde{A}_2}(x)/x \tag{12}$$

Complement of a type-2 fuzzy set

The complement of a set is another type-2 fuzzy set, just as the complement of type-1 fuzzy set A is another type-1 fuzzy set. More formally we have:

$$A' = \int_{X} \mu_{\tilde{A}'_1}(x)/x \tag{13}$$

where the prime denotes complement in the above equation. In this equation, $\mu_{\tilde{A}'1}$ is the secondary membership function [5, 6, 17, 20].

Type-2 fuzzy rules

Consider a type-2 FLS having r inputs $x_1 \in X_1, \ldots, x_r \in X_r$ and one output $y \in Y$. As in the type-1 case, we can assume that there are M rules; but, in the type-2 case the l th rule has the form:

$$R^1 : IF\ x_1\ is\ \tilde{A}_1^1\ and \ldots x_p\ is\ \tilde{A}_p^1,\ THEN\ y\ is\ \hat{Y}^1 \quad 1 = 1, \ldots M \tag{14}$$

The rules represent a type-2 rules relation between the input space $X_1 x \ldots x X_r$, and the space output set space Y, of the type-2 fuzzy system [30, 31].

2 Proposed Type-1 and Type-2 Fuzzy Classifier

In this section, we explain the creation of the type-1 and type-2 fuzzy system, as well as the particle for the type-1 and type-2 fuzzy system, rules and the objective function to optimize the structures of the fuzzy system type-1 and type-2.

2.1 *Blood Pressure*

The design of the new type-1 and type-2 fuzzy logic system for the classification of blood pressure consists of the two input variables of the fuzzy system have eight linguistic variables (Low, Optimal, Normal, High Normal, Grade 1, Grade 2, Grade 2 and ISH)., (as shown in Figs. 2 and 3) The output variable called Classification has eleven linguistic variables (Hypotension Low, Optimal, Normal, High Normal, Grade 1, Grade 2, Grade 2 and ISH1, ISH2 and ISH3).

Type-1 Memberships Functions:

$$gaussian = e^{-\frac{1}{2}\left(\frac{x-c}{\mathrm{r}}\right)^2} \tag{15}$$

$\boldsymbol{x}$, $\boldsymbol{c}$, r.

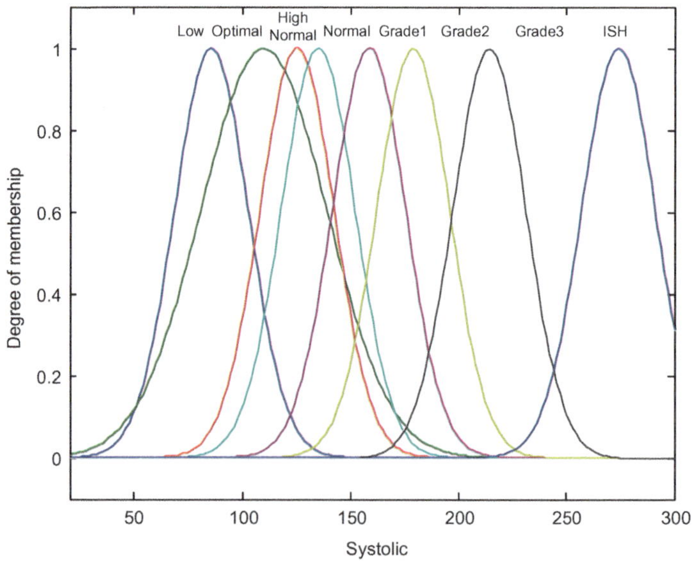

Fig. 2 Membership functions for input type-1 fuzzy system

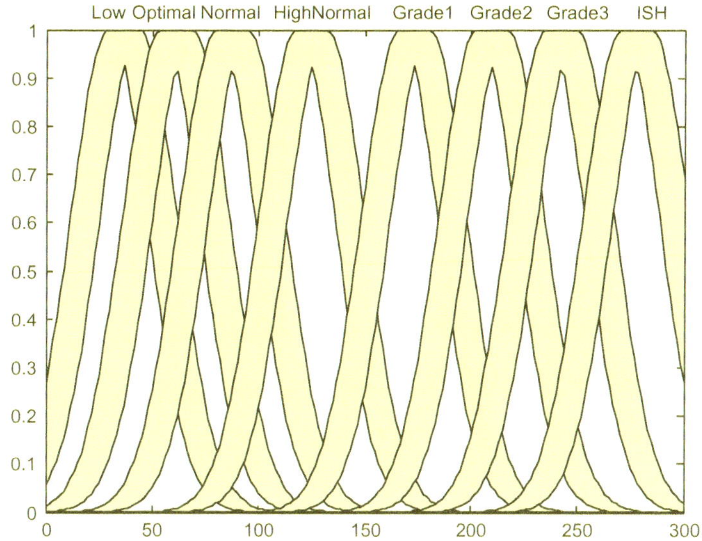

Fig. 3 Membership functions for input type-2 fuzzy system

where c is of the center of the function and r_i is the implication of the function.

MF too high:

$$MFToo\,high = e^{-\frac{1}{2}\left(\frac{x-234}{25.2}\right)^2} \tag{16}$$

where $c = 234$ and $r = 25.2$.

Gbell:

$$bell(a, b, c) = \frac{1}{1 + \left|\frac{x-c}{a}\right|2b} \tag{17}$$

MF Veryhigh:

$$\begin{aligned} MF\,Very\,high &= \frac{1}{1 + \left|\frac{x-169.9}{5.95}\right|2(2.479)} \\ &= \frac{1}{1 + \left|\frac{x-169.9}{5.95}\right|4.96} \end{aligned} \tag{18}$$

where $a = 5.9$, $b = 2.479$ and $c = 169.90$

Triangle:

$$triangle(x;\ a,\ b,\ c) = \begin{cases} 0, & x \le a \\ \frac{x-a}{x-b}, & a \le x \le b \\ \frac{c-x}{c-b}, & b \le x \le c \\ 0, & c \le x \end{cases} \tag{19}$$

0 is when there is no membership, in the corners of membership.

$$triangle(x; a, b, c) = \begin{cases} 0, x \le a \\ \frac{x-a}{x-b}, a \le x \le b \\ \frac{c-x}{c-b}, b \le x \le c \\ 0, c \le x \end{cases}$$

MF high:

$$MFHight(x) = \begin{cases} 0, x \le 138 \\ \frac{x-138}{11}, 138 \le x \le 149 \\ \frac{162-x}{13}, 149 \le x \le 162 \\ 0, 162 \le x \end{cases}$$

where $a = 138$, b = *149* and $c = 162$

The result of the denominators is the subtractions of the variables.

Type-2 Memberships Functions:

$$\tilde{\boldsymbol{u}}(\mathbf{x}) = \left[\mu(x),\ \tilde{\boldsymbol{u}}(\mathbf{x})\right] = \mathbf{gausstype2}(\mathbf{x},\ [\boldsymbol{\sigma}_x \boldsymbol{m}_1, \boldsymbol{m}_2]) \tag{20}$$

where "igaussstype2" stands for the Gaussian generalized type-2 membership function with uncertain mean:

$$\boldsymbol{m}_x = \frac{\boldsymbol{m}_1 + \boldsymbol{m}_2}{2} \tag{21}$$

$$\boldsymbol{p}_x = \mathbf{gaussmf}(\mathbf{x},\ [\boldsymbol{\sigma}_x \boldsymbol{m}_1, \boldsymbol{m}_2]) = \exp\left[\frac{1}{2}\left(\frac{x - \boldsymbol{m}_x}{\boldsymbol{\sigma}_x}\right)^2\right] \tag{22}$$

$$\boldsymbol{p}_x = \mathbf{gaussmf}(\mathbf{x},\ [\boldsymbol{\sigma}_x, \boldsymbol{m}_x])$$

$$\widetilde{\boldsymbol{u}}(\mathbf{x},\ \boldsymbol{\mu}) = \mathbf{gaussmf}\left(\boldsymbol{\mu}, \left[\boldsymbol{\sigma}_x,\ \boldsymbol{p}_x\right]\right)$$

$$= \exp\left[\frac{1}{2}\left(\frac{x - p_x}{\sigma_u}\right)^2\right] \tag{23}$$

where $= p_x =$ is of the center of the function and σ_u is the implication of the function.

MF too high:

$$MFToohigh = e^{-\frac{1}{2}}\left(\frac{x - 234}{25.2}\right)^2 \tag{24}$$

where $p_x = 234$ *and* $\sigma_u = 25.2$.

2.2 Rules of Fuzzy Systems

In this section the 64 possible rules of the fuzzy inference system are shown (Fig. 4).

2.3 Particle Swarm Optimization for the Fuzzy System

The particle structure is composed of 390 parameters, of which 1–328 are real numbers and represent parameters of the inputs and outputs of the membership functions used are Gaussian, generalized bell, triangular, and trapezoidal. The Parameters 325–388 represent the number of rules of type-1 fuzzy systems (Fig. 5).

The PSO structure is composed of 720 parameters, of which 1–712 are real numbers and represent parameters of the inputs and outputs of the membership functions used are Gaussian, generalized bell, triangular, and trapezoidal. The Parameters 649–722, represent the number of rules of type-2 fuzzy systems (Fig. 6).

Objective Function:

The main motivation for the optimize the structures of the fuzzy system type-1 and type-2, (as soon as parameters and type of membership functions, type of system, and a number of rules), to reduce the number of fuzzy rules and the classification error on the base patient data.

$$\textbf{Objective Function} = ep_c + \left(np_r / mp_r\right) \tag{25}$$

ep_c is the classification error of the PSO.

np_r is the rules number obtained for PSO.

$mp_{r\,\text{the}}$ maximum number of rules of the PSO.

1. If (Systolic is Low) and (Diastolic is Low) then (Levels is Hypotension) (1)
2. If (Systolic is Low) and (Diastolic is Optimal) then (Levels is Optimal) (1)
3. If (Systolic is Low) and (Diastolic is Normal) then (Levels is Normal) (1)
4. If (Systolic is Low) and (Diastolic is Grade1) then (Levels is HighNormal) (1)
5. If (Systolic is Low) and (Diastolic is Grade1) then (Levels is Grade1) (1)
6. If (Systolic is Low) and (Diastolic is Grade2) then (Levels is Grade2) (1)
7. If (Systolic is Low) and (Diastolic is Grade3) then (Levels is Grade3) (1)
8. If (Systolic is Low) and (Diastolic is ISH) then (Levels is ISH1) (1)
9. If (Systolic is Optimal) and (Diastolic is Low) then (Levels is Optimal) (1)
10. If (Systolic is Optimal) and (Diastolic is Optimal) then (Levels is Optimal) (1)
11. If (Systolic is Optimal) and (Diastolic is Normal) then (Levels is Normal) (1)
12. If (Systolic is Optimal) and (Diastolic is HighNormal) then (Levels is HighNormal) (1)
13. If (Systolic is Optimal) and (Diastolic is Grade1) then (Levels is Grade1) (1)
14. If (Systolic is Optimal) and (Diastolic is Grade2) then (Levels is Grade2) (1)
15. If (Systolic is Optimal) and (Diastolic is Grade3) then (Levels is Grade3) (1)
16. If (Systolic is Optimal) and (Diastolic is ISH) then (Levels is ISH1) (1)
17. If (Systolic is Normal) and (Diastolic is Low) then (Levels is Optimal) (1)
18. If (Systolic is Normal) and (Diastolic is Optimal) then (Levels is Optimal) (1)
19. If (Systolic is Normal) and (Diastolic is Normal) then (Levels is Normal) (1)
20. If (Systolic is Normal) and (Diastolic is HighNormal) then (Levels is HighNormal) (1)
21. If (Systolic is Normal) and (Diastolic is Grade1) then (Levels is Grade1) (1)
22. If (Systolic is Normal) and (Diastolic is Grade2) then (Levels is Grade2) (1)
23. If (Systolic is Normal) and (Diastolic is Grade3) then (Levels is Grade3) (1)
24. If (Systolic is Normal) and (Diastolic is ISH) then (Levels is ISH1) (1)
25. If (Systolic is HighNormal) and (Diastolic is Low) then (Levels is Optimal) (1)
26. If (Systolic is HighNormal) and (Diastolic is Optimal) then (Levels is Optimal) (1)
27. If (Systolic is HighNormal) and (Diastolic is Normal) then (Levels is Normal) (1)
28. If (Systolic is HighNormal) and (Diastolic is HighNormal) then (Levels is HighNormal) (1)
29. If (Systolic is HighNormal) and (Diastolic is Grade1) then (Levels is Grade1) (1)
30. If (Systolic is HighNormal) and (Diastolic is Grade2) then (Levels is Grade2) (1)
31. If (Systolic is HighNormal) and (Diastolic is Grade3) then (Levels is Grade3) (1)
32. If (Systolic is HighNormal) and (Diastolic is ISH) then (Levels is ISH1) (1)
33. If (Systolic is Grade1) and (Diastolic is Low) then (Levels is ISH1) (1)
34. If (Systolic is Grade1) and (Diastolic is Optimal) then (Levels is ISH1) (1)
35. If (Systolic is Grade1) and (Diastolic is Normal) then (Levels is ISH1) (1)
36. If (Systolic is Grade1) and (Diastolic is HighNormal) then (Levels is ISH1) (1)
37. If (Systolic is Grade1) and (Diastolic is Grade1) then (Levels is Grade1) (1)
38. If (Systolic is Grade1) and (Diastolic is Grade2) then (Levels is Grade2) (1)
39. If (Systolic is Grade1) and (Diastolic is Grade3) then (Levels is Grade3) (1)
40. If (Systolic is Grade1) and (Diastolic is Grade3) then (Levels is ISH1) (1)
41. If (Systolic is Grade2) and (Diastolic is Low) then (Levels is ISH2) (1)
42. If (Systolic is Grade2) and (Diastolic is Optimal) then (Levels is ISH2) (1)
43. If (Systolic is Grade2) and (Diastolic is Normal) then (Levels is ISH2) (1)
44. If (Systolic is Grade2) and (Diastolic is HighNormal) then (Levels is ISH2) (1)
45. If (Systolic is Grade2) and (Diastolic is Grade1) then (Levels is Grade2) (1)
46. If (Systolic is Grade2) and (Diastolic is Grade2) then (Levels is Grade2) (1)
47. If (Systolic is Grade2) and (Diastolic is Grade3) then (Levels is Grade3) (1)
48. If (Systolic is Grade2) and (Diastolic is ISH) then (Levels is ISH2) (1)
49. If (Systolic is Grade3) and (Diastolic is Low) then (Levels is ISH3) (1)
50. If (Systolic is Grade3) and (Diastolic is Optimal) then (Levels is ISH3) (1)
51. If (Systolic is Grade3) and (Diastolic is Normal) then (Levels is ISH3) (1)
52. If (Systolic is Grade3) and (Diastolic is HighNormal) then (Levels is ISH3) (1)
53. If (Systolic is Grade3) and (Diastolic is Grade1) then (Levels is Grade3) (1)
54. If (Systolic is Grade3) and (Diastolic is Grade2) then (Levels is Grade3) (1)
55. If (Systolic is Grade3) and (Diastolic is Grade3) then (Levels is Grade3) (1)
56. If (Systolic is Grade3) and (Diastolic is ISH) then (Levels is ISH3) (1)
57. If (Systolic is ISH) and (Diastolic is Low) then (Levels is ISH1) (1)
58. If (Systolic is ISH) and (Diastolic is Optimal) then (Levels is ISH2) (1)
59. If (Systolic is ISH) and (Diastolic is Normal) then (Levels is ISH3) (1)
60. If (Systolic is ISH) and (Diastolic is HighNormal) then (Levels is ISH3) (1)
61. If (Systolic is ISH) and (Diastolic is Grade1) then (Levels is Grade3) (1)
62. If (Systolic is ISH) and (Diastolic is Grade2) then (Levels is Grade3) (1)
63. If (Systolic is ISH) and (Diastolic is Grade3) then (Levels is Grade3) (1)

Fig. 4 Fuzzy rule base. Possible rules of the fuzzy system

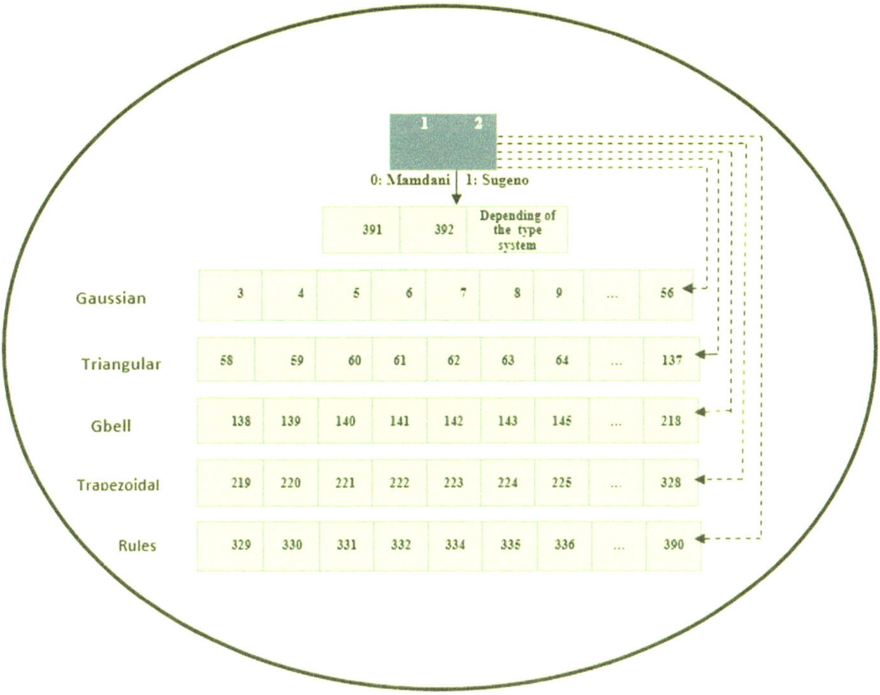

Fig. 5 PSO particle composition for the type 1 fuzzy system

3 Experimental Results

This part explains the experimental results of optimization of the type-1 and Type-2 classifiers.

Table 1 Summarizes all the experiments results of classifier type-1 where MFT (Membership Function Type), FST (Type of the Fuzzy system), Rules Number, Classification Error, and Time. Table 1 are also shown the best classification error 0.001045, where the best iteration was 8, in a time 01:17:15, with 22 fuzzy rules, the membership function type is Gaussian, and shows the results of the 30 Iterations.

Table 2 summarizes all the results of the experiments of classifier type-2 where MFT (Membership Function Type), FST (Type of the Fuzzy system), Rules Number, Classification Error, and Time. Table 2 are also shown best the classification error of 0.001272, where the best iteration was 18, in a time 02:21:18, with 24 fuzzy rules, the membership function type is Gaussian, and shows the results of the 30 Iterations.

Table 3 summarizes all the results of the 40 patients and from which to carry out the experiments we have 45 samples of diastolic and systolic pressure and as a result, we obtain the arterial classification with the type-1 fuzzy system.

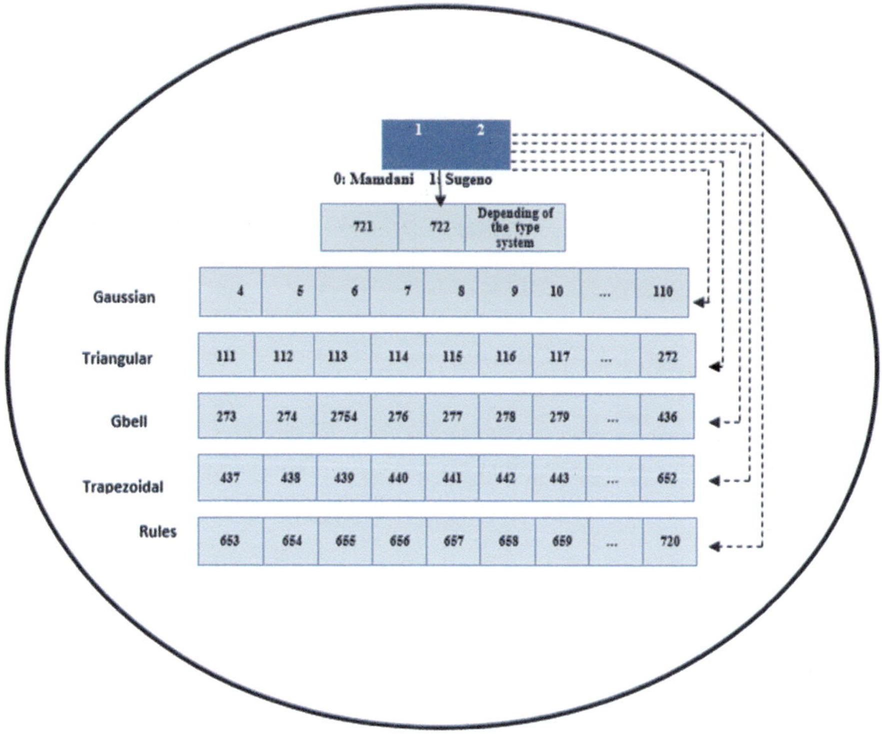

Fig. 6 PSO particle composition for the type-2 fuzzy system

Table 4 summarizes all the results of the 40 patients and from which to carry out the experiments we have 45 samples of diastolic and systolic pressure and as a result we obtain the arterial classification with the type-2 fuzzy system.

4 Conclusion

In this paper, a fuzzy classifier for systolic and diastolic hypertension was developed, with the type 1 fuzzy classifier optimization, with a classification error of 0.001045. We obtained with 22 rules, Mamdani fuzzy system with Gaussian-type membership functions. On the other hand, in the type-2 fuzzy classifier optimization, with a classification error of 0.001275, we obtained with 23 rules, Mamdani fuzzy system with Gaussian-type membership functions. Tests were carried out with 40 people and we obtained excellent results with the type-1 and type-2 fuzzy classifiers. Fuzzy systems and particle optimization algorithms are techniques that are effective and efficient in providing good classification solutions.

Table 1 Type-1 classifier results

Iterations	FST	TMF	Rules number	Time	Classification error
1	M	Gau	62	01:03:13	0.00221
2	M	Gau	18	00:54:26	0.002339
3	S	Gau	22	01:15:34	0.0013887
4	M	Gau	16	01:20:02	0.001425
5	M	Tri	23	01:22:10	0.002714
6	S	Gau	25	01:15:15	0.0023593
7	M	Tri	33	01:15:15	0.002651
8	M	Gau	22	01:17:15	0.001045
9	M	Gau	22	01:51:22	0.002665
10	M	Gau	35	01:50:24	0.001975
11	M	Tri	38	01:14:12	0.002265
12	S	Tri	22	00:54:26	0.002339
13	M	Tri	34	01:15:34	0.0013887
14	S	Gau	64	01:20:02	0.001295
15	S	Gau	33	01:22:10	0.003214
16	M	Gbell	26	01:15:15	0.003596
17	M	Tri	28	01:15:15	0.002651
18	S	Gau	39	01:17:15	0.001124
19	M	Gau	27	01:51:22	0.002665
20	M	Gbell	31	01:52:52	0.001978
21	M	Tri	44	01:23:13	0.00222
22	S	Gau	47	00:54:26	0.002339
23	M	Gau	58	01:15:34	0.002388
24	S	Tri	21	01:20:02	0.001555
25	M	Gau	36	01:22:10	0.002714
26	S	Gau	22	01:15:15	0.002366
27	M	Gau	29	01:15:15	0.001261
28	M	Tri	30	01:17:15	0.001735
29	M	Tri	31	01:31:22	0.002133
30	M	Tri	21	01:49:54	0.001836

Table 2 Result of the classification with type-2

Iterations	ST	TMF	Rules number	Time	Classification error
1	M	Gau	51	02:05:18	0.00346
2	M	Gau	63	02:46:27	0.00329
3	S	Gau	22	02:17:36	0.002387
4	M	Gau	16	01:20:02	0.002125
5	M	Tri	23	01:22:10	0.001896
6	S	Gau	25	01:15:15	0.002187
7	M	Tri	33	01:15:15	0.002651
8	S	Gau	44	01:17:15	0.001756
9	M	Gau	22	01:51:22	0.001366
10	M	Gau	36	01:49:54	0.002190
11	M	Tri	39	01:03:13	0.001887
12	S	Tri	22	00:54:26	0.002339
13	M	Tri	34	01:15:34	0.001638
14	S	Gau	64	01:20:02	0.002195
15	S	Gau	33	01:22:10	0.003214
16	M	Gbell	26	01:15:15	0.002387
17	M	Tri	28	01:15:15	0.002651
18	M	Gau	24	02:21:18	0.001272
19	M	Gau	27	01:51:22	0.002317
20	M	Gbell	33	01:49:54	0.001976
21	M	Tri	43	01:03:13	0.002123
22	S	Gau	47	00:54:26	0.001878
23	M	Gau	58	01:15:34	0.001977
24	S	Tri	22	01:20:02	0.001457
25	M	Gau	36	01:22:10	0.002156
26	S	Gau	22	01:15:15	0.001745
27	M	Gau	29	01:15:15	0.002261
28	M	Tri	30	01:17:15	0.001283
29	M	Tri	*31*	02:41:23	0.002178
30	M	Gau	27	02:29:57	0.002083

Table 3 Results with type-1 fuzzy classifier for the arterial hypertension

Persons	Systolic	Diastolic	Systolic classification	Diastolic classification	Classification
1	117	76	117	76	Optimal
2	118	77	115	75	Optimal
3	107	74	107	74	Optimal
4	122	75	122	75	Normal
5	114	66	115	61	Optimal
6	141	81	141	81	High
7	106	62	112	64	Optimal
8	120	81	120	81	Normal
9	107	61	112	58	Optimal
10	130	74	129	73	Normal
11	116	73	116	73	Optimal
12	134	62	130	62	Normal
13	135	82	136	81	Normal
14	121	77	120	72	Optimal
15	109	63	108	63	Optimal
16	123	71	124	71	Normal
17	125	77	126	76	Normal
18	106	65	106	65	Optimal
19	110	68	110	68	Optimal
20	123	76	123	76	Normal
21	115	72	112	71	Optimal
22	112	71	111	70	Optimal
23	122	76	122	73	Normal
24	117	68	116	67	Optimal
25	121	74	120	74	Optimal
26	129	82	129	80	Normal
27	121	63	120	61	Optimal
28	112	72	112	73	Optimal
29	123	82	121	82	Normal
30	95	61	95	61	Optimal
31	106	65	106	65	Optimal
32	110	69	116	75	Optimal
33	116	67	116	71	Normal
34	130	86	130	86	Normal
35	117	73	117	74	Optimal
36	117	54	117	58	Optimal

(continued)

Table 3 (continued)

Persons	Systolic	Diastolic	Systolic classification	Diastolic classification	Classification
37	113	72	113	70	Optimal
38	132	86	131	71	Normal
39	128	81	128	81	Normal
40	131	85	134	85	Normal

Table 4 Results with fuzzy classifier type-2 for the arterial hypertension

Persons	Systolic	Diastolic	Systolic classification	Diastolic classification	Classification
1	117	76	117	76	Optimal
2	118	77	115	75	Optimal
3	107	74	107	74	Optimal
4	122	75	122	75	Normal
5	114	66	114	61	Optimal
6	141	81	142	81	High
7	106	62	112	64	Optimal
8	120	81	120	81	Normal
9	107	61	112	58	Optimal
10	130	74	129	73	Normal
11	116	73	116	73	Optimal
12	134	62	130	77	Normal
13	135	82	136	81	Normal
14	121	77	120	72	Optimal
15	109	63	108	63	Optimal
16	123	71	124	70	Normal
17	125	77	126	76	Normal
18	106	65	106	65	Optimal
19	110	68	112	68	Optimal
20	123	76	123	76	Normal
21	115	72	114	72	Optimal
22	112	71	112	71	Optimal
23	122	76	122	76	Normal
24	117	68	116	68	Optimal
25	121	74	121	74	Optimal
26	129	82	130	82	Normal
27	121	63	121	63	Optimal
28	112	72	112	72	Optimal

(continued)

Table 4 (continued)

Persons	Systolic	Diastolic	Systolic classification	Diastolic classification	Classification
29	123	82	122	82	Normal
30	95	61	95	61	Optimal
31	106	65	109	77	Optimal
32	110	69	115	75	Optimal
33	116	67	127	69	Normal
34	130	86	125	80	Normal
35	117	73	119	78	Optimal
36	117	54	117	61	Optimal
37	113	72	108	70	Optimal
38	132	86	123	80	Normal
39	128	81	127	81	Normal
40	131	85	133	84	Normal

References

1. B. Akdemïr, S. Günes, B. Oran, S. Karaaslan, Expert Syst. Appl. **37**(8), 720–5727 (2010)
2. J.N. Booth III., K.M. Diaz, S.R. Seals, M. Sims, J. Ravenell, P. Muntner, D. Shimbo, Masked hypertension and cardiovascular disease events in a prospective cohort of blacks, the Jackson heart study. Hypertension **68**, 501–510 (2016)
3. H. Bustince, E. Barrenechea, M. Pagola, M. Fernandez, Interval valued fuzzy sets constructed from matrices, application to edge detection. Fuzzy Sets Syst. **160**(13), 1819–1840 (2009)
4. D.A. Calhoun, D. Jones, S. Textor, D.C. Goff, T.P. Murphy, R.D. Toto, A. White, W.C. Cushman, W. White, D. Sica, K. Ferdinand, T.D. Giles, B. Falkner, R.M. Carey, Resistant hypertension: diagnosis, evaluation, and treatment, a scientific statement from the American Heart Association Professional Education Committee of the Council for High Blood Pressure Research. Hypertension, 1403–1419 (2008)
5. O. Castillo, P. Melin, *Type-2 Fuzzy Logic Theory and Applications* (Springer, Berlin, Germany, 2008)
6. S. Chen, Y. Chang, J. Pan, Fuzzy rules interpolation for sparse fuzzy rule based systems based on interval type-2 Gaussian fuzzy sets and genetic algorithms. IEEE Trans. Fuzzy Syst. **21**(3), 412–425 (2013)
7. S. Das, P.K. Ghosh, S. Kar, Hypertension diagnosis: a comparative study using fuzzy expert system and neuro fuzzy system, in *IEEE International Conference on Fuzzy Systems, Durgapur, India*, vol. 148 (IEEE, 2013), pp. 113–120
8. X.Y. Djam, Y.H. Kimbi, Fuzzy expert system for the management of hypertension. Pac. J. Sci. Technol. **11**, 1 (2011)
9. S. Desch, T. Okon, D. Heinemann, K. Kulle, K. Röhnert, M. Sonnabend, M. Petzold, U. Müller, G. Schuler, I. Eitel, H. Thiele, P. Lurz, Randomized sham controlled trial of renal sympathetic denervation in mild resistant hypertension. Hypertension 1202–208 (2015)
10. B.M. Egan, Y. Zhao, J. Li, W.A. Brzezinski, T.M. Todoran, R.D. Brook, D.A. Calhoun, Prevalence of optimal treatment regimens in patients with apparent treatment resistant hypertension based on office blood pressure in a community-based practice network. Hypertension 691–697 (2013)
11. P. Engelbrech, Basic particle swarm optimization, in *Fundamentals of Computational of Swarm Intelligence* (Wiley, 2005), pp. 93–129

12. L. Golman, Ausiell, R.J. Johnson, I.B. Rodriguez, T. Nakagawa, D.H. Kang, D.I. Feig et al., Subtle renal injury is likely a common mechanism for salt-sensitive essential hypertension. Hypertension **45**, 326–330 (2001) (Diccionario de la Real Academia Española. 22ª ed. Tomo España)
13. B. González, F. Valdez, P. Melin, G. Prado-Arechiga, Fuzzy logic in the gravitational search algorithm for the optimization of modular neural network pattern recognition. Expert Syst. Appl. **42**(14), 5839–5847 (2015)
14. T.W. Hansen, Y. Li, J. Boggia, L. Thijs, T. Richart, J.A. Staessen, Predictive role of the nighttime blood pressure. Hypertension **57**, 3–10 (2011)
15. K. Heusser, J. Tank, J. Brinkmann, J. Menne, J. Kaufeld, S. Linnenweber-Held, J. Beige, M. Wilhelmi, A. Diedrich, H. Haller, J. Jordan, Acute response to unilateral unipolar electrical carotid sinus stimulation in patients with resistant arterial hypertension. Hypertension **67**, 585–591 (2016)
16. J.S.R. Jang, C.T. Sun, E. Mizutani, *Neuro-Fuzzy and Soft Computing* (Prentice Hall, 1996)
17. N. Karnik, M. Mendel, Applications of type-2 fuzzy logic systems to forecasting of time-series. 89–111 (1999)
18. J. Mendel, A quantitative comparison of interval type-2 and type1fuzzy logic systems: first results, in *Proceedings of the IEEE International Conference on Fuzzy Systems* (IEEE, 2010), pp. 1–8
19. J. Mendel, Comments on alpha-plane representation for type-2 fuzzy sets: theory and applications. IEEE Trans. Fuzzy Syst. **18**(1), 229–230 (2010)
20. J. Mendel, *Uncertain, Rule-Based Fuzzy Logic Systems: Introduction and New Directions* (Prentice-Hall, Englewood Cliffs, NJ, USA, 2001)
21. P. Muntner, R.C. Becker, D. Calhoun et al., Introduction to the American Heart Association's hypertension strategically focused research network. Hypertension **67**, 674–680 (2016)
22. S. Muthukaruppan, M.J. Er, A hybrid particle swarm optimization based fuzzy expert system for the diagnosis of coronary artery disease. Expert Syst. Appl. **39**(14), 11657–11665 (2012)
23. T. Noia, C. Ostuni, F. Pesce, G. Binetti, D. Naso, F.P. Schena, E. Sciascio, An end-stage kidney disease predictor based on artificial neural networks ensembles. Expert Syst. Appl. **40**(11), 4438–4445 (2013)
24. S.D. Persell, Prevalence of resistant hypertension in the United States. Hypertension **57**, 1076–1080 (2011)
25. S. Polak, M. Alesksander, Artificial neural networks based Internet hypertension prediction tool development and validation. Appl. Soft Comput. **8**(1), 734–739 (2008)
26. M. Rahman, T. Greene, R.A. Phillips et al., A trial of 2 strategies to reduce nocturnal blood pressure in blacks with chronic kidney disease. Hypertension **61**, 82–88 (2013)
27. J. Richards, A.N. Diaz, M.L. Gumz, Clock genes in hypertension: novel insights from rodent models. Blood Press. Monit. **19**, 249–254 (2014)
28. D.R. Roberie, W.J. Elliott, What is the prevalence of resistant hypertension in the United States. Curr. Opin. Cardiol. **27**, 86–391 (2012)
29. J. Sim, J. Handler, S.J. Jacobsen, M.H. Kanter, Systemic implementation strategies to improve hypertension: the Kaiser Permanente Southern California experience. Can. J. Cardio. **30**, 544–552 (2014)
30. C. Su, C. Yang, Feature selection for the SVM: an application to hypertension diagnosis. Expert Syst. Appl. **34**(1), 754–763 (2008)
31. M. Ture, I. Kurt, A. Kurum, K. Ozdamar, Comparing classification technique for predicting essential hypertension. Expert Syst. Appl. **29**(3), 583–588 (2005)
32. A. Wang, N. An, G. Chen, L. Li, G. Alterovitz, Predicting hypertension without measurement: a non-invasive, questionnaire-based approach. Expert Syst. Appl. **42**, 7601–7609 (2015)
33. J.T. Wright, L. Agodoa, G. Contreras, T. Greene, J.G. Douglas, J. Lash, O. Randall, N. Rogers, M.C. Smith, S. Massry, African American study of kidney disease and hypertension study group. Successful blood pressure control in the African American study of kidney disease and hypertension. Arch. Intern. Med. (2002)

34. L.A. Zadeh, R. Yager et al., *Fuzzy Sets and Applications: Selected* (Wiley, New York, 1987)
35. H. Zhang, F.C. Lin, Medical diagnosis by the virtual physician, in *Proceedings 12th IEEE Symposium on Computer-Based Medical Systems Computer based medical system*, 1999 (IEEE, 2009), pp. 296–302

A Survey of Models and Solution Methods for the Internet Shopping Optimization Problem

Miguel Ángel García Morales, Hector Joaquín Fraire Huacuja, Juan Frausto Solís, Laura Cruz Reyes, and Claudia Guadalupe Gómez Santillán

Abstract The Internet shopping optimization problem (IShOP) is an NP-hard combinatorial problem, which minimizes the total cost of shopping a list of products available in a set of shops on the Internet, considering the product price plus the shipping costs. With the advent of electronic commerce and the incredible popularity of Internet transactions, IShOP has become a problem of great relevance today in modern society with several variations of the practical problem application. This chapter reviews the different approaches applied to solve the problem. We review the used models, the solution methods, and the instances used to analyze the performance algorithms. Finally, we identify the main current and future research trends.

Keywords Internet shopping optimization problem · Heuristic algorithms · Metaheuristic algorithms

1 Introduction

Electronic commerce has revolutionized modern society today due to the important advancement of information technologies [1].

Undoubtedly, the large number of transactions carried out every day are primarily due to online purchases; users mainly make purchases of products and services that find meager prices. The main advantage of suppliers when offering their products on the Internet is that they are available to a broader public, and they do not have to worry about associated expenses such as maintenance, rent, among others, better prices, and a more comprehensive range of products and services [2].

The Internet shopping optimization problem (IShOP) [3] assumes a customer wants to purchase a shopping list in a set of online stores with the least possible expense. This problem is an Optimization problem formally modeled in [4]. However, IShOP is known to be an NP-hard combinatorial problem [5]. Therefore, achieving optimal solutions can be time-consuming, specifically for massive instances.

M. Á. G. Morales · H. J. F. Huacuja (✉) · J. F. Solís · L. C. Reyes · C. G. G. Santillán
Tecnológico Nacional de México/Instituto Tecnológico de Ciudad Madero, Cd. Madero, Tamaulipas, México
e-mail: automatas2002@yahoo.com.mx

O. Castillo and P. Melin (eds.), *Fuzzy Logic and Neural Networks for Hybrid Intelligent System Design*, Studies in Computational Intelligence 1061,
https://doi.org/10.1007/978-3-031-22042-5_6

Currently, there are many variants of IShOP, and this chapter reviews the different approaches applied to solve the problem, the solution methods, and the instances used to compare the algorithms and identify current research trends and future search topics.

2 Literature Review

This section presents the works in the state-of-the-art related to the Internet Shopping Optimization Problem. This section aims to diagnose the current situation and available knowledge regarding the solution to the problem. Furthermore, the works show a diversity of variants of the IShOP. Moreover, finally, an analysis is carried out to identify the problems or open areas of interest.

2.1 Internet Shopping Optimization Problem with Shipping Costs

Blazewicz et al. [4] formally define the IShOP problem for the first time. To define the problem consider that a customer needs to buy a set of n products N online, which he can buy in a set of m available stores M. The set N_i contains the products available in-store i, for each product $j \in N_i$ are given $c_{ij,}$ the cost of the product i in the store j and the shipping cost d_i from the store i. When a customer buys one or more products in a store, is add the shipping cost of this store i. Table 1, show the main parameters and variables used in the formal definition of the problem.

Formally, the IShOP problem consists in determining a partition of the products $X = (X_1, \ldots, X_m)$, such that $X_i \subseteq N_i$ and $U_{i=1}^{m} X_i = N$, and that minimizes the following total cost objective function:

$$F(X) = \sum_{i=1}^{m} \left(\sigma(|X_i|)d_i + \sum_{j \in X_i} c_{ij} \right) \tag{1}$$

where $|X_i|$ is the X_i cardinality, and $\sigma(X_i) = 0\ if\ i = 0$ and $\sigma(X_i) = 1\ if\ i > 0$.

Multiple items of the same type shopping are not supported; also it is assumed that all product is available in all stores. The instances used in the experiments consider five products and six stores.

Authors formally prove that the IShOP is NP-Hard in the strong sense even when the prices of the products are equal to zero and the shipping costs are equal to 1. Also they propose two polynomial algorithms to exactly solve small instances, the algorithm SHOP-ENUM has a $O(n2^m)$ time complexity and PRODUCT-ENUM has a $O(nm^n)$ time complexity.

Table 1 Variables and parameters

Variable/parameter	Explanation
M	Set of shops
N	Set of products
m	Number of shops, $\|M\|$
n	Number of products, $\|N\|$
i	Shop indicator
j	Product indicator
N_i	Multiset of products available from shop i
d_i	Delivery price of all products from shop i
c_{ij}	Cost of product j in shop i
x_{ij}	0–1 Usage indicator for product j in shop i
y_i	0–1 Usage indicator for shop i
T	Cumulative value of all products bought in all shops
T_i	Cumulative value of all products bought from shop i
$f_i(T_i)$	Piecewise function for all products (T_i) bought from shop i
$X = (X_1, \ldots, X_m)$	Sequence of selections of products from shops $1, \ldots, m$
$F(X)$	Sum of product and delivery costs
$\sigma(X)$	0–1 Indicator function for $x = 0$ and $x > 0$
X^*	Optimal sequence of selections of products
F^*	Optimal (minimum) total cost

The asterisk represents the optimal value

Lopez Locés et al. [2] uses a matrix to represent the candidate solutions and three sets of instances (small, medium and large) containing three subsets each one. The following subsets $3n20m$, $4n20m$ and $5n20m$ belong to the set of small instances, the medium instances includes the subsets $5n240m$, $5n400m$, $50n240m$, and finally, the large instances the subsets $50n400m$, $100n240m$, $100n400m$. They proposes the MinMin heuristic algorithm that goes through each store assigning a product and determines if it has the lowest total cost. Also they propose a metaheuristic cell processing algorithm which performs a simulation of multiple cell working in parallel. This work has the best results compared to the state-of-the-art algorithms. Multiple items of the same type shopping are not supported; also it is assumed that all product is available in all stores.

Table 2 Probability of occurrence

Occurrence (%)	$[a_{ij}]$	$[b_{ij}]$
8	$minimum$	$minimum + \frac{ref-minimum}{4}$
3	$minimum + \frac{ref-minimum}{4}$	$minimum + \frac{ref-minimum}{2}$
9	$minimum + \frac{ref-minimum}{2}$	$minimum + \frac{ref-minimum}{1.25}$
21	$minimum + \frac{ref-minimum}{1.25}$	ref
24	ref	$ref+\frac{maximum-ref}{4}$
9	$ref+\frac{maximum-ref}{4}$	$ref+\frac{maximum-ref}{2}$
10	$ref+\frac{maximum-ref}{2}$	$ref+\frac{maximum-ref}{1.25}$
16	$ref+\frac{maximum-ref}{1.25}$	$maximum$

The product prices are randomly generated as described below. Starting with an initial reference price (ref) for a given product $j : ref \in \{2, 4, \ldots, 100\}$, the price is randomly chosen with the probabilities of occurrence of 40% from 0 to 20, 16% from 22 to 30, 12% from 32 to 40, 16% from 42 to 60, and 16% from 62 to 100. The price for the product j from store i, is $p_{ij} \in [a_{ij}, b_{ij}]$, where $a_{ij} \geq 0.75ref_j$, $b_{ij} \leq 1.36ref_j$ and the intervals between $[a_{ij}, b_{ij}]$ are in Table 2 [6].

In this work authors propose an integer linear programming model (ILP) which can obtaining exact solutions in a reasonable time for the small instances subset. Table 1 show the main variables and parameters used in the ILP model.

In the ILP model the binary variable x_{ij} indicates whether a product j is bought at the store i, the binary variable y_i indicates if at least one product is purchased at the store i:

$$\min \sum_{i=1}^{m} \sum_{j=1}^{n} c_{ij} x_{ij} + \sum_{i=1}^{m} d_i y_i \tag{2}$$

$$s.t. \quad x_{i,j} \in \{0, 1\}, \forall i \in M, \forall j \in N, \tag{3}$$

$$y_i \in \{0, 1\}, \forall i \in M, \tag{4}$$

$$\sum_{i=1}^{m} \sum_{j=1}^{n} x_{ij} = n, \tag{5}$$

$$\sum_{i=1}^{m} x_{ij} = 1, \ \forall j \in N, \tag{6}$$

$$ny_i - \sum_{j=1}^{n} x_{ij} \geq 0, \ \forall i \in M. \tag{7}$$

The objective function is shown in (2), which is the total cost of purchasing a shopping list from the selected shops, including delivery costs, subject to the following constraints. The constraint (5) ensures that the number of purchased products is equal to the number of products on the shopping list, while constraint (6) guarantees that only one product of each kind is selected; the constraint (7) ensures that the variable y_i, takes the value of 1 when a product is purchased from shop i.

Wojciechowski and Musial [7] evaluated the best price comparison websites, creates a realistic model considering the actual conditions of internet purchases as possible and analyses the relationship between competitiveness, advertising, prices, and price dispersion in online stores. This model focuses on purchasing books as the product in most significant demand at that time and with a wide variety in virtual stores. Specific formulas are used to randomly calculate product prices and divide the price ranges into intervals. They propose a heuristic called 2-way basket optimization (2WBO), which orders the list of products in descending order and searches for the store with the lowest cost for each product. In this work, multiple items of the same type of shopping are supported; also, they assumed that all products are available in all stores. The instances used in the experiments consider 3, 5, 8, and 10 products and 20 stores.

López-Locés et al. [6] proposes two metaheuristics algorithms based on the trajectory: Tabu Search (TS) and Simulated Annealing (SA). The objective of these metaheuristics is to obtain an approximation of the optimal solution for the IShOP problem. For this, they created instances from actual conditions of online book stores such as Amazon, Barnes, and Noble. These instances are small (with ten instances with 20 instances) and large (with 30 instances). In this work multiple items of the same type shopping are not supported; also it is assumed that all product is available in all stores.

Verma et al. [8] propose a genetic algorithm (GA) to solve the IShOP, where a chromosome represents a solution and where each gene represents the index number of the store where they want to buy the product. The first generation of the solution is randomized; the fitness value is calculated by adding the costs of the products stored in the cost matrix of the stores and the products. The chromosome with the lowest total cost is the most suitable. The computational experiments consider instances created using the cost of 10 different products in 20 different virtual stores. They carry out an extension of the proposed work dedicated mainly to users who do not use smartphones. Multiple items of the same type shopping are not supported; also it is assumed that all product is available in all stores.

Sayyaadi et al. [9] propose a discretization approach and the use of the water cycle algorithm (WCA) to solve instances of the IShOP problem with a size of $10n30m$ (10 products and 30 stores). In this work multiple items of the same type shopping are not supported; also it is assumed that all product is available in all stores.

Huacuja et al. [10] propose a new metaheuristic algorithm based on the memetic algorithm methodology. The memetic algorithm uses a vector representation of the solutions and incorporate a mechanism that speeds up the calculation of the objective function. The computational experiments consider instances created randomly, three sets of instances (small, medium and large) containing three subsets each one. The following subsets $3n20m$, $4n20m$ and $5n20m$ belong to the set of small instances, the medium instances includes the subsets $5n240m$, $5n400m$, $50n240m$, and finally, the large instances the subsets $50n400m$, $100n240m$, $100n400m$. Each subgroup contains 30 cases. In this work multiple items of the same type shopping are not supported; also it is assumed that all product is available in all stores. The computational experiments show that the memetic algorithm performance outperforms to the cell processing algorithm.

2.2 *Internet Shopping Optimization Problem with Shipping Costs and Discounts*

The notation used in the Eq. 8 is given in Table 3. Musial et al. [11] for the first time approaches the IShOP with shipping costs and discounts and propose the following model:

$$\min \sum_{i=1}^{m} \sum_{j=1}^{n} f_j(p_{ij}x_{ij}) + \sum_{j=1}^{n} d_j y_j, \tag{8}$$

$$s.t. \sum_{j=1}^{n} x_{ij} = 1, i = 1, \ldots, m \tag{9}$$

$$0 \leq x_{ij} \leq y_j, i = 1, \ldots, m, j = 1, \ldots, n, \tag{10}$$

$$x_{ij} \in \{0, 1\}, y_j \in \{0, 1\}, i = 1, \ldots, m, j = 1, \ldots, n, \tag{11}$$

where a customer wants to buy products from a given set $M = \{1, \ldots, m\}$ in a given set of Internet shops $N = \{1, \ldots, n\}$ at the minimum total final price. These are the following given parameters and decision variables:

d_j—delivery price of all products from shop j,

y_j—usage indicator for shop j,

p_{ij}—standard price of product i in shop j,

x_{ij}—usage indicator for product i in shop j,

$f_j(T_j)$—piecewise function (discounting) for final price of all products T bought in shop j.

Błażewicz et al. [12] associate problem location facilities to IShOP with discounts and propose and develop simple heuristics to solve the IShOP with discounts. This

Table 3 Variables and parameters

Variable/parameter	Explanation
M	Set of products
N	Set of shops
m	Number of products
n	Number of shops
i	Product indicator
j	Shop indicator
M_j	Multiset of products available in shop j
d_j	Delivery price of all products from shop j
y_j	Usage indicator for shop j
p_{ij}	Cost of product i in shop j
x_{ij}	Usage indicator for product i in shop j
T	Cumulative value of all products bought in all shops
T_j	Cumulative value of all products bought from shop j
$f_j(T_j)$	Piecewise function for all products bought in shop j
$X = (X_1, \ldots, X_n)$	Sequence of selections of products in shops $1, \ldots, n$
$F(X)$	Sum of product and delivery costs
$\delta(X)$	0–1 Indicator function for $x = 0$ and $x > 0$
X^*	Optimal sequence of selections of products
F^*	Optimal (minimum) total cost

The asterisk represents the optimal value

heuristics obtains better results than the book price comparison sites. In this work, multiple items of the same type of shopping they supported also assumed that all products are available in all stores. Ten instances used in the experiments are generated for each pair (n, m).

Józefczyk and Ławrynowicz [13] consider the price-sensitive discounts of each product (ISOPwD), and propose two metaheuristic algorithms: tabu search (TS) and the simulated annealing (SA). Other possible discounts connected, for example, with bundles of products and (or) coupons, are not considered. The main contributions of this work are: First of all, a more overall version of IShOPwD they considered with an arbitrary number of purchased products, the possibility to buy an identical product in different stores when its quantity in the selected store is not enough and considering the particular parameters in the decision making processes such as product weight, amount and availability. In this work, multiple items of the same type shopping are supported; also, are assumed that all products are available in all stores. The instances used in the experiments consider 5, 10, 15, 20, 25, 30, 35, 40, 45, 50 and 55 products, and 50, 100, 150, 200, 250, 300, 350, 400, 450, 500, 550 stores, and ordered amount of product 10, 20, 30, 40, 50, 60, 70, 80, 90, 100, 110.

Musial et al. [11] consider the additional shipping costs to the prices of the products for each store; in addition, the IShOP model is extended including discounts on the prices of the products, a comparison of several optimization algorithms are made (Greedy, Forecasting, Cellular, MinMin, B&B) to fix the IShOP issue with discounts. They assume that there are no differences between the quality of the goods the web stores sell besides their prices for the different products. They introduce a new set of heuristic approaches to solve the problem. The heuristic comprise a new lightweight metaheuristic based on a cellular optimization process, an extended greedy algorithm, and two state-of-the-art algorithms. The instances used in the experiments consider 5, 10, 15, 20, 25, 30, 35, 40, 45, 50, 55, 60, 65, 70, 75, 80, 85, 90, 95 products and 20, 40 stores. In this work, multiple items of the same type shopping do not supported them; also, is assumed that all product are available in all stores.

Blazewicz et al. [14] consider the shipping costs and discounts sensitive to the price of each product and in each store to solve the IShOP problem, a comparative evaluation is carried out between several algorithms (BB, PCS, PCS+, Greedy, Forecasting and a new implementation) to solve the IShOP problem with discounts and shipping costs. In this work, multiple items of the same type shopping do not support them; also, they assumed that all the products are available in all stores. The instances used in the experiments consider 2, 3, 4, 5, 6, 7 products and 20, 30, 40 stores.

Sadollah et al. [15] assign a maximum budget per client for the IShOP problem; in addition, discounts offered and the selection of multiple elements within the candidate solutions is added. In this work, the ISOP formulation is enhanced by considering several extra assumptions and constraints such as the maximum affordable budget, discounts offered by Internet shops, and permission for multiple-item selections. An ISOP experiment with 300 Internet shops and 400 products is investigated. Three metaheuristic optimization methods, including the genetic algorithm (GA), the harmony search (HS), and the water cycle algorithm (WCA) have been utilized to find the better optimal solutions; also, is assumed that all products are available in all stores.

Orciuoli et al. [16] consider non-cumulative discount coupons in the IShOP solution, a new type of coupon that only applies to a single product, and a new way of collecting discount coupons will be incorporated. This work aimed at systematically comparing three bio-inspired optimization approaches, genetic algorithms, memetic ones, and ant colony optimization, to detect the best performer for solving the shopping plan problem in a blended shopping scenario. The instances used in the experiments consider 10, 15, 20 products and 10, 40, 80 number of items in the wish list, and 15, 30 percentage of products subject to a coupon, also it is assumed that all product are available in all stores.

D'Aniello et al. [17] propose using a framework based on AmI to solve the IShOP problem, and mobile devices will be incorporated in scenarios with discount coupons to obtain candidate solutions. Moreover, the work defines a genetic algorithm to face the shopping plan problem with a heuristic approach. A genetic algorithm is used to solve various instances of the IShOP with discount coupons. The dataset for the evaluation has been generated randomly with the following characteristics: number of items in the wish list = 10, average number of product for each set of products = 100 and variance = 10 (generated by using a Gaussian probability distribution), the probability to use a coupon for a single product = 0.5, the probability of generating a coupon for a single product = 0.5, also it is assumed that all product is available in all stores.

D'Aniello et al. [18] use a framework to build scenarios that support the user to establish a purchasing plan to solve the IShOP problem, discount coupons they have taken into account, and a memetic algorithm is used to solve the problem. The algorithm uses a stochastic local search within the IShOP solution establishing a purchasing plan for each user. The dataset for the evaluation has been generated randomly with the following characteristics: number of items in the wish list = {10, 15, 20}, the average number of product for each wish = {10, 40, 80}, and number of involved coupons = {15, 30}, the probability to generating a coupon for a single product = 0.5, also it is assumed that all product is available in all stores.

Gaeta et al. [19] propose using a framework to solve the IShOP problem under an intelligent environment, which defines purchase plans that allow minimizing the total costs of purchases, taking into account discount coupons, also using a Genetic algorithm for solving the IShOP problem with discount coupons. The dataset for the evaluation has been generated randomly with the following characteristics: number of items in the wish list = 20, average number of products for each set of products = 1000 and variance = 100 (generated by using a Gaussian probability distribution), the probability to use a coupon for a single product = 0.06, the probability of generating a coupon for a single product = 0.04, also it is assumed that all product are available in all stores.

2.3 *Internet Shopping Optimization Problem with a Budget (B-ISOP)*

The notation used in the Eq. 12 is given in Table 4. Marszaªkowski [20] for the first time formulates the Internet shopping optimization problem with budget as follows. The customer wishes to acquire a set M of m products, where for each product I the user assigns its perceived value v_i. Said purchase is limited by a budget P. From the data collected in a Database we have the set N of n stores: for each store j, the standard delivery price is d_j and for each product i available in store j, your cost in this store is p_j. The objective is to maximize the amount of products that the customer can buy in the stores considering the budget limitation. This can be stated as follows:

Table 4 Variables and parameters

Variable/parameter	Explanation
M	Set of products
N	Set of shops
m	Number of products
n	Number of shops
i	Product indicator
j	Shop indicator
d_j	Delivery price of all products from shop j
y_j	Indicator variable for shop j
p_{ij}	Cost of product i in shop j
x_{ij}	Indicator variable for product i in shop j
v_i	User perceived value of product i
P	Budget, limit of total cost

$$\max \sum_{i=1}^{m} \sum_{j=1}^{n} x_{ij} v_i \tag{12}$$

$$s.t. \quad \sum_{i=1}^{m} \sum_{j=1}^{n} p_{ij} x_{ij} + \sum_{j=1}^{n} d_j y_j \leq P, \tag{13}$$

where there are additional restrictions so that each product is chosen at least once and that if any product is selected from a store j, this shops delivery cost will paid. Finally, the indicators are binary.

$$\begin{aligned}
&\sum_{j=1}^{n} x_{ij} \leq 1, i = 1, \ldots, m, \\
&0 \leq x_{ij} \leq y_j, i = 1, \ldots, m, j = 1, \ldots, n, \\
&x_{ij} \in \{0, 1\}, y_j \in \{0, 1\}, i = 1, \ldots, m, j = 1, \ldots, n.
\end{aligned} \tag{14}$$

Marsza[a]kowski [20], for the first time, B-ISOP is defined as a problem in which a customer wants to buy a list of products considering not exceeding an established budget. Additional restrictions limit the selection of a product only once and, if so, add the shipping cost from the store. Therefore, a mathematical formulation he carried out for the B-ISOP problem in the case of budget limitations, and he considered to maximizing the number of products that he can purchase with the established budget. The BKP problem he used to test the NP-Completes of this problem, however, for the particular case where the customer wants to maximize the number of products he receives, a new transformation of this problem to the MC problem is carried out. They include a relationship of the B-ISOP with other problems such as BKP, MCKP, WMC, BMC and GMC.

2.4 Internet Shopping Optimization Problem with Price Sensitive Discounts

Blazewicz et al. [17] conducted a study and optimization aspect to Internet shopping with price-sensitive discounts from customer perspective. Specifically, they consider a problem in which a customer would like to buy products of a given set $N = \{1, \ldots, n\}$ in a given set of Internet shops $M = \{1, \ldots, m\}$ at the final minimum price. The show parameters and decision variables are in Table 5.

They denote the above problem as IShOPwD, where the abbreviation stands for Internet Shopping with price sensitive discounts. The mathematical formulation of the model is as follows:

$$\min \sum_{i=1}^{m} f_i \left(d_i y_i + \sum_{j \in N_i} p_{ij} x_{ij} \right) \tag{15}$$

$$s.t. \sum_{i \in M_j} x_{ij} = 1, j = 1, \ldots, n \tag{16}$$

$$0 \leq x_{ij} \leq y_i, i = 1, \ldots, m, j = 1, \ldots, n \tag{17}$$

$$x_{ij} \in \{0, 1\}, y_i \in \{0, 1\}, i = 1, \ldots, m, j = 1, \ldots, n \tag{18}$$

Blazewicz et al. [17] consider the IShOP study with price-sensitive discounts from the customer's perspective. Two sub-problems are defined and both consider

Table 5 Variables and parameters

Variable/parameter	Explanation
d_i	Delivery price of all products from shop i to the customer
p_{ij}	Standard price of product j in shop i, $p_{ij} = p_j$ if Standard prices of product j are the same in all shops
N_i	Subset of products of the set N in shop i (eligible products for shop i),$N_i \subseteq N$
M_j	Subset of shops in which product j can be bought (eligible shops for product j), $M_j \subseteq M$
S_i	Subset of products selected by the customer in shop i (basket of shop i, decision variable), $N = \cup_{i=1}^{m} S_i$ and $S_i \cap S_j = \emptyset$, $i \neq j$, for a feasible solution
$T_i(S_i) = d_i + \sum_{j \in S_i} p_{ij}$	Total delivery and standard price in shop I for a given set of products $S_i \subseteq N_i$; if there is no ambiguity, notation S_i in $T_i(S_i)$ can be omitted
$f_i(T)$	Discounting function for final price, a concave increasing differentiable or concave piecewise linear function of total delivery and standard price T in shop i at all points $T > 0$, $f_i(0) = 0$

that all the stores have all the products and the opposite case where not all the stores sell them. They implemented a branch and bound algorithm to calculate the optimal solutions of the evaluated instances; for the experiments carried out, 50 small-size instances are used.

Musial et al. [21] investigate an extended version of the IShOP, considering price-sensitive discounts. A new set of heuristic approaches to solve the problem is introduced. The heuristics are composed of a new lightweight metaheuristic based on a cellular optimization process, a new greedy algorithm, and two state-of-the-art ones. They have designed different heuristics to consider a different solution quality regarding computational time and results close to the optimum solution. The optimal solutions for small problem instances are solved using a branch and bound algorithm; they also assumed that all products are available in all stores. In computational experiments assume that the number of stores is $\{20, 40\}$ and the number of products is $\{2, 3, \ldots, 10, 15, \ldots, 100\}$.

2.5 Trusted Internet Shopping Optimization Problem (T-ISOP)

Musial and López-Locés [22] propose for the first time a new, more sophisticate model where all the variables and symbols are described in Table 6, which reflects real shopping situations more accurately. Trusted Internet Shopping Optimization Problem (T-ISOP) is presented in the following way:

$$\min \sum_{i=1}^{m} \sum_{j=1}^{n} x_{ij} \frac{p_{ij}}{OPay_j} v_j + \sum_{j=1}^{n} y_j d_j$$

$$s.t. \tag{19}$$

$$\sum_{j=1}^{n} x_{ij} = 1 \forall i \in M \tag{20}$$

$$\sum_{i=1}^{m} \sum_{j=1}^{n} x_{ij} = m \tag{21}$$

$$m * y_j - \sum_{i=1}^{m} x_{ij} \geq 0 \forall j \in N \tag{22}$$

$$x_{ij} \in \{0, 1\} \quad \forall i \in M, \forall j \in N \tag{23}$$

$$y_j \in \{0, 1\} \quad \forall j \in N \tag{24}$$

Table 6 Variables and parameters

Variable/parameter	Explanation
m	Number of products to buy
n	Number of shops
i	Product indicator
j	Shop indicator
d_j	Delivery price for all products from shop j
y_j	Usage indicator for delivery price from shop j
p_{ij}	Cost of product i in shop j
x_{ij}	Usage indicator for product i in shop j
$OPay_j$	Overpay trust function for shop j
v_j	Trust veto factor for shop j

$$OPay_j = (1, \ldots, \max OP) \forall j \in N \tag{25}$$

$$v_j \in \{1, \infty\} \quad \forall j \in N \tag{26}$$

Musial and López-Locés [22] formulate a new model to solve the IShOP problem that includes trust and reputation for each of the stores to be evaluated; this model named Trusted Internet Shopping Optimization Problem (T-ISOP), a genetic algorithm (GA) is used to solve the IShOP problem that includes trust and reputation of stores. In this work, multiple items of the same type shopping are not supported; also, they assume that all products are available in all stores. The instances used in the experiments consider two sets of 10 products and 20 stores, and 20 products and 20 stores.

Musial et al. [1] propose an expansion of the IShOP problem model considering discounts among other variables; the recommendations are used so that the user can make decisions in the selection of stores where each product should be purchased. A mapping of the IShOP problem is carried out to the Cloud Brokering type and a framework they also developed that allows optimizing the IShOP problem.

2.6 Bi-objective Internet Shopping Optimization Problem

The notation used in Eq. 27 is in Table 7. Chung [23], for the first time, formulated the bi-objective Internet shopping optimization problem. The model considers the purchase cost and delivery time limitation as targets at the same time. The bi-objective model is the following:

Table 7 Variables and parameters

Variable/parameter	Explanation
M	Set of shops
N	Set of products
n	Number of products
m	Number of shops
p_{ij}	Price of product i at shop j
f_j	Delivery cost at shop j
d_{ij}	Expected delivery time of product i from shop j
x_{ij}	Binary decision variable of product i is selected from shop j
y_j	Binary decision variable of delivery cost at shop j

$$\text{Min} \sum_i \sum_j p_{ij}x_{ij} + \sum_j f_j y_j \tag{27}$$

$$\text{Min}\ \max_{i,j}\left(d_{ij}x_{ij}\right) \tag{28}$$

$$s.t. \sum_j x_{ij} = 1, \forall i = 1, \ldots, n \tag{29}$$

$$\sum_i x_{ij} \le n y_j, j = 1, \ldots, m \tag{30}$$

$$x_{ij} = 0/1, y_j = 0/1 \tag{31}$$

The objective function (27) means that you want to minimize the purchase cost, including price of products and delivery cost. The objective function (28) means that you want to minimize the delivery time of all products. Constraints (29) means that all products to buy must be selected from available shops and constraints (30) means that fixed delivery cost incurs whenever there is any product selection from shop. Constraints (31) means binary decision variables.

Chung [23] proposes a bi-objective solution to the IShOP problem using the costs of the products plus the shipping cost and adds the restriction of the product availability time. To solve the problem, two heuristics are used to satisfy the two objective functions to obtain the Pareto optimal set.

2.7 Research Open Issues

In this section, the main research issues for each variant of IShOP are described.

2.7.1 Internet Shopping Optimization Problem with Shipping Costs

This variant of the IShOP problem is the most studied [2, 4, 6–10]. The literature available can find several solution methods like linear programming algorithm, heuristic algorithms, and various metaheuristics algorithms. This model considers that the customers buy only one product of each type, and the stores have all the products in the shopping list. Practically all the reviewed works in this overview make this consideration which is a limitation for real application.

2.7.2 Internet Shopping Optimization Problem with Shipping Costs and Discounts

Błażewicz et al. [12] propose the IShOP with discounts model and develop simple heuristics to solve the problem. The only work with this IShOP variant where multiple items of the same type shopping are supported, and they assumed that all product are available in all stores. Solution methods are reported only for the variant that considers that the customers buy only one product of each type and include: Simple heuristics, Tabu search algorithm, and simulated annealing.

2.7.3 Internet Shopping Optimization Problem with a Budget (B-ISOP)

Future work could focus first on algorithms for the problem described. Including at least: an analysis of the usability of the greedy approximation algorithm for a more general GMC problem, a proposal for heuristic and metaheuristic algorithms. Therefore an approach is also required that provides optimal solutions at least to evaluate an optimization gap of the other algorithms in small instances. The ILP model should be able to be developed in existing resolution software. Second, the presented model simplifies reality that could they expanded to capture more real-world situations. Some other ideas such as price-sensitive discounts or dual-discount features have already are proposed above.

The areas of possible application of the presented problem are not limited to optimizing purchases on the Internet. Similar problems can already are found, where the user wants to buy the best set of cloud resources. It could be helpful in the future in energy markets, where the user wants to buy electricity from different sources, valuing even clean energy more than dirty energy.

2.7.4 Internet Shopping Optimization Problem with Sensitivity Discounts

The direction of future studies could focus on optimizing purchases from a supplier perspective or from both points of view as customers and suppliers strive to maximize their profits. The latter case assumes that, at the same time, stores are trying to sell their supplies as soon as possible because the value of most products decreases during the heuristic algorithms and potential customers delay buying while waiting for some discounts. Finally, other versions of ISOPwD may require the use of metaheuristics that have not they applied so far, e.g., ant colony optimization or particle swarm optimization.

2.7.5 Trusted Internet Shopping Optimization Problem (T-ISOP)

Future work will involve a deep analysis of the trust factor and combine it with optimization of the problem from a more technical point of view. In addition, the intermediation problem in the cloud could enrich with the analysis of the trust factor. Said topic is vibrant and, based on the knowledge base, it lacks and suffers from adequate (or even absolute) influence from factors of trust and reputation in the market. A work future research idea is to prepare computational experiments based on actual data. A simple decision aid simulation tool could be another exciting research point.

Moreover, it could be beneficial to use biology-inspired approaches, models, and computations to tackle the problem. Furthermore, both OPay and functions could are personalized to the user. This personalization process will are realized as a decision-aided questioner tool. Allows to prepare experimental analysis dedicated to every single user accordingly to his detailed needs and requirements.

2.7.6 Bi-objective Internet Shopping Optimization Problem

Currently, Internet search engine recommends a product based on a single criterion such as price or reputation. However, more advanced search engines or shopping robots will recommend multiple products based on multiple criteria such as purchasing cost, delivery time, and various delivery options. Therefore, extending the model to consider other delivery options is one of the other research topics.

3 Conclusions

This chapter reviews the most recent and relevant works related to the different variants of the Internet Shopping Optimization Problem. First, are identified the characteristics of the used model, the proposed solution methods, and the used instances for each variant. Once the works are analyzed, a series of open research areas are determined to give an overview to the researchers that want to work with the variants of the IShOP problem.

Acknowledgements Authors thanks to CONACYT for supporting the projects from (a) Cátedras CONACYT Program with Number 3058. (b) CONACYT Project with Number A1-S-11012 from Convocatoria de Investigación Científica Básica 2017–2018 and CONACYT Project with Number 312397 from Programa de Apoyo para Actividades Científicas, Tecnológicas y de Innovación (PAACTI), a efecto de participar en la Convocatoria 2020-1 Apoyo para Proyectos de Investigación Científica, Desarrollo Tecnológico e Innovación en Salud ante la Contingencia por COVID-19. (c) M.A. García Morales would like to thank CONACYT for the support number 658787. Also H. Fraire thank to Tecnológico Nacional de México for the support to the project 10362.21-P.

References

1. J. Musial, J. Pecero, B. Dorronsoro, J. Blazewicz, Internet shopping optimization project (IShOP), in *European IST Projects*, p. 16 (2014)
2. M.C. López-Locés, J. Musial, J.E. Pecero, H.J. Fraire-Huacuja, J. Blacewicz, P. Bouvry, Exact and heuristic approaches to solve the Internet shopping optimization problem with delivery costs. Int. J. Appl. Math. Comput. Sci. **26**(2), 391–406 (2016)
3. J. Musial, *Applications of Combinatorial Optimization for Online Shopping*. NAKOM, Poznán (2012)
4. J. Błażewicz, M.Y. Kovalyov, J. Musiał, A.P. Urbański, A. Wojciechowski, Internet shopping optimization problem. Int. J. Appl. Math. Comput. Sci. **20**(2), 385–390 (2010). https://doi.org/10.2478/v10006-010-0028-0
5. M. Gen, R. Cheng, *Genetic Algorithms & Engineering Optimization* (Wiley, New York, 2000)
6. M.C. López-Locés, K. Rege, J.E. Pecero, P. Bouvry, H.J.F. Huacuja, Trajectory metaheuristics for the internet shopping optimization problem, in *Design of Intelligent Systems Based on Fuzzy Logic, Neural Networks and Nature-Inspired Optimization* (Springer, Cham. 2015), pp. 527–536
7. A. Wojciechowski, J. Musial, Towards optimal multi-item shopping basket management: heuristic approach, in *On the Move to Meaningful Internet Systems: OTM 2010 Workshops*, ed by R. Meersman et al. Lecture Notes in Computer Science, Vol. 6428. (Springer, Berlin, 2010), pp. 349–357
8. S. Verma, A. Sinha, P. Kumar, A. Maitin, Optimizing online shopping using genetic algorithm, in *2020 3rd International Conference on Information and Computer Technologies (ICICT)* (IEEE, 2020), pp. 271–275
9. H. Sayyaadi, A. Sadollah, A. Yadav, N. Yadav, Stability and iterative convergence of water cycle algorithm for computationally expensive and combinatorial Internet shopping optimization problems. J. Exp. Theor. Artif. Intell. 1–21 (2018)
10. H.J.F. Huacuja, M.Á.G. Morales, M.C.L. Locés, C.G.G. Santillán, L.C. Reyes, M.L.M. Rodríguez, Optimization of the internet shopping problem with shipping costs, in *Fuzzy Logic Hybrid Extensions of Neural and Optimization Algorithms: Theory and Applications* (Springer, Cham, 2021), pp. 249–255
11. J. Musial, J.E. Pecero, M.C. Lopez-Loces, H.J. Fraire-Huacuja, P. Bouvry, J. Blazewicz, Algorithms solving the Internet shopping optimization problem with price discounts. Bull. Pol. Acad. Sci. Tech. Sci. 505–516 (2016)
12. J. Błażewicz, J. Musiał, E-commerce evaluation–multi-item Internet shopping. Optimization and heuristic algorithms, in *Operations Research Proceedings 2010*, ed. by B. Hu, et al. (Springer, Berlin, 2011), pp. 149–154

13. J. Józefczyk, M. Ławrynowicz, Heuristic algorithms for the Internet shopping optimization problem with price sensitivity discounts. Kybernetes **47**(4), 831–852 (2018)
14. J. Blazewicz, N. Cheriere, P.F. Dutot, J. Musial, D. Trystram, Novel dual discounting functions for the Internet shopping optimization problem: new algorithms. J. Sched. **19**(3), 245–255 (2016)
15. A. Sadollah, K. Gao, A. Barzegar, R. Su, Improved model of combinatorial Internet shopping optimization problem using evolutionary algorithms, in *2016 14th International Conference on Control, Automation, Robotics and Vision (ICARCV)* (IEEE, 2016), pp. 1–5
16. F. Orciuoli, M. Parente, A. Vitiello, Solving the shopping plan problem through bio-inspired approaches. Soft. Comput. **20**(5), 2077–2089 (2016)
17. G. D'Aniello, M. Gaeta, V. Loia, F. Orciuoli, An ami-based software architecture enabling evolutionary computation in blended commerce: the shopping plan application. Mobile Inf. Syst. (2015)
18. G. D'Aniello, F. Orciuoli, M. Parente, A. Vitiello, Enhancing an AmI-based framework for u-commerce by applying memetic algorithms to plan shopping, in *2014 International Conference on Intelligent Networking and Collaborative Systems* (IEEE, 2014), pp. 169–175
19. M. Gaeta, V. Loia, F. Orciuoli, M. Parmentola, A genetic approach to plan shopping in the AmI-based blended commerce, in *2013 IEEE International Symposium on Industrial Electronics* (IEEE, 2013), pp. 1–6
20. J. Marsza[a]kowski, *Budgeted Internet Shopping Optimization Problem* (2015)
21. J. Blazewicz, P. Bouvry, M.Y. Kovalyov, J. Musial, Internet shopping with price sensitive discounts. 4OR-A Q. J. Oper. Res. **12**(1), 35–48 (2014a)
22. J. Musial, M.C. Lopez-Loces, Trustworthy online shopping with price impact. Found. Comput. Decis. Sci. **42**(2), 121–136 (2017)
23. J.B. Chung, Internet shopping optimization problem with delivery constraints. J. Distrib. Sci. **15**(2), 15–20 (2017)

A Comparison Between MFCC and MSE Features for Text-Independent Speaker Recognition Using Machine Learning Algorithms

Joseph Isaac Ramírez-Hernández, Alain Manzo-Martínez, Fernando Gaxiola, Luis C. González-Gurrola, Vania C. Álvarez-Oliva, and Roberto López-Santillán

Abstract Speaker identification is the process through which a person can be identified by using the physical characteristics of his voice. In recent years, this identification is done by using artificial intelligence algorithms along to feature extract methods from the speech signal such as the MFCCs (Mel Frequency Cepstral Coefficients). Text-independent speaker recognition consists of identifying a person by training and testing of the model with voice signals where one speaker will say different phrases. In the present research two types of audio features were extracted, on one hand the MFCC and on the other hand the MSE (Multiband Spectral Entropy). For the classification stage we use machine learning algorithms, such as k Nearest Neighbors, Random Forest, Deep Neural Networks, and Decision Trees. Two important databases of the literature were used in our experiments, LibriSpeech and ELSDSR. Four different experiments were defined: speaker identification in a group of 20 participants, speaker identification between men, speaker identification between women, and speaker identification by gender. Significant results were obtained when using the ELSDSR database, obtaining 93.99% precision in the experiment for speaker recognition by gender.

Keywords Text-independent speaker identification · MFCC · Multiband spectral entropy

1 Introduction

A computational model is a mathematical model applied in computer science. Usually, this model does not has a well-defined analytical solution but is based on the idea of looking for parameters and favorable condition for the solution of a problem.

J. I. Ramírez-Hernández · A. Manzo-Martínez (✉) · F. Gaxiola · L. C. González-Gurrola · V. C. Álvarez-Oliva · R. López-Santillán
Universidad Autónoma de Chihuahua, Chihuahua 31125, México
e-mail: amanzo@uach.mx

F. Gaxiola
e-mail: lgaxiola@uach.mx

O. Castillo and P. Melin (eds.), *Fuzzy Logic and Neural Networks for Hybrid Intelligent System Design*, Studies in Computational Intelligence 1061,
https://doi.org/10.1007/978-3-031-22042-5_7

An Automatic Speaker Recognition System (ASR) consists of a stage which studies the speech and then it can interpret the speaker identity based on its features. Speech signals contain useful information of the speaker like emotions, gender, and accent [1].

In general, speakers can be identified using two different methodologies: text-dependent and text-independent. In the first hand, the text used for training and testing must be the same. In the second hand, the text used for training and testing is generally different.

Speech signals generally can be considered as non-stationary time series. For this reason, an ASR system faces various challenges as the following: speakers with different language, human interaction is not limited to speech and it also uses corporal movements and gestures, the vocal tract varies according to gender, accent, emotions, and age.

The importance of voice signal analysis is observed in the technological tools that work on these signals. ASR system has matured to the point of being robust and reliable. However, a general rule has not been found that solves the problem of identifying the speaker from the characteristics of his voice. Improving these voice recognition systems will allow this technology to be applied in automotive environments (autopilot systems, navigation systems, etc.) [2, 3], home automation (smart home, energy savings, etc.) [4, 5] and banking systems (data security, access to mobile applications, etc.) [6, 7].

Currently the ASR systems transform the message said by the speaker into text, this implies that the way the system recognizes is according to what the speaker says, namely, this is a text-dependent classification. Such is the case with Siri, Apple's personal assistant, which was introduced in 2010 by the brand. This personal assistant is activated and works by recognizing the text mentioned by a speaker, this implies that if any person gives instructions to the device, it will carry out those actions [8].

On the other hand, text-independent speech recognition analyzes the properties of the voice beyond the text emitted by the speaker and even more so, regardless of what the speaker says. For this reason, the need arises to use qualities and characteristics of the voice signal that mutually contrast the speakers and thus define a pattern for each of them.

1.1 Database Description

This section describes the different databases used in this research work, in order to better understand the information that constitutes them for later use in the methodology and results obtained.

ELSDSR

The Database ELSDSR (English Language Speech Database for Speaker Recognition) is a set of speech signals recorded in Denmark Technic University (DTU). This database contains samples from 30 Danish participants, 1 Icelander and 1 Canadian. All the samples were recorded at 16,000 Hz of sampling frequency [9].

LibriSpeech

This database was proposed by Vassil Panayotov and it is based on LibriBox project, an audiobooks project. It contains more of 1000 h of speech signals. All data were recorded using 16,000 Hz of sampling frequency. For this work, a developer database is used, and it contains 40 participants divided in 20 women and 20 men [10].

1.2 Literature Reviewed

Speaker recognition models can be divided in two phases: the feature extraction and a classification model. Feature extraction plays an important role, because it affects the performance of the classification model. For this research work, a search for previous work in the-state-of-art was carried out, all of them based on the text-independent methodology.

Ittichaichareon et al. [11] use MFCCs as a feature extraction method and a support vector machine classifier on an own database obtaining 44% of precision. MFCCs are powerful features for speaker recognition since they use the spectrum of the speech signal and a filterbank to generate frequency sub-bands.

Camarena et al. [12] use formants as a feature extraction method on speech signals with vocalized sound, namely, sub-bands, where there is speaker voice. The authors use two different datasets: a Spanish dataset and ELSDSR database obtaining a precision of 90% using a k-nearest neighbors (kNN) classifier.

Luque Suárez et al. [13] use Multiband Spectral Entropy (MSE) as a feature extraction model. The authors use a kNN classifier using a ball tree proximity index. They obtain a 97% of precision on an uncontrolled environment and a 99% of precision on a controlled environment. MSE are a novel methodology for speaker recognition, it was proposed firstly for Misra et al. [14] and it's based on the theory of information.

Jahangir et al. [1] use MFCCT (Mel Frequency Cepstral Coefficients and Time-Based Algorithms) as feature extraction model. They use three different datasets: LibriSpeech, VCTK and ELSDSR. As classifiers they use Support Vector Machine, Random Forest, k-nearest neighbors, decision tree J48 and a deep neural network. The more significant results are obtained in LibriSpeech, achieving an error of 0.06 in men and 0.08 in women.

In this work, it will be compared the efficiency of machine learning algorithms in the classification task using MFCC and MSE as feature extraction methodology. As a hypothesis, it is thought that multiband spectral entropy can generate better precision in speech recognition task. The first part of methodology is the same for MFCC and MSE, but in the last part there is a variation for MSE computing proposed. The machine learning algorithms that were used are random forest, decision trees, k-nearest neighbors and deep neural networks.

2 Proposed Methodology and Feature Extraction

This section will review the different techniques used for feature extraction of multi-band spectral entropy (MSE) and MFCCs. In addition, there is a description of machine learning algorithms that were used during the work.

2.1 Preprocessing

Speech signals can be modeled as discrete functions sampled with a sampling frequency f_s. The ELSDSR and LibriSpeech databases use a sampling rate of 16 kHz. All this speech signals have different length. This is not a problem in the feature extraction phase, but in the classification.

Pre-emphasis: A pre-emphasis filter is a highpass type filter, which is used in speech recognition. It can enhance the hidden patters in the human voice. A pre-emphasis filter can enhance high frequencies of speech signals. The difference function of pre-emphasis filter is showed in Eq. 1.

$$y[n] = x[n] - \alpha x[n-1] \tag{1}$$

where α is defined as an equalization number. This value is normally defined in a closed set [0.9, 1]. In this work, a value of 0.95 is used to filter all speech signals. Figure 1 shows a 3-secong-long speech signal, where red plot shows the filtered signal. As can be seen, the amplitude of the signal is modified reducing its magnitude.

Framing: In the framing process, speech signals can be considered time series of N points. This process consists of generating a group of subseries (also called frames) of length n, which are overlapped in certain percentage. Framing results in a $n \times m$ matrix, where m is the total number of frames. Figure 2 shows the framing process in a speech signal of 1-second-long using frames of 250 ms and an overlapping of 50% between frames.

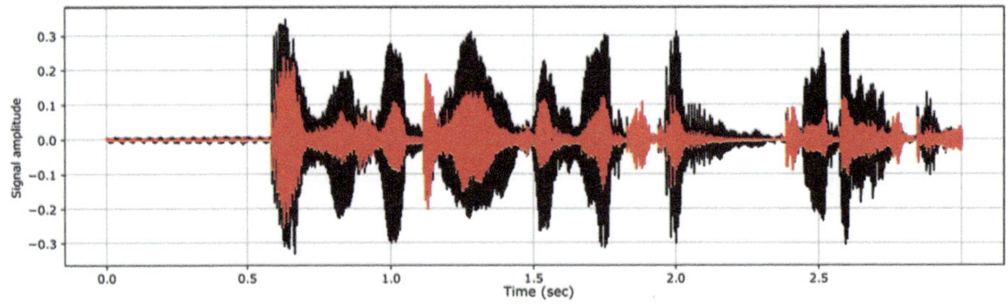

Fig. 1 Filtered signal using a pre-emphasis filter

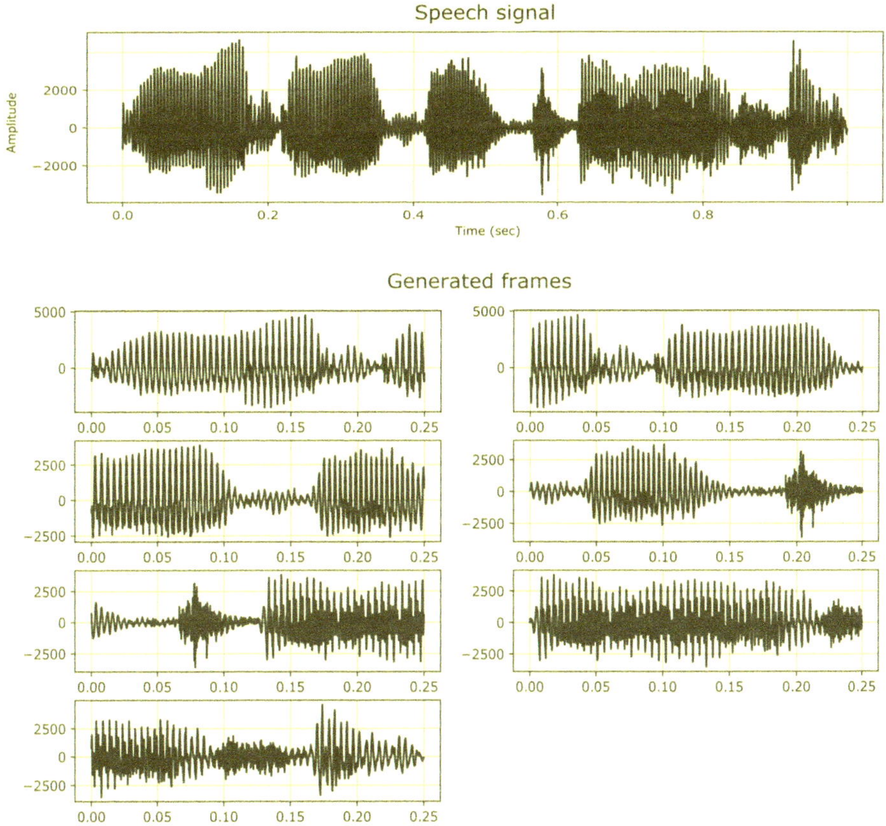

Fig. 2 Framing process of a 1 second long speech signal using frames of 250 ms

Vocalized sound: Vocalized sound is defined as parts of the signal where there is human voice (voiced sounds), namely where the vocal cords vibrate. Generally, a speech signal contains certain intervals where the participant does not emit a sound. These intervals can be found using an autocorrelation function. An autocorrelation function estimates the periodicity of a signal and with it the vocalized sound [15]. The Short-Time Autocorrelation Function (STAF) $R_l(k)$ of a frame starting in a sample l of a signal x is defined as

$$R_l[k] = \sum_{m=0}^{n-1} x[l+m]x[l+m+k] \tag{2}$$

where n is the frame length. The autocorrelation function generates a new discrete function. If the maximum value of this function is greater than 0.1 then the frame contains vocalized sound. The selection of this value was the result of empirical experiments using different values of threshold.

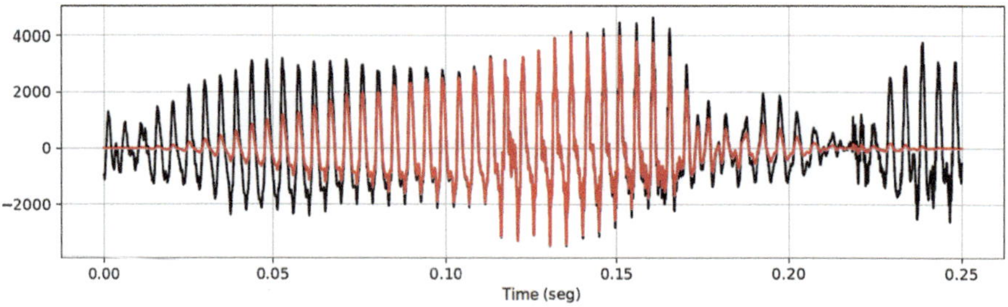

Fig. 3 Audio signal obtained from the windowing process using a frame of 250 ms

Windowing: The windowing process is used to convert every frame into a pseudo periodic function which facilitates the Fourier transform analysis. There are a lot of different window functions, but in this work the Hanning window is used. This window was proposed by Von Hann and it is defined as a cycle of a cosine function. The Hanning window is defined in Eq. 3

$$H(k) = 0.5 - 0.5\cos\left(\frac{2\pi k}{n-1}\right) \tag{3}$$

where n is the frame length. Figure 3 shows the result of using the Hanning window on a frame.

Fast Fourier Transform (FFT): The Fourier transform of a function is defined as a mapping to a new space of functions in terms of frequency. The discrete form of Fourier transform is defined in Eq. 4.

$$X[k] = \sum_{n=0}^{N-1} x[n]e^{\frac{-2i\pi kn}{N}} \tag{4}$$

The fast Fourier transform (FFT) is an optimized form of the discrete Fourier transform. FFT is used in several physics and mathematics applications.

Following the proposed methodology, the FFT was used in each frame. This results in a matrix called spectrogram of the signal. Figure 4 shows the spectrogram of a speech signal sampled at 16 kHz and frames of 25.6 ms.

Sub-bands generation: The crucial part of the methodology is create frequency sub-bands using a Mel filter bank. A Mel filter bank is defined as an array of triangular bandpass overlapped filters. This filterbank uses a frequency scale named Mel scale, defined by Stevens, Volkman and Newman in 1937.

In the experiments, 40 triangular filters were used, which were multiplied by each of the frames of the signal, thus obtaining each frequency sub-band.

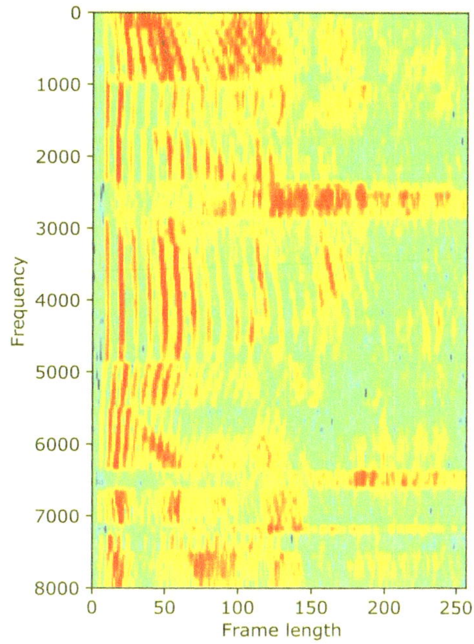

Fig. 4 Spectrogram of a speech signa using frames of 25.6 ms

2.2 Mel Frequency Cepstral Coefficients

MFCCs were proposed by David and Mermelstein in the eighties. These coefficients are used widely as a feature extraction method for a speech signal [16]. Each coefficient can be defined as a sum of energies of each sub-band. Symbolically we have,

$$X'[m] = \log\left(\sum_{k=0}^{N-1} X[k]H[k,m]\right) \tag{5}$$

for $m = 1, 2, \ldots, M$, where M is the number of filters. The Mel filter bank, $H(k, m)$, is a set of triangular filters. This new result is a set of points in the frequency space. Usually, discrete cosine transform (DCT) is used to map this points to the temporal space. DCT is defined as

$$c[l] = \sum_{m=1}^{M} X'[m] \cos\left[l \frac{\pi}{M}\left(m - \frac{1}{2}\right)\right] \tag{6}$$

for $l = 1, 2, \ldots, L$, where $c[l]$ is the lth MFCC. After DCT certain coefficients can be used. In this work, 40 coefficients were used in our experiments. Each group of coefficients can be used as a 40-dimensional feature vector, assigning a label (target) to it.

2.3 Multiband Spectral Entropy

This methodology presents a novel feature extraction method for speech recognition. Unlike MFCCs the entropy can measure the information quantity contained in every sub-band.

Shannon's entropy is a real number that measures uncertainty of an information source. This measure provides useful criteria to analyze and compare distinct probability distributions. Shannon's entropy is defined as

$$E = -\sum_{i=1}^{n} p_i \cdot \log(p_i) \tag{7}$$

where p_i is the occurrence probability of an event i with a probability distribution P. Spectral entropy was presented originally as an additional feature for speech recognition. Mirsa et al. [14] consider every point of sub-band as a distinct event with a probability defined as

$$p_k = \frac{X[k]}{\sum_{i=1}^{N} X[k]} \tag{8}$$

where $X[k]$ is the energy of k-th component from spectrum and N is the total points in the spectrum.

On the other hand, Camarena et al. [17] use a variation in the entropy computing. They use the real and imaginary part of FFT to calculate a covariance matrix Σ and then obtain the entropy as

$$E = \ln(2\pi e) + \frac{1}{2}\ln[det(\Sigma)] \tag{9}$$

where e is the Euler's number and $\det(\Sigma)$ is the determinant of covariance matrix, i.e. $\sigma_{xx}\sigma_{yy} - \sigma_{xy}^2$ where σ_{xx} and σ_{yy} are the variances of the real and imaginary parts, respectively, and σ_{xy} is the covariance between the real and imaginary parts. Figure 5 shows an entropygram obtained from MSE method. As same as MFCCs this feature vectors can be used on classification assigning a label to every vector.

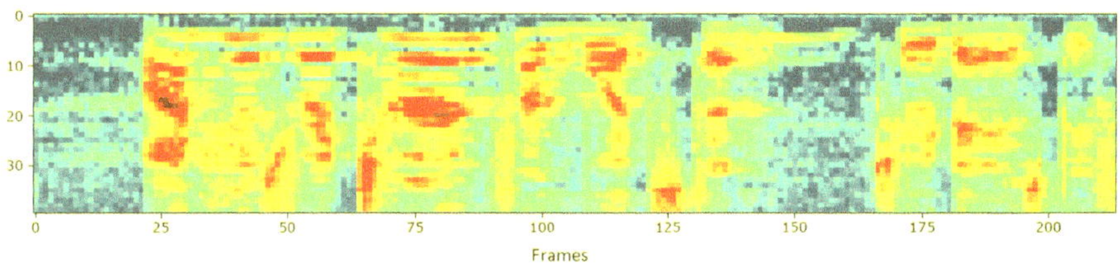

Fig. 5 Entropygram of a speech signal using frames of 25.6 ms

2.4 *Machine Learning Algorithms*

Machine learning is defined as the search for a general rule that allows explaining the behavior or distribution of certain data (called variables or characteristics) given a limited sample size.

Machine learning can be studied from different perspectives: supervised learning and unsupervised learning. In supervised learning the characteristics have a target vector, which is commonly called a class vector. This class vector allows to identify each point of the set of characteristics in a different cluster. A cluster is an agglomeration of points with similar characteristics in a n-dimensional space. On the other hand, if the data do not have a class vector, the learning is of the unsupervised type.

2.5 *Decision Trees*

A decision tree is a classification algorithm that is based on the structure of a tree. The algorithm builds nodes (leaves) of the tree by performing different partitions of the input variables. The decision tree is made up of the following elements: a) A root node that has no input edges and has zero or more output edges. b) Internal nodes which have an entry edge and two or more exit edges. c) Leaves or terminal nodes that have exactly one input edge and no output edges.

Figure 6 shows the basic structure of a decision tree. In this algorithm, each leaf node is assigned a class label. On the other hand, the internal nodes contain conditions that separate each vector of characteristics and methods to measure the efficiency of separation of classes such as entropy or Gini [18].

2.6 *Random Forest*

The random forest is an assemblage algorithm, composed of a specified number of decision trees. The classification of this algorithm is based on the wisdom of the crowd ideology that basically considers that the opinion of a large group of participants is

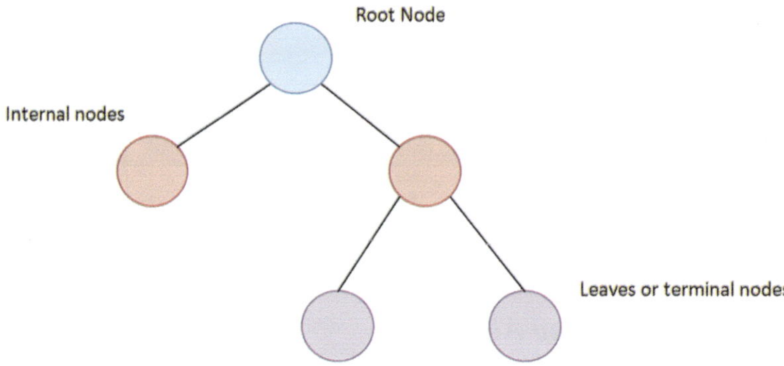

Fig. 6 Basic structure of a decision tree

the correct answer. The output of the algorithm is determined from the vote of each of the decision trees.

The assembly is initialized with N decision trees, which are defined as Tree (X; q_i) with $i = 1, 2, 3, \ldots, N$. Each tree works as a base classifier, where q_i is a sequence of random variables that are determined as follows:

1. From the set X, training sets of the same size as X are selected and a decision tree is generated for each training set. The number of sets is usually selected as an odd number, this to avoid a tie in the vote of each tree.
2. Each node of the decision tree is divided, and a subset of characteristics is extracted with the same probability. The generation of each decision tree is defined as follows: (a) The training set is given the number of samples, K, and the number of features m. (b) Then M characteristics of the set are randomly selected, with $M < m$. (c) n samples are drawn from the given set, to form a new set of samples to train the expansion of the tree. (d) For the division of the node, the best characteristic is calculated, based on the set of M characteristics defined previously. (e) All decision trees are expanded.

The random forest algorithm adjusts according to the number of characteristics and the number of levels in the decision tree. Therefore, the greater the depth of the tree (levels of nodes), the better the separation of data in space [18].

2.7 K-nearest Neighbors

The k-nearest neighbors algorithm was first proposed by Cover and Hant in 1967. It is an algorithm that classifies feature vectors according to the spatial proximity of the vectors. It is an unsupervised and non-parametric algorithm, which basically estimates the value of the probability density function or directly the probability that a vector x_i belongs to a class C_j for $j = 1, 2, 3, \ldots, m$.

For each vector in the test set the distance $d(x_i, x_k)$ to each vector in the training set is calculated, forming a distance matrix D. This distance matrix is square and symmetric. The distance metric is usually the Euclidean distance, although different ones such as the Manhatan distance can be used [19].

2.8 Artificial Neural Networks

An artificial neural network (ANN) is a mathematical model used for classification and regression. This model simulates the biological behavior of the human nervous system and its neurons. The artificial neural network uses a transfer function that transforms input variables and their weights into an output variable. Let x_i be input variables and let w_i be weights assigned to each variable, the output variable is calculated as

$$y = f\left(\sum_{i=1}^{n} x_i w_i\right) \tag{10}$$

where f is the transfer function. Some frequently used transfer functions are: softmax, ReLu, tanh, and sigmoid. ANN can have hidden layers and various output units. Figure 7 shows the basic structure of a neural network. As can be seen, this algorithm emulates the connection between the dendrites and the axons in its analogous human neural net [20].

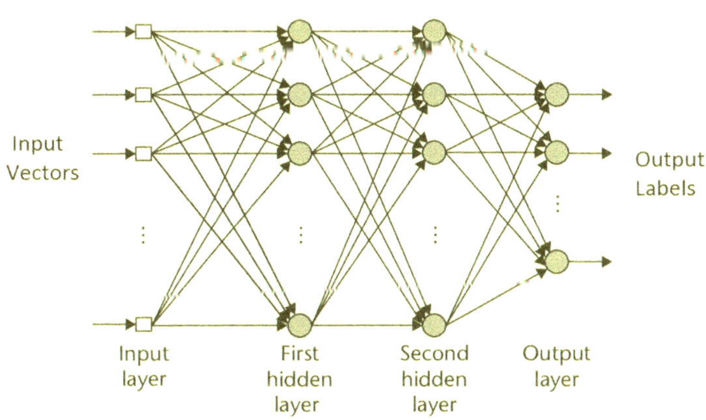

Fig. 7 Structural graph of a neural network model

2.9 Preprocessing for Classification

Regardless of the type of feature extraction method used, it is important to assign a label to each vector before starting the classification. In this work, 40-dimesional feature vectors are used for classification (MFCCs and MSE). To extract both MFCC and MSE features, the next procedure was implemented. (a) First, stereo signals are changed to monoaural by averaging both channels. (b) Frames of 30 ms are used to divide the monoaural signal (we use a sampling frequency of 16 kHz). (c) Consecutive frames have an overlap of 50%. (d) A Hann window function is applied to each frame. (e) The FFT is computed for each frame. (f) A bank of 40 Mel filters is used to split the full band spectrum with $f_{min} = 0$ Hz, $f_{max} = 8$ kHz. (g) Before computing FFT, we rule out unvoiced sounds from the speech signals by using the autocorrelation function. With the FFTs, we are ready to compute Mel Frequency Cepstral Coeffcients (Sect. 2.2), and the Multiband Spectral Entropy Signature (Sect. 2.3).

Normalization: Is the process of reduce the magnitude of the audio dataset and can be based on statistical methods. In this work, min–max feature scaling was used. This normalization technique gets a new value using

$$x^* = \frac{x - \min(X)}{\max(X) - \min(X)} \tag{11}$$

where x represents the i-th coefficient of the feature vector and X is the feature value. This normalization converts the feature set to a closed set [0, 1]. After normalizing the data, stratification is used to separate the set into two different sets, a training set and a testing set. Stratification helps in cases where the data is unbalanced, such as in speech recognition. In addition, 80% of the data was used for training each of the models and 20% for testing.

2.10 Setup for Machine Learning Algorithms

In this work, 4 different machine learning algorithms were used: random forest, k-nearest neighbors, ID3 decision tree and a deep neural network. The objective of using these algorithms is to test their operation given the nature of each one of them and how this can help in the task of speech recognition. Table 1 shows the parameters for every machine learning algorithm, they were varied using a grid search algorithm.

To measure the efficiency and performance of the above classifiers, it is necessary to use an evaluation metric that provides relevant information to evaluate and compare the algorithms with each other.

Table 1 Parameters for every machine learning algorithm

Machine learning algorithm	Parameters
k-nearest neighbors	Neighbors, starting in a defined value
Decision tree	Criteria for a partition (gini or entropy) Strategy for splitting (best, random)
Random forest	Criteria for a partition (gini or entropy) Estimators (number of trees)
Deep neural network	Number of hidden layers, number of hidden neurons and activation functions

2.11 Defined Experiments

The first of the experiments consists of determining whether a speaker is male or female, that is, the gender of the speaker. This experiment is of the binary type and can be considered the simplest of all. In the ELSDSR database there are 10 voice recordings of women and 12 for men, while in the LibriSpeech database there are 20 recordings of men and 20 for women.

The second and third experiment consists of identifying a man or a woman from the data sets of men and women respectively. For the ELSDSR database there are 12 classes for men and 10 of women. In LibriSpeech there are 20 classes for men and 20 classes for women.

The last experiment is the most complicated since it consists of identifying a speaker from the total set of speakers. ELSDSR has a total of 22 classes and LibriSpeech a total of 40 classes.

3 Results

This section presents the results obtained by implementing the machine learning algorithms in the speech recognition problem. The results are presented according to each of the experiments carried out and defined in the methodology section. Tables are used to better understand the results obtained.

3.1 Speaker Recognition

The recognition of all speakers is the most complex task of this research work, since it contains a greater number of classes (labels) to be classified. There are 22 different classes in the ELSDSR database and 40 classes in the LibriSpeech database.

The results of the decision tree, random forest, artificial neural network, and k-nearest neighbors algorithms are evaluated using the accuracy metric. The results

Table 2 Obtained results in the experiment of recognition of all speakers in the ELSDSR database

Audio feature	Deep neural network (%)	k-nearest neighbors (%)	Random forest (%)	Decision tree (%)
MFCC	**66.07**	64.68	65.17	34.71
MSE	77.13	72.58	**78.92**	53.22

Table 3 Obtained results in the experiment of recognition of all speakers in the LibriSpeech database

Audio feature	Deep neural network (%)	k-nearest neighbors (%)	Random forest (%)	Decision tree (%)
MFCC	59.69	64.62	**67.52**	34.86
MSE	63.91	68.18	**74.73**	46.81

obtained in each of the algorithms in this experiment are shown in Table 2. The best results in accuracy are shown in bold, noting that for MFCC feature the best performance was 66.07% by using a deep neural network classifier, on the other hand, the best performance for multiband spectral entropy was 78.92% by using a random forest classifier, above the Mel coefficients by more than 12 percentage points.

There are 40 speakers in the LibriSpeech database, so for this experiment 40 different classes (labels) will be obtained. Likewise, the evaluation metric to be used in this experiment was accuracy. The results obtained in this experiment are shown in Table 3 As can be seen, the best result was obtained in the random forest classifier by using spectral entropy, with 74.73% in accuracy, whereas MFCC obtained 67.52% with the same classifier. This represents 7 percentage points above Mel's coefficients using random forest. Something important to mention is that it was expected that the accuracy results would decrease with respect to those obtained in ELSDSR. This may be due to the number of classes that are held.

3.2 Speaker Recognition in Men

The results of this experiment are based on the voice identification task in men. This experiment uses only 12 different classes in ELSDSR and 20 classes on LibriSpeech, for which the algorithms previously described were used. Table 4 shows the accuracy results for this experiment. With fewer classes than in the previous experiment, it would expect the ranking results to be higher. However, as can be seen, the results are lower than those found in the previous experiment. The best result was obtained with the random forest algorithm using multiband spectral entropy with a percentage of 70.6%, whereas for MFCC was of 66.88%, 4 percentage points above the Mel coefficients. Table 5 shows the results of the experiment in LibriSpeech using the metric accuracy. As can be seen, the best result is obtained when using the random forest

Table 4 Obtained results in the speech recognition experiment in men in the ELSDSR database

Audio feature	Deep neural network (%)	k-nearest neighbors (%)	Random forest (%)	Decision tree (%)
MFCC	66.83	67.63	**66.88**	37.58
MSE	68.65	61.97	**70.6**	41.53

Table 5 Obtained results in the speech recognition experiment in men in the LibriSpeech database

Audio feature	Deep neural network (%)	k-nearest neighbors (%)	Random forest (%)	Decision tree (%)
MFCC	71.23	75.13	**75.55**	45.04
MSE	75.15	76.33	**82.15**	58.33

algorithm using multiband spectral entropy, with 82.15%, whereas that the couple MFCC and random forest algorithm achieved the 75.55%. When comparing both features using the same algorithm, a difference of 7 percentage points is observed. This result is higher than that obtained in the ELSDSR database in the same experiment, this may be since there is a greater amount of information from each of the speakers.

3.3 Speaker Recognition in Women

The results of this experiment consist of recognizing female speakers. This experiment uses a total of 10 different classes in ELSDSR and 20 for LibriSpeech, using the algorithms described above. Table 6 shows the accuracy results of this experiment. It is observed that the classification was higher than that obtained in the two previous experiments, this may be since there were fewer classes, or else the algorithms measure the characteristics of the female voice in greater detail. The best result was obtained in the random forest algorithm using multiband spectral entropy, obtaining a 79.03%. The result was very closed compared to the artificial neural network. In respect to MFCC feature, a percentage of 75.11% is obtained by using deep neural network. Table 7 shows the results obtained in this experiment with LibriSpeech database. As can be seen, the best result was obtained using the random forest algorithm and multiband spectral entropy with 90.79%. Comparing this result with its analogue in the Mel coefficients, a difference of 2 percentage points is obtained since MFCC had an accuracy of 88.78% by using deep neural network classifier. Something important to mention is that this result is higher than the one obtained in the ELSDSR database. This is not necessarily good, as there may be a more contrasting pattern in the LibriSpeech database.

Table 6 Obtained results in the speech recognition experiment in women in the ELSDSR database

Audio feature	Deep neural network (%)	k-nearest neighbors (%)	Random forest (%)	Decision tree (%)
MFCC	**75.11**	70.94	72.58	49.90
MSE	78.32	72.58	**79.03**	53.06

Table 7 Obtained results in the speech recognition experiment in women in the LibriSpeech database

Audio feature	Deep neural network (%)	k-nearest neighbors (%)	Random forest (%)	Decision tree (%)
MFCC	**88.78**	88.26	86.67	77.36
MSE	87.42	86.62	**90.79**	82.49

3.4 Genre Recognition

The results of this experiment consist of the identification of the gender of the speaker, so there are only two different classes. Table 8 shows the results obtained using the implemented algorithms in ELSDSR. As can be seen, the results are superior to previous experiments. This is because the experiment is binary, that is, of two classes. The best result is obtained using the artificial neural network algorithm with multiband spectral entropy extraction (93.75% and 93.99% for MFCC and MSES respectively), although there is no significant difference to that obtained in the random forest. One of the points that could break this little difference is the computational cost of training the model, which could be analyzed in more detail. The results of this experiment applied in the LibriSpeech database are shown in Table 9. As can be seen, the best result was obtained using the random forest with multiband spectral entropy, obtaining 90.79% in accuracy. For the case of MFCC feature, a accuracy of 88.78% by using deep neural network. This result is superior to its analogue in the Mel coefficients by 2 percentage points. This difference is greater than that observed in the ELSDSR database, where the difference in accuracy between the extractions was smaller.

As in the ELSDSR database, the best results were obtained using multiband spectral entropy. Also, the random forest algorithm is generally the best of the four numerically speaking. The results obtained in these experiments show that the algorithms find a contrasting pattern that allows them to reach that precision. To choose

Table 8 Obtained results in the gender recognition experiment in the ELSDSR database

Audio feature	Deep neural network (%)	k-nearest neighbors (%)	Random forest (%)	Decision tree (%)
MFCC	**93.75**	91.86	91.14	84.92
MSE	**93.99**	91.3	93.29	88.54

Table 9 Obtained results in the gender recognition experiment in the LibriSpeech database

Audio feature	Deep neural network (%)	k-nearest neighbors (%)	Random forest (%)	Decision tree (%)
MFCC	**88.78**	88.26	86.67	77.36
MSE	87.42	86.62	**90.79**	82.49

one of these algorithms, other factors such as computational time or computational cost would have to be considered.

4 Conclusions

In this research work, the extraction of features based on multiband spectral entropy shows superior results than the extraction of Mel coefficients. Each of the experiments carried out shows a higher precision for multiband spectral entropy, even in some cases 12% above the Mel coefficients. In this research, four classification algorithms were considered, which are: random forest, decision tree, k-nearest neighbors and a deep neural network. The most significant results are shown by the neural network and the random forest algorithm, this may be due to the operation and the mathematical foundation of each of these algorithms.

On the other hand, when comparing the results obtained with related work or the state of the art, competitive results are shown in terms of accuracy. Jahanhir et al. [1] obtained MFCCT (temporal spectrum characteristics) and obtained 92.9% in their gender experiment applied in the ELSDSR database, while using multiband spectral entropy an accuracy of 93.99% was obtained. Likewise, in the male voice identification experiment in the ELSDSR database, the authors obtained 82% in accuracy, while the multiband spectral entropy obtained 70.6%. Also, in this same database, but in the voice identification experiment in women, the authors obtain an accuracy of 78%, while the spectral entropy shows an accuracy of 79.03%. Something important to mention is that for these last two experiments the authors use a double check. They first use a pre-trained model that indicates the gender of the speaker to later classify it according to the experiment of men or women. This has shown better results according to the-state-of-the-art. On the other hand, in the LibriSpeech database the authors show 94% in accuracy in men, which is above the 82.15% achieved by multiband spectral entropy. In turn, in the female experiment, the authors obtained 92%, which is also higher than the 79.03% obtained by multiband spectral entropy. Although the extractions of characteristics of both works are different, it is statistically comparable since the same databases are used.

References

1. R. Jahangir, Y.W. Teh, N.A. Memon, G. Mujtaba, M. Zareei, U. Ishtiaq, M.Z. Akhtar, I. Ali, Text-independent speaker identification through feature fusion and deep neural network. IEEE Access **8**, 32 187–32 202 (2020)
2. A. Ali, M. Siregar, T. Taryo, Analysis voice recognition Pada system autopilot. Humanities Manage. Sci. Proc. **1**(2), 91–102 (2019)
3. S. Priyanayana, A.G. Buddhika, P. Jayasekara, Developing a voice-controlled wheelchair with enhanced safety through multimodal approach, in *IEEE Region 10 Humanitarian Technology Conference R10-HTC*, vol. 2018-December, pp. 1–6 (2019)
4. S. Venkatraman, A. Overmars, M. Thong, Smart home automation—use cases of a secure and integrated voice-control system. Systems **9**(4), 77 (2021)
5. C.Y. Peng, R.C. Chen, Voice recognition by google home and raspberry Pi for smart socket control, in *Proceedings of the 2018 Tenth International Conference on Advanced Computational Intelligence (ICACI)* (2020), pp. 324–329
6. N. Zhang, X. Mi, X. Feng, X.F. Wang, Y. Tian, F. Qian, Understanding and mitigating the security risks of voice controlled third-party skills on amazon Alexa and google home," arXiv preprint arXiv:1805.01525 (2018)
7. S. Mayer, G. Laput, C. Harrison, 2020 Enhancing Mobile Voice Assistants with WorldGaze, in *Conference on Human Factors in Computing Systems—Proceedings* (Association for Computing Machinery, 2020)
8. S. Team, Hey Siri: an on-Device DNN-Powered voice trigger for Apple's personal assistant. Apple Mach. Learn. J. **1**(6) (2017)
9. L. Feng, L.K. Hansen, A new database for speaker recognition, tech. rep. (2005)
10. V. Panayotov, G. Chen, D. Povey, S. Khudanpur, Librispeech: an ASR corpus based on public domain audio books, in *2015 IEEE International Conference on Acoustics, Speech and Signal Processing (ICASSP)* (2015), pp. 5206–5210
11. C. Ittichaichareon, S. Suksri, T. Yingthawornsuk, Speech recognition using MFCC, in *International Conference on Computer Graphics, Simulation and Modeling (ICGSM'2012)* (2012), pp. 28–29
12. A. Camarena Ibarrola, M. Castro Coria, K. Figueroa, Cloud point matching for text- independent speaker identification, in *2018 IEEE International Autumn Meeting on Power, Electronics and Computing (ROPEC)* (IEEE, 2018), pp. 1–6
13. F. Luque Suárez, A. Camarena Ibarrola, E.Chávez, Efficient speaker identification using spectral entropy. Multimedia Tools Appl. **78**(12), 16 803–16 815 (2019)
14. H. Misra, S. Ikbal, H. Bourlard, H. Hermansky, Spectral entropy based feature for robust ASR. in *2004 IEEE International Conference on Acoustics, Speech, and Signal Processing*, vol. 1 (IEEE, 2004), pp. I–193
15. A. Camarena-Ibarrola, M. Castro-Coria, K. Figueroa, Cloud point matching for text-independent speaker identification, in *2018 IEEE International Autumn Meeting on Power, Electronics and Computing (ROPEC 2018)* (2018), pp. 1–6
16. M.S. Likitha, S.R.R. Gupta, K. Hasitha, A.U. Raju, Speech based human emotion recognition using MFCC, in *2017 International Conference on Wireless Communications, Signal Processing and Networking (WiSPNET)* (2017), pp. 2257–2260
17. J.A. Camarena-Ibarrola, E. Chávez, On musical performances identification, entropy and string matching, in *2006 Mexican International Conference on Artificial Intelligence*, pp. 952–962 (2006)
18. J. López-Rentería, Análisis de Señales Electroencefalográficas para Clasificar Emociones Utilizando el Modelo Bidimensional Valencia Excitación. (2020)
19. T. Cover, P. Hart, Nearest neighbor pattern classification. IEEE Trans. Inf. Theory **13**(1), 21–27 (1967)
20. A.A. Rodríguez-Miranda, Modeling and analysis of the air quality in the city of Oviedo (Northern Spain) using the PSO-SVM-based approach, MLP neural network, and M5 model tree. In Doctoral Thesis, León University, p. 334 (2018)

Forecasting Based on Fuzzy Logic of the Level of Epidemiological Risk for the Mexican State of Tamaulipas

Paula Hernández-Hernández and Norberto Castillo-García

Abstract Nowadays, in Mexico there exists a traffic light monitoring system to regulate the use of public space according to the risk level of infection with SARS–CoV–2. The monitoring system is applied to each state in Mexico and consists of four levels of risk encoded with four colors: green, yellow, orange and red. In this chapter we propose a Fuzzy Time Series Model to forecast the next color to be assigned to the Mexican state of Tamaulipas based on historical data from the monitoring system. We conducted a computational experiment to measure the accuracy of the model. The model accuracy was measured by the well–known Root Mean Square Error (RMSE) index.

Keywords Fuzzy time series model · COVID 19 · SARS CoV-2 · Mexican traffic · Light monitoring system · Public health

1 Introduction

Mexico has implemented a traffic–light monitoring system to indicate the risk level of infection with the novel coronavirus SARS–CoV–2 [1]. A similar color–based monitoring system is also used in other geographic regions [2]. The Mexican monitoring system is based on four different levels of risk, namely, maximal, high, medium and low. Each risk level is represented by a specific color. More precisely, maximal risk is represented by red, high risk is represented by orange, medium risk is represented by yellow and low risk is represented by green. The risk level is determined by several indicators such as the hospital occupancy rate, the COVID–19 test positivity rate and the tendency of patients who require hospitalization, among many others [3]. The goal of the monitoring system is to regulate the use of public space in order to mitigate the virus dissemination. Since the risk level varies over time, the validity period of the monitoring system color is 15 days. According to the Mexican government

P. Hernández-Hernández · N. Castillo-García (✉)
Department of Engineering, Tecnológico Nacional de México/I.T. Altamira, Altamira, Mexico
e-mail: norberto_castillo15@hotmail.com

O. Castillo and P. Melin (eds.), *Fuzzy Logic and Neural Networks for Hybrid Intelligent System Design*, Studies in Computational Intelligence 1061,
https://doi.org/10.1007/978-3-031-22042-5_8

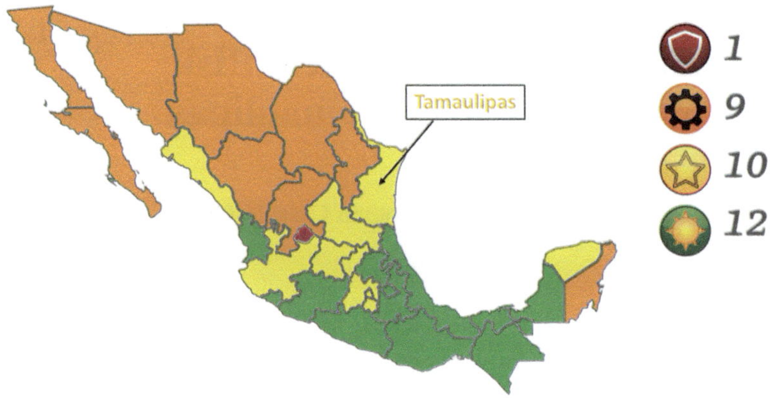

Fig. 1 Colors assigned to the Mexican states by the traffic–light monitoring system valid from January 24, 2022 to February 6, 2022 [4]

policies, the monitoring system is independently applied to each autonomous administrative division [4]. In Mexico there are 32 different autonomous divisions that are called states. Figure 1 depicts the Mexican map with the states colored according to traffic–light monitoring system valid from January 24, 2022 to February 6, 2022.

In this chapter we propose a fuzzy time series model (FTSM) to forecast the color of the Mexican traffic–light monitoring system of COVID–19 for the state of Tamaulipas (see Fig. 1). As far as we know, this is the first time that the color assigned to Tamaulipas by the Mexican monitoring system is forecasted. Our FTSM is a first–order model that uses the well–known average–based partition method to divide the universe of discourse and the three principles proposed in [5] to perform the defuzzification process. Since the colors are qualitative values, we assign a number to each color in order for our FTSM to be able to process them. The time series consists of 43 observations collected from June 8, 2020 to January 24, 2022 [6]. We conducted a computational implementation to assess the accuracy of the model through the Root Mean Square Error index. The experimental results show that the accuracy of FTSM is about 6.73 units. This means that the predicted values fit relatively well to the real values.

The remainder of this chapter is organized as follows. In Sect. 2 we briefly review the fundamental concepts of fuzzy time series models. In Sect. 3 we describe our proposed fuzzy time series model as well as the implementation. Finally, in Sect. 4 we discuss the main conclusions and further work of this research.

2 Fuzzy Time Series Definitions

A time series $Y(t)$ (for $t = \ldots, 0, 1, 2, \ldots$) is a set of real numbers (a.k.a. observations) chronologically ordered which describes the behavior of a variable over time.

Let $U = \{u_1, u_2, \ldots, u_n\}$ be the universe of discourse defined from the observations of the time series $Y(t)$. A fuzzy set A of the universe of discourse U can be formally defined as follows [5]:

$$A = f_A(u_1)/u_1 + f_A(u_2)/u_2 + \cdots + f_A(u_n)/u_n,$$

where $f_A : U \rightarrow [0, 1]$ is known as the membership function of A and its purpose is to associate an element of the universe of discourse U with a real number between zero and one. Thus, $f_A(u_i)$ indicates the level of membership of u_i in fuzzy set A, where $f_A(u_i) \in [0, 1]$ and $i = 1, 2, \ldots, n$. If $F(t)$ represents a collection of fuzzy sets $f_1, f_2, \ldots$ defined on the universe of discourse U, then $F(t)$ is called a fuzzy time series defined on $Y(t)$ [7].

Now suppose that $F(t)$ is exclusively caused by $F(t-1)$. This causal relationship between $F(t)$ and $F(t-1)$ can be formally expressed as:

$$F(t) = F(t-1) \circ R(t, t-1),$$

where $R(t, t-1)$ is known as the fuzzy logical relationship between $F(t)$ and $F(t-1)$, and $\circ$ represents a mathematical operator. If $F(t-1) = A_i$ and $F(t) = A_j$ then the fuzzy logical relationship can be denoted by $A_i \rightarrow A_j$. In this notation, A_i is the left–hand side and A_j is the right–hand side of the fuzzy logical relationship [7]. The previous fuzzy logical relationship can be interpreted as follows. If the value at time $t-1$ is A_i then the value at time t is A_j. Thus, $A_i \rightarrow A_j$ is a fuzzy statement in which A_i is the antecedent and A_j is the consequent. Frequently, fuzzy time series models have several fuzzy logical relationships with the same antecedent and different consequent. For example:

$$A_i \rightarrow A_{j1}, A_i \rightarrow A_{j2}, \ldots, A_i \rightarrow A_{jk}.$$

In the previous example, there are k fuzzy logical relationships with exactly the same antecedent: the fuzzy set A_i. In these cases, it is convenient to group the k relationships into one single fuzzy logical relationship group:

$$A_i \rightarrow A_{j1}, A_{j2}, \ldots, A_{jk}.$$

In this group, the antecedent has only one fuzzy set (A_i) and the consequent has k fuzzy sets ($A_{j1}, A_{j2}, \ldots, A_{jk}$). The grouping process is performed on all the fuzzy logical relationships with the same antecedent in order to obtain the fuzzy logical relationship groups.

3 Proposed Forecasting Model

As mentioned previously, the goal of our fuzzy time series model (FTSM) is to forecast the color of the Mexican traffic–light monitoring system for the state of Tamaulipas. Section 3.1 presents the first step of our FTSM, which consists in defining the universe of discourse. Then, in Sect. 3.2 we describe the method used to partition the universe of discourse. The fuzzification process is explained in Sect. 3.3. Section 3.4 describes how to compute the fuzzy logical relationships. Section 3.5 explains the defuzzification process. Finally, in Sect. 3.6 we report the computational implementation and the model accuracy through the RMSE index.

3.1 Universe of Discourse

In order to define the universe of discourse, we collect the colors assigned to the state of Tamaulipas by the traffic–light monitoring system from June 8, 2020 to January 24, 2022. This gives us a time series consisting of 43 observations. However, since the elements in the time series are not numeric, we assign a number to each color in order to be able to compute the data. Thus, we arbitrarily assign the number 10 to color green (low risk), the number 20 to color yellow (medium risk), the number 30 to color orange (high risk) and the number 40 to color red (maximal risk). All the observations of the time series are shown in Table 1. This table has four sections and each section has two headings: the time (t) and the observation at time t ($(Y(t))$).

The universe of discourse is defined by the minimum ($D_{\min}$) and maximum ($D_{\max}$) values from the time series $Y(t)$. Clearly, in this research $D_{\min} = 10$ (observed at $t = 22$) and $D_{\max} = 40$ (observed at $t = 1$). The domain of the universe of discourse

Table 1 Time series of the encoded colors assigned by the traffic–light monitoring system of COVID–19 for the state of Tamaulipas, Mexico

t	$(Y(t))$	t	$(Y(t))$	t	$(Y(t))$	t	$(Y(t))$
1	40	12	20	23	10	34	30
2	30	13	20	24	20	35	20
3	40	14	30	25	20	36	10
4	40	15	20	26	20	37	10
5	40	16	20	27	20	38	10
6	30	17	30	28	30	39	10
7	20	18	30	29	30	40	10
8	20	19	20	30	30	41	10
9	20	20	20	31	30	42	30
10	20	21	20	32	40	43	20
11	20	22	10	33	30	–	–

is computed as $U = [D_{\min} - D_1, D_{\max} + D_2]$, where D_1 and D_2 are two positive numbers conveniently determined [5]. In this study we set $D_1 = D_2 = 0$. Therefore, the domain of the universe of discourse is $U = [10, 40]$.

3.2 Partition of the Universe of Discourse

The partition process consists in computing a collection of n sub–intervals $u_1, \ldots, u_n$ from the universe of discourse U. In this research we use the well–known average–based partition method. This method is deterministic and computes equally–sized intervals. The first step is to compute the average of the absolute difference between of each pair of consecutive observations. For our time series this value is computed as follows:

$$avg = \left\lfloor \frac{|30-40| + |40-30| + |40-40| + \cdots + |20-30|}{42} \right\rfloor = \left\lfloor \frac{180}{42} \right\rfloor = 4.$$

Once the average has been determined, the next step is to compute the half of $avg = 4$, which is $half_{avg} = 2$. Since the value of $half_{avg}$ has one digit, the partition length ℓ is precisely $half_{avg}$, i.e., $\ell = half_{avg} = 2$. With this information, we proceed to compute the number of sub–intervals and the domain of each sub–interval as follows. The number of sub–intervals is:

$$n = \left\lfloor \frac{D_{\max} - D_{\min}}{\ell} \right\rfloor = \left\lfloor \frac{40 - 10}{2} \right\rfloor = 15.$$

Now the domain of each sub–interval can be computed by:

$$u_i = [D_{\min} + (i-1) \times \ell, \quad D_{\min} + i \times \ell] \quad \forall i = 1, \ldots, n.$$

Thus, the $n = 15$ sub–intervals are: $u_1 = [10, 12]$, $u_2 = [12, 14]$, $u_3 = [14, 16]$, $u_4 = [16, 18]$, $u_5 = [18, 20]$, $u_6 = [20, 22]$, $u_7 = [22, 24]$, $u_8 = [24, 26]$, $u_9 = [26, 28]$, $u_{10} = [28, 30]$, $u_{11} = [30, 32]$, $u_{12} = [32, 34]$, $u_{13} = [34, 36]$, $u_{14} = [36, 38]$, $u_{15} = [38, 40]$. Notice that each sub–interval u_i has a length $\ell = 2$ and their lengths are all equal.

3.3 Fuzzification

The goal of the fuzzification process is to determine the fuzzy sets and their membership functions that constitute the fuzzification model defined on the universe of discourse U. In this study we follow the approach proposed in [5] to construct the fuzzification model. This model exclusively uses triangular membership functions to cover the universe of discourse.

Every triangular function can be fully described by three parameters: $a < b < c$. On the one hand, parameters a and c represent the points where the function reaches its lowest value: zero. On the other hand, parameter b is the point where the function reaches its largest value: one. Typically, this parameter is located in the middle of the function, i.e., $b = (a + c)/2$. Given a real number x and a triangular membership function T defined by parameters a, b and c, the fuzzified value of x can be computed by Eq. (1):

$$T(x, a, b, c) = \begin{cases} 0 & x \leq a \\ \dfrac{x-a}{b-a} & a \leq x \leq b \\ \dfrac{c-x}{c-b} & b \leq x \leq c \\ 0 & x \geq c \end{cases} \tag{1}$$

The fuzzification model followed in this research considers $n + 1$ fuzzy sets to cover the universe of discourse. Since there are $n = 15$ sub–intervals, there will be 16 fuzzy sets A_1, ..., A_{16} defined as follows:

$$A_1 = \frac{1}{u_1} + \frac{0.5}{u_2} + \frac{0}{u_3} + \frac{0}{u_4} + \frac{0}{u_5} + \frac{0}{u_6} + \frac{0}{u_7} + \frac{0}{u_8} + \frac{0}{u_9} + \frac{0}{u_{10}} + \frac{0}{u_{11}} + \frac{0}{u_{12}} + \frac{0}{u_{13}} + \frac{0}{u_{14}} + \frac{0}{u_{15}} + \frac{0}{u_{16}}$$

$$A_2 = \frac{0.5}{u_1} + \frac{1}{u_2} + \frac{0.5}{u_3} + \frac{0}{u_4} + \frac{0}{u_5} + \frac{0}{u_6} + \frac{0}{u_7} + \frac{0}{u_8} + \frac{0}{u_9} + \frac{0}{u_{10}} + \frac{0}{u_{11}} + \frac{0}{u_{12}} + \frac{0}{u_{13}} + \frac{0}{u_{14}} + \frac{0}{u_{15}} + \frac{0}{u_{16}}$$

$$A_3 = \frac{0}{u_1} + \frac{0.5}{u_2} + \frac{1}{u_3} + \frac{0.5}{u_4} + \frac{0}{u_5} + \frac{0}{u_6} + \frac{0}{u_7} + \frac{0}{u_8} + \frac{0}{u_9} + \frac{0}{u_{10}} + \frac{0}{u_{11}} + \frac{0}{u_{12}} + \frac{0}{u_{13}} + \frac{0}{u_{14}} + \frac{0}{u_{15}} + \frac{0}{u_{16}}$$

$$A_4 = \frac{0}{u_1} + \frac{0}{u_2} + \frac{0.5}{u_3} + \frac{1}{u_4} + \frac{0.5}{u_5} + \frac{0}{u_6} + \frac{0}{u_7} + \frac{0}{u_8} + \frac{0}{u_9} + \frac{0}{u_{10}} + \frac{0}{u_{11}} + \frac{0}{u_{12}} + \frac{0}{u_{13}} + \frac{0}{u_{14}} + \frac{0}{u_{15}} + \frac{0}{u_{16}}$$

$$A_5 = \frac{0}{u_1} + \frac{0}{u_2} + \frac{0}{u_3} + \frac{0.5}{u_4} + \frac{1}{u_5} + \frac{0.5}{u_6} + \frac{0}{u_7} + \frac{0}{u_8} + \frac{0}{u_9} + \frac{0}{u_{10}} + \frac{0}{u_{11}} + \frac{0}{u_{12}} + \frac{0}{u_{13}} + \frac{0}{u_{14}} + \frac{0}{u_{15}} + \frac{0}{u_{16}}$$

$$A_6 = \frac{0}{u_1} + \frac{0}{u_2} + \frac{0}{u_3} + \frac{0}{u_4} + \frac{0.5}{u_5} + \frac{1}{u_6} + \frac{0.5}{u_7} + \frac{0}{u_8} + \frac{0}{u_9} + \frac{0}{u_{10}} + \frac{0}{u_{11}} + \frac{0}{u_{12}} + \frac{0}{u_{13}} + \frac{0}{u_{14}} + \frac{0}{u_{15}} + \frac{0}{u_{16}}$$

$$A_7 = \frac{0}{u_1} + \frac{0}{u_2} + \frac{0}{u_3} + \frac{0}{u_4} + \frac{0}{u_5} + \frac{0.5}{u_6} + \frac{1}{u_7} + \frac{0.5}{u_8} + \frac{0}{u_9} + \frac{0}{u_{10}} + \frac{0}{u_{11}} + \frac{0}{u_{12}} + \frac{0}{u_{13}} + \frac{0}{u_{14}} + \frac{0}{u_{15}} + \frac{0}{u_{16}}$$

$$A_8 = \frac{0}{u_1} + \frac{0}{u_2} + \frac{0}{u_3} + \frac{0}{u_4} + \frac{0}{u_5} + \frac{0}{u_6} + \frac{0.5}{u_7} + \frac{1}{u_8} + \frac{0.5}{u_9} + \frac{0}{u_{10}} + \frac{0}{u_{11}} + \frac{0}{u_{12}} + \frac{0}{u_{13}} + \frac{0}{u_{14}} + \frac{0}{u_{15}} + \frac{0}{u_{16}}$$

$$A_9 = \frac{0}{u_1} + \frac{0}{u_2} + \frac{0}{u_3} + \frac{0}{u_4} + \frac{0}{u_5} + \frac{0}{u_6} + \frac{0}{u_7} + \frac{0.5}{u_8} + \frac{1}{u_9} + \frac{0.5}{u_{10}} + \frac{0}{u_{11}} + \frac{0}{u_{12}} + \frac{0}{u_{13}} + \frac{0}{u_{14}} + \frac{0}{u_{15}} + \frac{0}{u_{16}}$$

$$A_{10} = \frac{0}{u_1} + \frac{0}{u_2} + \frac{0}{u_3} + \frac{0}{u_4} + \frac{0}{u_5} + \frac{0}{u_6} + \frac{0}{u_7} + \frac{0}{u_8} + \frac{0.5}{u_9} + \frac{1}{u_{10}} + \frac{0.5}{u_{11}} + \frac{0}{u_{12}} + \frac{0}{u_{13}} + \frac{0}{u_{14}} + \frac{0}{u_{15}} + \frac{0}{u_{16}}$$

$$A_{11} = \frac{0}{u_1} + \frac{0}{u_2} + \frac{0}{u_3} + \frac{0}{u_4} + \frac{0}{u_5} + \frac{0}{u_6} + \frac{0}{u_7} + \frac{0}{u_8} + \frac{0}{u_9} + \frac{0.5}{u_{10}} + \frac{1}{u_{11}} + \frac{0.5}{u_{12}} + \frac{0}{u_{13}} + \frac{0}{u_{14}} + \frac{0}{u_{15}} + \frac{0}{u_{16}}$$

$$A_{12} = \frac{0}{u_1} + \frac{0}{u_2} + \frac{0}{u_3} + \frac{0}{u_4} + \frac{0}{u_5} + \frac{0}{u_6} + \frac{0}{u_7} + \frac{0}{u_8} + \frac{0}{u_9} + \frac{0}{u_{10}} + \frac{0.5}{u_{11}} + \frac{1}{u_{12}} + \frac{0.5}{u_{13}} + \frac{0}{u_{14}} + \frac{0}{u_{15}} + \frac{0}{u_{16}}$$

$$A_{13} = \frac{0}{u_1} + \frac{0}{u_2} + \frac{0}{u_3} + \frac{0}{u_4} + \frac{0}{u_5} + \frac{0}{u_6} + \frac{0}{u_7} + \frac{0}{u_8} + \frac{0}{u_9} + \frac{0}{u_{10}} + \frac{0}{u_{11}} + \frac{0.5}{u_{12}} + \frac{1}{u_{13}} + \frac{0.5}{u_{14}} + \frac{0}{u_{15}} + \frac{0}{u_{16}}$$

$$A_{14} = \frac{0}{u_1} + \frac{0}{u_2} + \frac{0}{u_3} + \frac{0}{u_4} + \frac{0}{u_5} + \frac{0}{u_6} + \frac{0}{u_7} + \frac{0}{u_8} + \frac{0}{u_9} + \frac{0}{u_{10}} + \frac{0}{u_{11}} + \frac{0}{u_{12}} + \frac{0.5}{u_{13}} + \frac{1}{u_{14}} + \frac{0.5}{u_{15}} + \frac{0}{u_{16}}$$

$$A_{15} = \frac{0}{u_1} + \frac{0}{u_2} + \frac{0}{u_3} + \frac{0}{u_4} + \frac{0}{u_5} + \frac{0}{u_6} + \frac{0}{u_7} + \frac{0}{u_8} + \frac{0}{u_9} + \frac{0}{u_{10}} + \frac{0}{u_{11}} + \frac{0}{u_{12}} + \frac{0}{u_{13}} + \frac{0.5}{u_{14}} + \frac{1}{u_{15}} + \frac{0.5}{u_{16}}$$

$$A_{16} = \frac{0}{u_1} + \frac{0}{u_2} + \frac{0}{u_3} + \frac{0}{u_4} + \frac{0}{u_5} + \frac{0}{u_6} + \frac{0}{u_7} + \frac{0}{u_8} + \frac{0}{u_9} + \frac{0}{u_{10}} + \frac{0}{u_{11}} + \frac{0}{u_{12}} + \frac{0}{u_{13}} + \frac{0}{u_{14}} + \frac{0.5}{u_{15}} + \frac{1}{u_{16}}$$

The graphical representation of the fuzzification model is depicted in Fig. 2.

From Fig. 2 we can observe that the universe of discourse U is entirely covered by triangular membership functions that represent fuzzy sets $A_1, \ldots, A_{16}$. Table 2 shows the parameter values for each triangular membership function. In this table, T_{A_i} stands for the triangular membership function of fuzzy set A_i ($\forall i = 1, \ldots 16$). Table 2 has two sections and each one has four headings: the membership function, the value of parameter a, the value of parameter b and the value of parameter c.

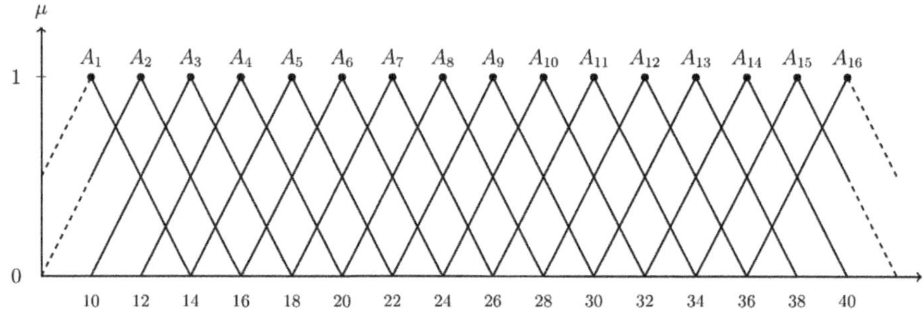

Fig. 2 Graphical representation of the fuzzification model defined on the universe of discourse U

Table 2 Parameter values for the triangular membership functions associated to fuzzy sets $A_1, \ldots, A_{16}$

Membership function	a	b	c	Membership function	a	b	c
T_{A_1}	6	10	14	T_{A_9}	22	26	30
T_{A_2}	8	12	16	$T_{A_{10}}$	24	28	32
T_{A_3}	10	14	18	$T_{A_{11}}$	26	30	34
T_{A_4}	12	16	20	$T_{A_{12}}$	28	32	36
T_{A_5}	14	18	22	$T_{A_{13}}$	30	34	38
T_{A_6}	16	20	24	$T_{A_{14}}$	32	36	40
T_{A_7}	18	22	26	$T_{A_{15}}$	34	38	42
T_{A_8}	20	24	28	$T_{A_{16}}$	36	40	44

Once defined the fuzzification model, each element of the time series $Y(t)$ must be fuzzified. Specifically, each observation in $Y(t)$ must be mapped to the fuzzy set for which its membership value is maximal [7]. This is performed by computing the degree of membership of all the observations to each fuzzy set by applying Eq. (1) with their corresponding parameter values from Table 2. In order to illustrate this process, let us consider the first observation of the time series, $Y(1) = 40$. The membership levels of this observation for fuzzy sets $A_1, \ldots, A_{14}$ are zero. The membership level of $x = 40$ to A_{15} is:

$$\mu_{A_{15}} = T(x = 40, a = 34, b = 38, c = 42) = \frac{c - x}{c - b} = \frac{42 - 40}{42 - 38} = \frac{2}{4} = 0.5.$$

Similarly, the membership level of $x = 40$ to A_{16} is:

$$\mu_{A_{15}} = T(x = 40, a = 36, b = 40, c = 44) = \frac{x - a}{b - a} = \frac{40 - 36}{40 - 36} = \frac{4}{4} = 1.$$

Since $\mu_{A_1}, \mu_{A_2}, \ldots, \mu_{A_{14}} = 0$, $\mu_{A_{15}} = 0.5$ and $\mu_{A_{16}} = 1$, observation $Y(1) = 40$ is associated to fuzzy set for which its membership level is the largest, that is, A_{16}. As mentioned previously, this process must be performed on every observation of the time series. Given the particular nature of the time series, only four fuzzy sets are used. Specifically, value 10 (green) is mapped to A_1, value 20 (yellow) is mapped to A_6, value 30 (orange) is mapped to A_{11} and value 40 (red) is mapped to A_{16}. Table 3 shows the result of the fuzzification process.

Table 3 Fuzzified values of the time series under study

Crisp value	Fuzzy set	Crisp value	Fuzzy set	Crisp value	Fuzzy set	Crisp value	Fuzzy set
40	A_{16}	20	A_6	10	A_1	30	A_{11}
30	A_{11}	20	A_6	20	A_6	20	A_6
40	A_{16}	30	A_{11}	20	A_6	10	A_1
40	A_{16}	20	A_6	20	A_6	10	A_1
40	A_{16}	20	A_6	20	A_6	10	A_1
30	A_{11}	30	A_{11}	30	A_{11}	10	A_1
20	A_6	30	A_{11}	30	A_{11}	10	A_1
20	A_6	20	A_6	30	A_{11}	10	A_1
20	A_6	20	A_6	30	A_{11}	30	A_{11}
20	A_6	20	A_6	40	A_{16}	20	A_6
20	A_6	10	A_1	30	A_{11}	–	–

3.4 Fuzzy Logical Relationships

Once all the observations of the time series $Y(t)$ have been fuzzified, the next step is to formulate the fuzzy logical relationships. This is achieved by iteratively comparing two consecutive fuzzified values A_i and A_j and stablishing the fuzzy logical relationship $A_i \rightarrow A_j$. Thus, the fuzzy logical relationships for the fuzzified values of our time series (see Table 3) are reported in Table 4.

Fuzzy logical relationship groups are obtained from the fuzzy logical relationships reported in Table 4 by grouping those with the same antecedent. Table 5 shows the fuzzy logical relationship groups for the time series under study.

Table 4 Fuzzy logical relationships of the time series under study

$A_{16} \rightarrow A_{11}$	$A_{11} \rightarrow A_{16}$	$A_{16} \rightarrow A_{16}$	$A_{16} \rightarrow A_{16}$	$A_{16} \rightarrow A_{11}$
$A_{11} \rightarrow A_6$	$A_6 \rightarrow A_6$	$A_6 \rightarrow A_6$	$A_6 \rightarrow A_6$	$A_6 \rightarrow A_6$
$A_6 \rightarrow A_6$	$A_6 \rightarrow A_6$	$A_6 \rightarrow A_{11}$	$A_{11} \rightarrow A_6$	$A_6 \rightarrow A_6$
$A_6 \rightarrow A_{11}$	$A_{11} \rightarrow A_{11}$	$A_{11} \rightarrow A_6$	$A_6 \rightarrow A_6$	$A_6 \rightarrow A_6$
$A_6 \rightarrow A_1$	$A_1 \rightarrow A_1$	$A_1 \rightarrow A_6$	$A_6 \rightarrow A_6$	$A_6 \rightarrow A_6$
$A_6 \rightarrow A_6$	$A_6 \rightarrow A_{11}$	$A_{11} \rightarrow A_{11}$	$A_{11} \rightarrow A_{11}$	$A_{11} \rightarrow A_{11}$
$A_{11} \rightarrow A_{16}$	$A_{16} \rightarrow A_{11}$	$A_{11} \rightarrow A_{11}$	$A_{11} \rightarrow A_6$	$A_6 \rightarrow A_1$
$A_1 \rightarrow A_1$	$A_1 \rightarrow A_1$	$A_1 \rightarrow A_1$	$A_1 \rightarrow A_1$	$A_1 \rightarrow A_1$
$A_1 \rightarrow A_{11}$	$A_{11} \rightarrow A_6$	–	–	–

Table 5 Fuzzy logical relationship groups for the time series under study

Group 1:	$A_1 \rightarrow A_1, A_6, A_1, A_1, A_1, A_1, A_1, A_{11}$
Group 2:	$A_2 \rightarrow \emptyset$
Group 3:	$A_3 \rightarrow \emptyset$
Group 4:	$A_4 \rightarrow \emptyset$
Group 5:	$A_5 \rightarrow \emptyset$
Group 6:	$A_6 \rightarrow A_6, A_6, A_6, A_6, A_6, A_6, A_{11}, A_6, A_{11}, A_6, A_6, A_1, A_6, A_6, A_6, A_{11}, A_1$
Group 7:	$A_7 \rightarrow \emptyset$
Group 8:	$A_8 \rightarrow \emptyset$
Group 9:	$A_9 \rightarrow \emptyset$
Group 10:	$A_{10} \rightarrow \emptyset$
Group 11:	$A_{11} \rightarrow A_{16}, A_6, A_6, A_{11}, A_6, A_{11}, A_{11}, A_{11} A_{16}, A_{11}, A_6, A_6$
Group 12:	$A_{12} \rightarrow \emptyset$
Group 13:	$A_{13} \rightarrow \emptyset$
Group 14:	$A_{14} \rightarrow \emptyset$
Group 15:	$A_{15} \rightarrow \emptyset$
Group 16:	$A_{16} \rightarrow A_{11}, A_{16}, A_{16}, A_{11}, A_{11}$

3.5 Defuzzification

The goal of the defuzzification process is to compute the forecasted output. In this research we follow the principles proposed by Chen to perform this process on the time series of the colors assigned to Tamaulipas by the Mexican traffic–light monitoring system. These principles state the following [5]:

1. If the fuzzy logical relationship is $A_i \rightarrow A_j$ then the output value is the middle point of A_j.
2. If the fuzzy logical relationship has more than one fuzzy set in the consequent, i.e., $A_i \rightarrow A_{j1}, A_{j2}, \ldots, A_{jk}$, then the output value is computed as the average of the maximal values of the fuzzy sets involved.
3. If the fuzzy logical relationship does not have any fuzzy set in the consequent, i.e., $A_i \rightarrow \emptyset$, then the output value is the middle point of A_i.

Frequently, the output value computed by the Chen principles does not correspond to any numerical code of color. For example, suppose that we want to forecast the next color given that the current color is orange. Since the numeric code of orange is 30, the fuzzy set associated with this observation is A_{11} (see Table 3). Moreover, from Table 5 we know that the fuzzy logical relationship of A_{11} has more than one fuzzy set in the consequent, and hence, the second principle must be applied. The output value is computed as follows:

$$Y(t+1) = \frac{40+20+20+30+20+30+30+30+40+30+20+20}{12} = 27.5$$

As we can observe, the output value (forecast) is $Y(t+1) = 27.5$, which does not correspond to any color code (10, 20, 30 or 40). In order to overcome this drawback, we propose to round the output to the nearest ten as follows:

$$\tilde{Y}(t+1) = \left\lfloor \frac{Y(t+1)}{10} + 0.5 \right\rfloor \times 10.$$

Thus, $Y(t+1) = 27.5$ is rounded to $\tilde{Y}(t+1) = 27.5/10 + 0.5 \times 10 = 30$. Therefore, the forecasted color is orange.

3.6 Computational Implementation and Model Accuracy

The proposed fuzzy time series model (FTSM) was implemented in Java in order to measure its accuracy. The accuracy of our FTSM can be obtained by the Root Mean Square Error index (RMSE), which requires the actual values of the time series and the forecasted values computed by the FTSM. Figure 3 shows the original time series from the observed data (black line) and the forecasted time series by our FTSM (red line).

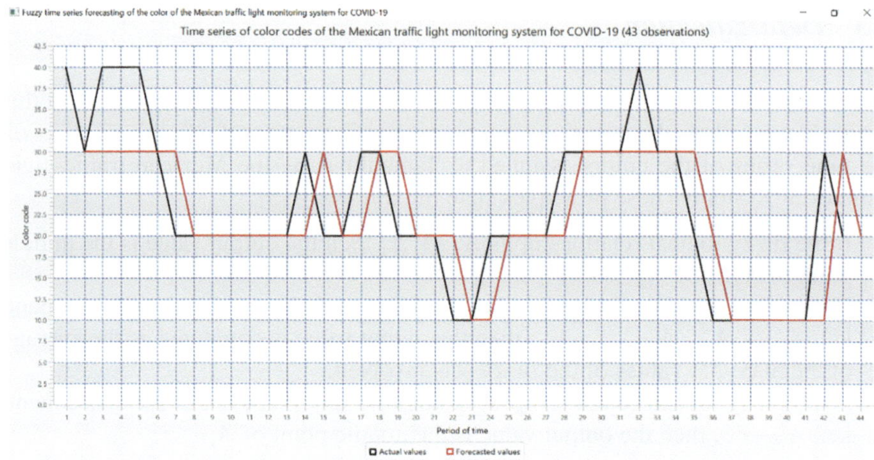

Fig. 3 Original time series (black line) and time series forecasted by the FTSM (red line)

As can be observed from Fig. 3, the forecasted time series fits relatively well to the real data. In order to quantitatively measure the performance of our proposed FTSM, we compute the RMSE index as follows:

$$RMSE = \sqrt{\frac{\sum_{t=1}^{T}\left[Y(t) - \hat{Y}(t)\right]^2}{T}},$$

where $Y(t)$ stands for the actual observation at time t, $\hat{Y}(t)$ is the value forecasted by our FTSM at time t and T is the number of paired values, both observed and forecasted. The RMSE value computed for our FTSM is 6.73. This value is relatively low. Therefore, we conclude that the accuracy of the proposed FTSM is high.

4 Conclusions

In this chapter we propose a fuzzy time series model (FTSM) to forecast the color assigned to the Mexican state of Tamaulipas by traffic–light monitoring system of COVID–19. The proposed FTSM divides the universe of discourse by using the well–known average–based partition method. Furthermore, it uses the three principles of Chen along with a rounding process to forecast the next color of the monitoring system.

We implement the FTSM in Java to quantitatively evaluate the accuracy. We use the RMSE index and the obtained value was 6.73, which is relatively low. According to the computational results, we conclude that the model proposed here can be used to forecast the color of the traffic–light monitoring system for Tamaulipas. The proposed fuzzy time series model can be easily applied to other states, the entire country or other geographical regions worldwide.

Acknowledgements The authors would like to thank *Tecnológico Nacional de México*, and especially the authorities of *Instituto Tecnológico de Altamira* for their support in this research. Besides, the authors also thank the Mexican Council for Science and Technology (CONACYT) for its support through the Mexican National System of Researchers (SNI).

References

1. Mexican Government. Coronavirus official web site. COVID–19 (2022). https://coronavirus.gob.mx/covid-19/
2. F. He, X. Shang, F. Ling, Z. Chen, T. Fu, J. Lin, Z. Wang, A practice of using five-colour chart to guide the control of COVID-19 and resumption of work in Zhejiang Province China. Sci. Rep. **11**(1), 1–7 (2021)
3. Coronavirus official web site, Lineamientos para la estimación de riesgos del semáforo por regiones COVID-19. https://coronavirus.gob.mx/wp-content/uploads/2022/01/Metodo_semaforo_COVID_actualiz_220124-20_19hrs.pdf (2022)
4. Coronavirus official web site, Mexican traffic–light monitoring system. https://coronavirus.gob.mx/semaforo/ (2022)
5. S.M. Chen, Forecasting enrollments based on fuzzy time series. Fuzzy Sets Syst. **81**(3), 311–319 (1996)
6. Mexican Government, Coronavirus COVID19 Comunicados Técnicos Diarios Históricos. https://www.gob.mx/salud/acciones-y-programas/covid19-informacion-relevante (2022)
7. H.K. Yu, Weighted fuzzy time series models for TAIEX forecasting. Physica A **349**(3–4), 609–624 (2005)

Bio-inspired Flower Pollination Algorithm for the Optimization of a Monolithic Neural Network

Hector Carreon-Ortiz, Patricia Melin, and Fevrier Valdez

Abstract The Flower Pollination Algorithm (FPA) is a metaheuristic that is inspired by nature and is based on the process of pollination of flowering plants and the constancy of flowers associated with pollinating agents that can be insects, water or the wind. This paper provides a study of the optimization of a multilayer monolithic neural network for pattern recognition with the ORL face dataset, comparisons are also made with other studies that used the ORL dataset and performed optimization of the neural network with other metaheuristics such as PSO (Particle Swarm Optimization) and GA (Genetic Algorithms).

Keywords Neural networks · Flower pollination algorithm · Optimization · Metaheuristic · Bio-inspired algorithm

1 Introduction

Nowadays many people seem to be worried about the future arrival of machines with artificial intelligence. Some people even propose to take measures to avoid it, including the prohibition of certain developments, something that has historically been shown not to work, since the advance in the research of intelligent computing cannot be avoided, such as the utilization of Artificial Neural Networks (ANNs) and Algorithms Inspired by Nature (Bioinspired Algorithms) [1].

ANNs have received considerable attention due to their powerful capacity in image processing, speech recognition, natural language processing, etc. The performance of ANN models depends to a large extent on the quantity and quality of the data, the calculation power and the efficiency of the algorithms [2].

In Evolutionary Computation (EC), stochastic optimization methods that have been developed to obtain almost optimal solutions in complex optimization problems are used, for which classical mathematical techniques usually fail.

H. Carreon-Ortiz · P. Melin (✉) · F. Valdez
Tijuana Institute of Technology/TecNM, Tijuana, BC, Mexico
e-mail: pmelin@tectijuana.mx

O. Castillo and P. Melin (eds.), *Fuzzy Logic and Neural Networks for Hybrid Intelligent System Design*, Studies in Computational Intelligence 1061,
https://doi.org/10.1007/978-3-031-22042-5_9

The methods of EC use as inspiration our scientific understanding of biological, natural or social systems, which at some level of abstraction can be represented as optimization processes.

In their operation, the search agents imitate a group of biological or social entities that interact with each other on the basis of specialized operators that make up a certain biological or social behavior.

These operators are applied to a population (or several subpopulations) of candidate solutions (individuals) that are evaluated with respect to their physical fitness. Thus, in the evolutionary process, individual positions are successively close to the optimal response of the system to be solved [3].

The Flower Pollination Algorithm (FPA) is a metaheuristic inspired by the pollination of flowering plants, whose pollinating agents can be insects such as bees, but also wind and water [4].

The goal of this work is the optimization of a monolithic multilayer neural network using the metaheuristics represented by the Flower Pollination Algorithm, by optimizing the parameters of the number of neurons in the hidden layers and the number of epochs.

The motivation for this work is to obtain the maximum recognition of images from the ORL database and compare the performance with other works where the neural network is optimized with different metaheuristics.

This paper is organized as follows:

In Sect. 2 we described the Artificial Neural Networks (ANN) and its biological inspiration, perceptron architecture, Sect. 3 is about FPA Algorithm, its inspiration and structure, Sect. 4 is about Optimization Monolithic Neural Network using FPA, network architecture and its integration with the FPA, Sect. 5 is about the simulation with the proposal method, description of ORL database used in this work, experiments with manual method and proposal method and statistical test between methods, and Sect. 6 is about conclusions.

2 Artificial Neural Networks (ANNs)

ANNs are traditional machine learning techniques that mimic the learning mechanism in biological organisms. The human nervous system contains cells, which are known as neurons. Neurons are connected to each other by the use of axons and dendrites, and the regions of connection between axons and dendrites are called synapses Fig. 1. The strength of synaptic connections continually changes in response to external stimuli; this change is how learning takes place in living organisms [5, 6].

This biological mechanism is simulated in artificial neural networks, which contain units of computation that are called neurons. Neurons are connected to each other through weights, which fulfill the same function as the strengths of synaptic connections in biological organisms. Each input to a neuron is scaled with a weight, which affects the function calculated in that unit and this is illustrated in Fig. 2. An

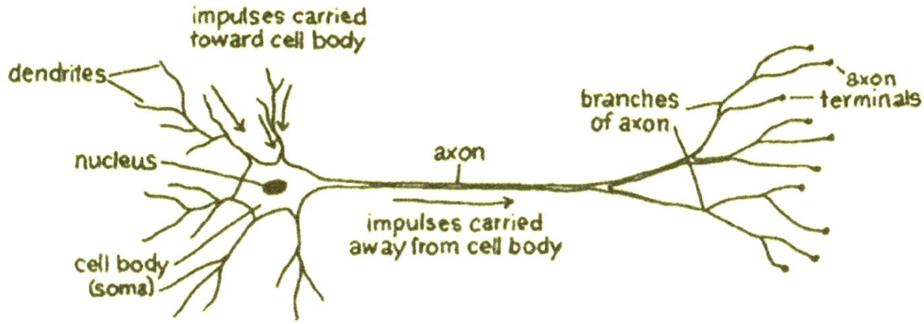

Fig. 1 Structure of a neuron

Fig. 2 Structure of an artificial neuron

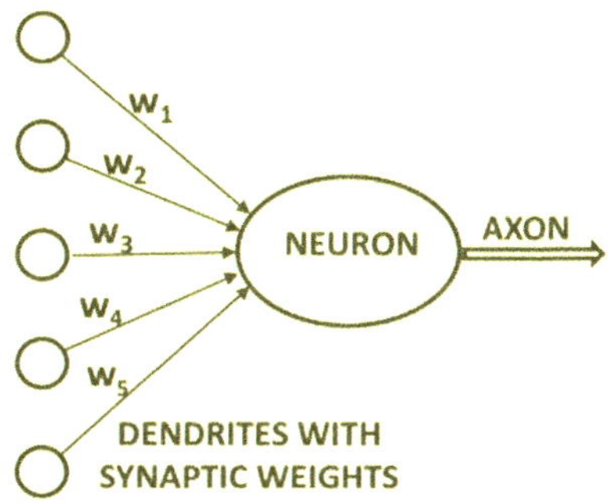

artificial neural network calculates a function of the inputs by propagating the calculated values from the input neurons to the output neurons and using the weights as intermediate measures [7]. Learning occurs by changing the weights that connect the neurons. Just as external stimuli are necessary for learning in biological organisms, the external stimulus in artificial neural networks is facilitated by the training data that contain the input–output data pairs of the function to be learned [8, 9].

2.1 *Multilayer Perceptron (MLP)*

The neuron (node) is the fundamental unit of a neural network. In the case of the MLP, it includes an input layer, an output layer and at least one hidden layer.

The perceptron is a simple neuron model that takes input signals (patterns) coded as (real) *input* vectors $\overline{x} = (x_1, x_2, \ldots, x_{n+1})$ through the associated (real) vector of synaptic *weights* $\overline{w} = (w_1, w_2, \ldots, w_{n+1})$ [10].

The layers consist of a set of nodes; in the case of the hidden layer, its inputs come from units in the previous layer and send their outputs to the next layer [11]. The input and output layers indicate the information traffic during the training phase where the learning algorithm is carried out [12].

The MLP usually learns by means of the backpropagation algorithm that in itself it is a gradient technique. Variants of the algorithm are also implemented to work on the problem of slow convergence (momentum). Once the training process is carried out, the weights of the network are frozen and can be used to calculate the output values for the new input samples. Next, a brief explanation of the backpropagation algorithm is provided [13, 14].

The learning of the network is a process in which the weights, w, are adapted through a continuous interaction (k) with the environment, in such a way that (1):

$$w_{nj}(k+1) = w_{nj}(k) + \Delta w_{nj}(k) \tag{1}$$

where $w(k)$ is the previous value of the weighting vector and $w(x+1)$ is the updated value. The learning algorithm consists of a set of rules to solve the learning problem and determine the values $w_{nj}(k)$.

One of the most important algorithms is that of error correction. Consider the nth neuron in the iteration.

Let y_n be the response of this neuron; $x(k)$ is the vector of environmental stimuli, and $\{x(k), d_n(k)\}$ is the training pair. Error signal Eq. (2):

$$e_n(k) = d_n(k) - y_n(k) \tag{2}$$

The goal is to minimize the objective function that takes this error into account. After selecting the criteria, the problem of correction of errors in learning becomes one of optimization. Consider a function $\varepsilon(w)$, which is a continuously differentiable function of a weight vector. The function $\varepsilon(w)$ transforms the elements of w into real numbers. We need to find an optimal solution $w*$ that satisfies the condition (3):

$$\epsilon\left(w^*\right) \leq \epsilon(w) \tag{3}$$

Then, it is necessary to solve an optimization problem without restrictions posed as: the minimization of the cost function $e(w)$ with respect to the weight vector. The necessary condition for optimization is given by (4):

$$\nabla\epsilon\left(w^*\right) = 0 \tag{4}$$

where ∇ is the gradient operator. An important class of unrestricted optimization algorithms is based on the idea of iterative descent (gradient descent method and Newton method). Starting with an initial condition $w(0)$, it generates a sequence $w(1)$, $w(2)$, …, in such a way that the cost function $\varepsilon(w)$ decreases in each iteration of the algorithm. It is desirable that the algorithm eventually converge on the optimal solution (5) in such a way that:

$$\epsilon(w(k+1)) < \epsilon(w(k)) \tag{5}$$

In the gradient descent method, successive adjustments are applied to the weight vector, in the direction of gradient descent. For convenience, we will use the following notation (6):

$$g = \nabla \in (w) \tag{6}$$

The gradient descent algorithm can be formally written as (7):

$$w(k+1) = w(k) - \eta g(k) \tag{7}$$

where η is a positive constant called the learning rate, and $g(k)$ is the gradient vector evaluated in $w(k)$.

Therefore, the correction applied to the weight vector can be written as (8):

$$\Delta w(k) = w(k+1) - w(k) = -ng(k) \tag{8}$$

This method slowly converges to an optimal solution w^*. However, the learning rate has a greater impact on this convergent behavior. When η is small, the trajectory of $w(k)$ on the plane W is smooth. When η is large, the trajectory of $w(k)$ on the plane W is oscillatory, and when η exceeds a certain critical value, the trajectory $w(k)$ on the plane W becomes unstable. Therefore, the backward propagation algorithm is a technique for implementing the gradient descent method in a weight space for a multilayer network. The basic idea is to efficiently calculate the partial derivatives of an approximate function of the behavior of the neural network with respect to all the elements of the adjustable vector of parameters w for a given value of the input vector x [13].

3 Flower Pollination Algorithm (FPA)

The FPA was developed by Xin She Yang in the year 2012 [15–17], inspired by the pollination of flowering plants, pollination of the flower involves the transfer of pollen between the flowers. This is basically done in two ways:

1. Autopollination or local pollination, which is a biotic form and contributes 10% of the global pollination, where pollinators are not required.
2. Cross-pollination or global pollination, which is an abiotic form and contributes 90% of global pollination, where pollinators such as insects, birds, bats and other animals, wind and water are required [18, 19].

The FPA algorithm is governed by the following 4 rules:

1. Cross pollination occurs from the pollen of a flower of different plants [20]. Pollinators obey the rules of a Levy Distribution [21, 22] (10) by jumping or flying distant passes, this is known as global pollination (9).

$$x_i^{t+1} = x_i^t + \gamma L(\lambda)\left(g^* - x_i^t\right) \tag{9}$$

$$L \sim \frac{\lambda \Gamma(\lambda) \sin(\pi \lambda/2)}{\pi} \frac{1}{s^{1+\lambda}}, \quad (s \gg s_0 > 0) \tag{10}$$

2. Self-pollination occurs when pollen from the same flower or other flowers of the same plant, this is known as local pollination [20] (11).

$$x_i^{t+1} = x_i^t + \epsilon\left(x_j^t - x_k^t\right) \tag{11}$$

3. The constancy of the flower is associated between flowers and pollinators; this improves the pollination process of the flowers.
4. Global and local pollination are controlled by a probability between 0 and 1, known as the probability of change [16, 23–25].

4 Monolithic Neural Network Optimization with FPA

Architecture of the Neural Network: A monolithic neural network Feed-Forward Backpropagation Network [26–30] of 3 hidden layers with 300 neurons each is being proposed, and according to the literature this is a good model for image recognition. Figure 3 show the architecture of a Monolithic Neural Network.

Figure 4 refers to the flow diagram of the FPA Algorithm including the Neural Network, it is the method that was used for the dynamic optimization of the Neural

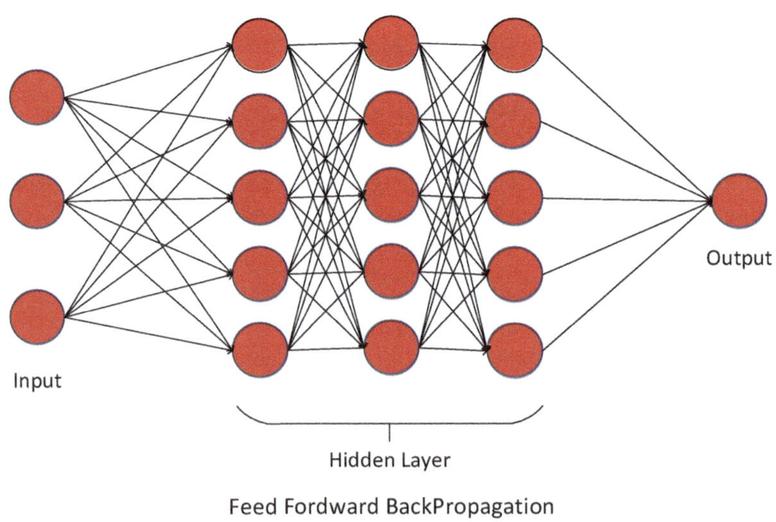

Fig. 3 Architecture of the monolithic neural network

Network, the parameters that were optimized were the neurons of the 3 hidden layers and the number of epochs [18].

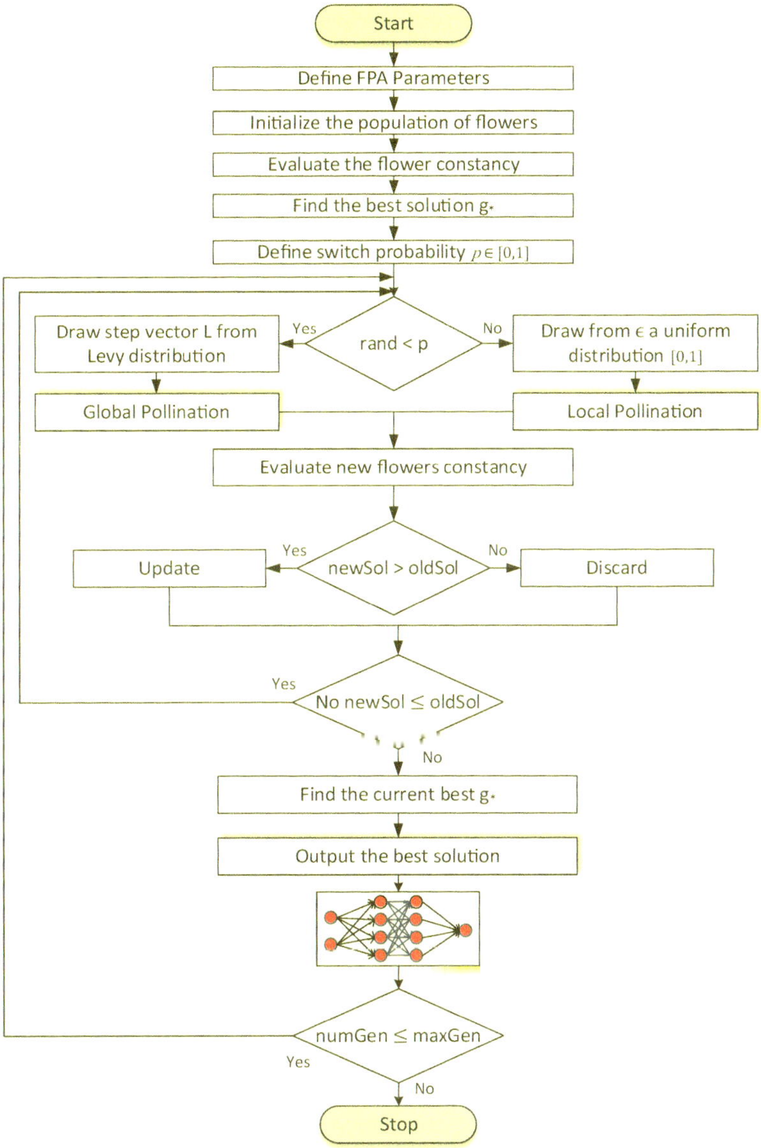

Fig. 4 Flowchart of the proposal of the FPA-NN method

Table 1 Characteristics of the ORL DB

ORL data base	
No. persons	40
No. images for person	10
Total of images	400
Image size	112 × 92
Size of the vector by image	10,304

5 Simulation

In this section we summarize the simulation results with the proposed approach.

5.1 ORL Database

This is the Database of Data (BD) of faces that is being used for the realization of the experiments, this BD belongs to the AT&T laboratories of Cambridge, and it is formed by 400 images of the faces of 40 different people, 10 images for each person, the images were taken at different times, varying the lighting, the facial expressions (eyes opened/closed, smiling/not smiling) and the facial details (glasses/without glasses). All images were taken on a dark and homogeneous background with people in a vertical position, (with tolerance for some lateral movements). Table 1 shows the basic characteristics of the ORL BD, such as the number of people, number of images per person, the total number of images that make up the BD, among others [31]. Figure 5 shows some examples of the ORL database images.

5.2 Method: Manual Adjustment of Parameters (MAN-NN)

Table 2 shows the setting of the proposed Neural Network Architecture for the method of manual adjustment of parameters.

Table 3 shows the number of hidden layers, neurons per layer of the different Architectures of the Neural Networks that were used with the method of manual adjustment of parameters, each of these Architectures has an input layer and an output layer.

Table 4 refers to the best four experiments of more than 100 that were performed with the method of manual adjustment of parameters, the minimum and maximum value, as well as the Mean and Standard Deviation obtained from each of the Architectures are also shown of every one of them.

Fig. 5 Sample of the faces of the ORL DB

Table 2 Neural network parameters used with the manual adjustment method

Parameters	
Modules	1
Hidden layers	2
Neurons hidden layer	100–400
Neurons output layer	40
Transfer function	tansig
Training method	trainscg
Epochs	1000
Meta error	1.00E−05
Learning rate	0.05

Table 3 Parameters of the different architectures used in the experiments

Neural network architecture			
Architecture	Hidden layers	Neurons per layer	Epochs
1	2	200	200
2	2	300	350
3	2	400	500
4	2	270	250

Table 4 Experimental results with the manual parameter adjustment method

Exp/arch	1	2	3	4
1	68.75	68.75	63.75	83.75
2	78.75	82.50	77.50	65.00
3	63.75	72.50	68.75	80.00
4	81.25	76.25	87.50	67.50
5	88.75	80.00	85.00	63.75
6	71.25	60.00	71.25	81.25
7	77.50	57.50	76.25	73.75
8	71.25	81.25	72.50	63.75
9	33.75	70.00	53.75	76.25
10	73.75	70.00	71.25	71.25
11	82.50	75.00	70.00	72.50
12	32.50	81.25	68.75	83.75
13	72.50	83.75	77.50	63.75
14	78.75	67.50	66.25	62.50
15	67.50	72.50	68.75	62.50
16	85.00	72.50	70.00	83.75
17	72.50	71.25	62.50	76.25
18	82.50	85.00	68.75	77.50
19	76.25	76.25	73.75	77.50
20	77.50	76.25	52.50	65.00
21	63.75	66.25	65.00	61.25
22	73.75	76.25	71.25	73.75
23	76.25	71.25	70.00	82.50
24	77.50	68.75	76.25	52.50
25	75.00	53.75	70.00	83.75
26	72.50	68.75	68.75	70.00
27	85.00	76.25	77.50	70.00
28	68.75	71.25	85.00	76.25
29	71.25	72.50	61.25	86.25
30	80.00	73.75	73.75	80.00
$\sum$	2180.00	2178.75	2125.00	2187.50
Minimum	32.50	53.75	52.50	52.50
Maximum	88.75	85.00	87.50	86.25
Mean	72.67	72.63	70.83	72.92
σ	12.33	7.25	7.90	8.67

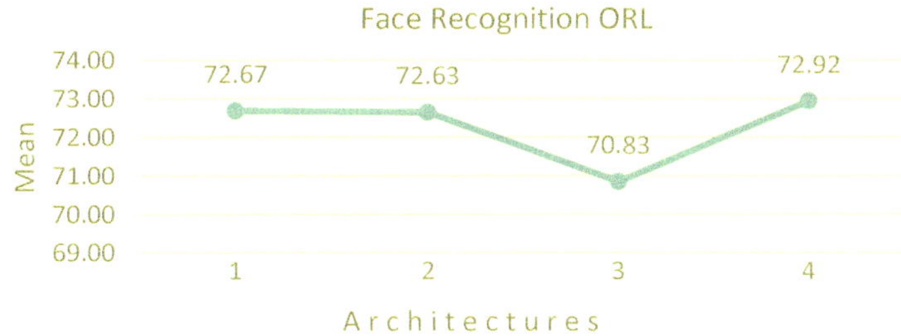

Fig. 6 Average of the experiments of the method manual adjustment of parameters

Figures 6, 7 and 8 show the graphical representation of the linear behavior of the Mean and Standard Deviation [32, 33] respectively of the 4 Architectures of the best Neural Networks obtained by the manual parameter adjustment method.

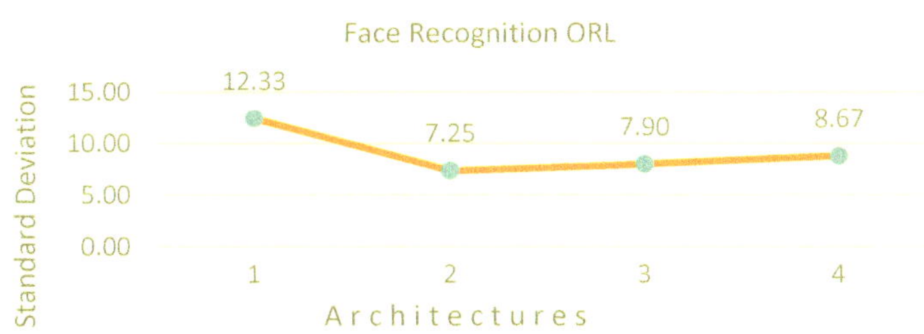

Fig. 7 Standard deviation of the experiments of the method manual adjustment of parameters

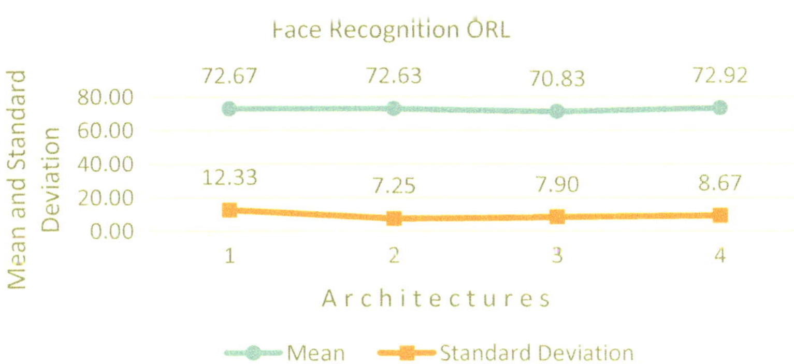

Fig. 8 Mean and standard deviation of the experiments of the method manual adjustment of parameters

5.3 *Neural Network Optimized with FPA (FPA-NN)*

Table 5 shows the number of hidden layers, neurons per layer of the different Architectures of the Neural Networks that were used with the FPA method, each of these Architectures has an input layer and an output layer.

Table 6 shows the number of hidden layers, neurons per layer of the 13 different Architectures of the Neural Networks that were used with the FPA method, each of these Architectures has an input layer and an output layer.

Table 5 Parameters of the neural network that were used with the FPA method

Parameters	
Modules	1
Hidden layers	3
Neurons hidden layer	100–400
Neurons output layer	40
Transfer function	tansig
Training method	trainscg
Epochs	1000
Meta error	1.00E−05
Learning rate	0.05

Table 6 Architectures of the neural networks that were used with the FPA method

Neural network architecture					
Architecture	Hidden layers	Neurons per layer			Epochs
		1	2	3	
1	3	361	362	361	800
2	3	391	390	343	1000
3	3	400	400	400	700
4	3	390	350	361	711
5	3	388	374	380	747
6	3	388	304	311	721
7	3	385	344	313	729
8	3	399	342	334	1000
9	3	369	322	342	714
10	3	359	343	379	711
11	3	400	306	329	733
12	3	370	311	301	1000
13	3	359	311	217	755

Table 7 refers to the 13 best experiments of more than 1000 that were carried out with the FPA method, the minimum and maximum value, as well as the Mean and Standard Deviation obtained from each of the Architectures are also shown from each of the Architectures.

Figures 9, 10 and 11 show the graphical representation of the linear behavior of the Mean and Standard Deviation respectively of the 4 Architectures of the best Neural Networks obtained by the FPA method.

After performing the statistical tests among all the experiments of the methods of manual adjustment of parameters and FPA, Architecture 5 of the FPA method obtained the best result, Figs. 12 and 13 show the probability and the dispersion of the winning neural network.

5.4 Statistical Test Between Methods

Statistical tests [34, 35] were performed among the best experiments performed with the FPA-NN, MAN-NN and SD + T1FIS methods [36–38].

SD + T1FIS is a diffuse edge detection method optimized for image recognition using a modular neural network [39] and the PSO algorithms [40–43] and GA [44], in addition to using the ORL database [36].

Table 8 shows the best experiments done with the different methods described above with which the statistical tests were performed, in the lower part we can find the summation, the maximum and minimum values, the mean and the standard deviation of each of the different methods.

Table 9 presents the results of the statistical tests between the different methods and the proposed FPA-NN method, the result is that there was significant evidence to accept that the FPA-NN method obtained better results (highlighted in bold) than the other methods.

Figure 14 is a linear representation of the face recognition of the ORL database that was obtained with the methods FPA-NN, MAN-NN and SD + T1FIS.

Figures 15 and 16 represent the linear graphs of the Mean and Standard Deviation respectively of the methods FPA-NN, MAN-NN and SD + T1FIS, in the face recognition of the ORL database.

6 Conclusions

We performed more than 100 experiments with the MAN-NN method and more than 1000 experiments with the FPA-NN method. In addition, statistical tests were also performed among these methods and the two best experiments were obtained, that is, with the manual method and the FPA method, statistical tests were performed with

Table 7 Experiments performed with the FPA method

Exp/arch	1	2	3	4	5	6	7	8	9	10	11	12	13
1	93.75	91.25	92.50	80.00	90.00	88.75	88.75	88.75	88.75	88.75	90.00	86.25	86.25
2	86.25	85.00	86.25	88.75	87.50	85.00	88.75	88.75	88.75	90.00	93.75	90.00	85.00
3	87.50	90.00	88.75	91.25	87.50	91.25	92.50	95.00	91.25	85.00	90.00	90.00	92.50
4	92.50	91.25	86.25	90.00	83.75	88.75	91.25	87.50	88.75	92.50	87.50	86.25	85.00
5	90.00	93.75	87.50	91.25	88.75	90.00	86.25	86.25	85.00	92.50	88.75	88.75	88.75
6	87.50	87.50	85.00	88.75	88.75	90.00	90.00	86.25	90.00	90.00	86.25	88.75	88.75
7	88.75	90.00	95.00	88.75	88.75	92.50	87.50	91.25	91.25	91.25	85.00	87.50	90.00
8	88.75	88.75	90.00	91.25	92.50	90.00	88.75	92.50	87.50	86.25	90.00	88.75	86.25
9	91.25	90.00	92.50	90.00	88.75	86.25	85.00	88.75	90.00	86.25	88.75	88.75	91.25
10	90.00	85.00	92.50	88.75	91.25	91.25	91.25	87.50	88.75	88.75	90.00	82.50	87.50
11	86.25	85.00	86.25	88.75	85.00	87.50	86.25	90.00	86.25	90.00	88.75	85.00	88.75
12	90.00	92.50	90.00	92.50	87.50	90.00	81.25	87.50	86.25	91.25	88.75	90.00	86.25
13	91.25	87.50	86.25	88.75	86.25	90.00	90.00	87.50	93.75	90.00	88.75	90.00	88.75
14	82.50	92.50	92.50	91.25	88.75	88.75	92.50	90.00	90.00	90.00	90.00	86.25	91.25
15	85.00	86.25	87.50	90.00	90.00	88.75	91.25	93.75	85.00	88.75	83.75	88.75	92.50
16	87.50	88.75	90.00	92.50	87.50	87.50	95.00	87.50	92.50	85.00	90.00	92.50	87.50
17	88.75	87.50	90.00	86.25	90.00	91.25	87.50	86.25	91.25	90.00	91.25	92.50	88.75
18	86.25	82.50	88.75	88.75	87.50	90.00	87.50	91.25	83.75	90.00	88.75	91.25	87.50
19	93.75	91.25	91.25	91.25	87.50	87.50	92.50	83.75	87.50	87.50	88.75	88.75	87.50
20	92.50	90.00	86.25	86.25	87.50	86.25	81.25	87.50	92.50	85.00	88.75	83.75	90.00
21	88.75	86.25	82.50	85.00	90.00	92.50	88.75	87.50	83.75	86.25	85.00	85.00	87.50

(continued)

Table 7 (continued)

Exp/arch	1	2	3	4	5	6	7	8	9	10	11	12	13
22	87.50	86.25	88.75	90.00	88.75	90.00	90.00	87.50	87.50	87.50	85.00	86.25	80.00
23	90.00	92.50	86.25	90.00	91.25	86.25	91.25	87.50	92.50	85.00	81.25	86.25	90.00
24	88.75	86.25	90.00	93.75	90.00	90.00	91.25	91.25	87.50	87.50	88.75	92.50	87.50
25	86.25	88.75	90.00	86.25	90.00	82.50	91.25	86.25	87.50	90.00	90.00	85.00	87.50
26	90.00	88.75	90.00	93.75	90.00	85.00	86.25	86.25	91.25	88.75	87.50	83.75	87.50
27	91.25	91.25	93.75	92.50	91.25	86.25	91.25	87.50	90.00	90.00	86.25	92.50	87.50
28	86.25	91.25	88.75	82.50	92.50	90.00	81.25	88.75	90.00	85.00	90.00	88.75	88.75
29	87.50	83.75	90.00	90.00	88.75	90.00	85.00	92.50	81.25	92.50	92.50	95.00	92.50
30	87.50	95.00	88.75	87.50	87.50	88.75	90.00	87.50	87.50	85.00	90.00	90.00	86.25
∑	2663.75	2666.25	2673.75	2676.25	2665.00	2662.50	2661.25	2660.00	2657.50	2656.25	2653.75	2651.25	2645.00
Min	82.50	82.50	82.50	80.00	83.75	82.50	81.25	83.75	81.25	85.00	81.25	82.50	80.00
Max	93.75	95.00	95.00	93.75	92.50	92.50	95.00	95.00	93.75	92.50	93.75	95.00	92.50
Mean	88.79	88.88	89.13	89.21	88.[illegible]3	88.75	88.71	88.67	88.58	88.54	88.46	88.38	88.17
σ	2.62	3.10	2.81	3.09	1.[illegible]9	2.34	3.48	2.54	2.97	2.46	2.60	3.05	2.58

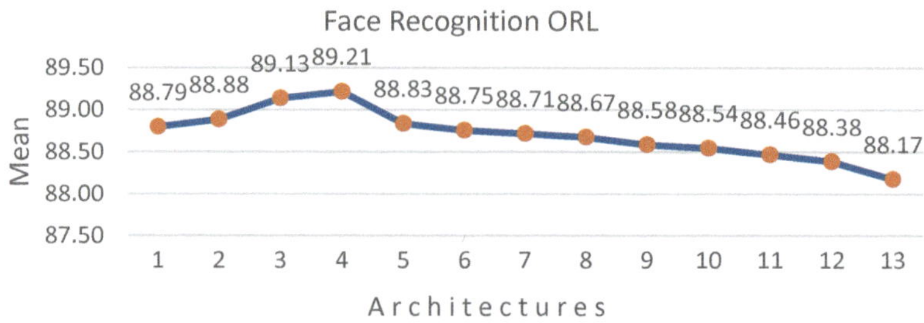

Fig. 9 Average of the experiments with the FPA method

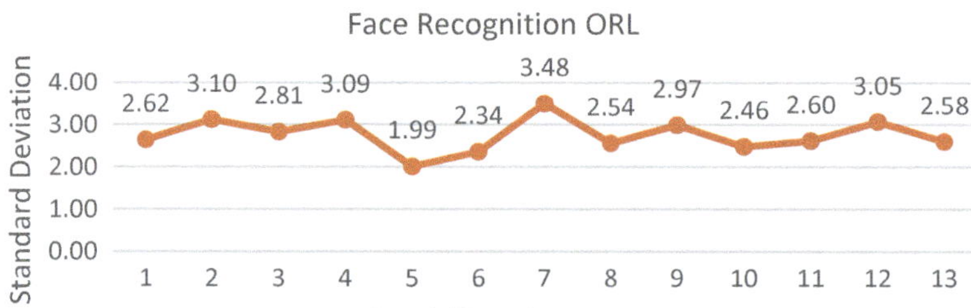

Fig. 10 Standard deviation of the experiments with the FPA method

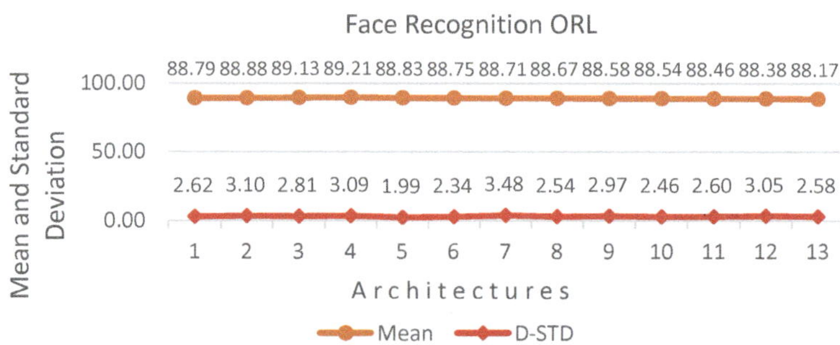

Fig. 11 Mean and standard deviation of the experiments with the FPA method

these two best winning methods and the SD + T1FIS method, the method with the best result was FPA-NN. In all the experiments of the three different methods the ORL face database was used for training and neural network test was used 70 and 30 percent of the database respectively.

Although the SD + T1FIS method used a 3-module neural network architecture and the bio-inspired algorithms PSO and GA, and with the FPA-NN method, a multilayer monolithic neural network was used, as well as the bio-inspired FPA

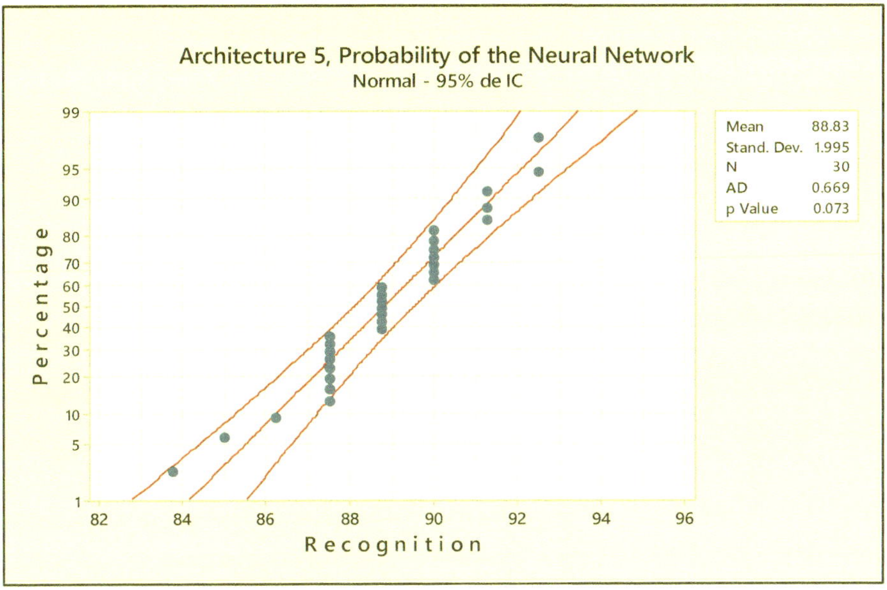

Fig. 12 Probability of architecture A5

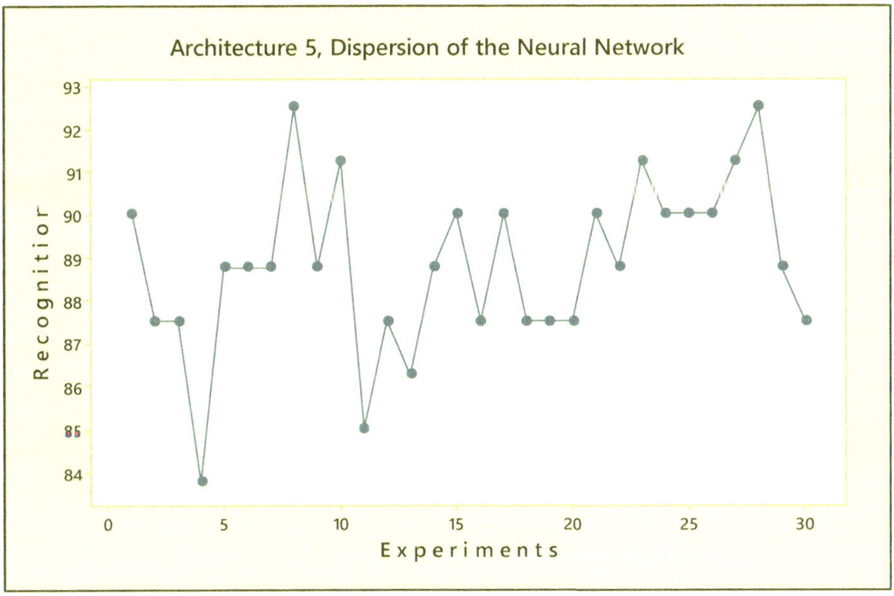

Fig. 13 Dispersion of the A5 architecture experiments

Table 8 Experiments of the different methods

Exp/arch	FPA-NN	MAN-NN	SD + T1FIS
1	90.00	68.75	82.86
2	87.50	82.50	83.93
3	87.50	72.50	77.86
4	83.75	76.25	85.36
5	88.75	80.00	87.14
6	88.75	60.00	81.07
7	88.75	57.50	81.07
8	92.50	81.25	86.79
9	88.75	70.00	86.79
10	91.25	70.00	82.86
11	85.00	75.00	80.71
12	87.50	81.25	73.93
13	86.25	83.75	80.00
14	88.75	67.50	83.21
15	90.00	72.50	84.29
16	87.50	72.50	85.00
17	90.00	71.25	83.93
18	87.50	85.00	88.57
19	87.50	76.25	81.43
20	87.50	76.25	87.50
21	90.00	66.25	90.00
22	88.75	76.25	80.00
23	91.25	71.25	83.21
24	90.00	68.75	75.00
25	90.00	53.75	86.43
26	90.00	68.75	77.50
27	91.25	76.25	81.43
28	92.50	71.25	77.50
29	88.75	72.50	84.64
30	87.50	73.75	86.43
$\sum$	2665.00	2178.75	2486.44
Min	83.75	53.75	73.93
Max	92.50	85.00	90.00
Mean	88.83	72.63	82.88
σ	1.99	7.25	3.96

Table 9 Statistical test between the different methods

	MAN-NN	SD + T1FIS
Difference	16.208	5.952
Z (observed value)	**11.800**	**7.358**
Z (critical value)	1.645	1.645
Valor-p (unilateral)	1.000	1.000
Alfa	0.050	0.050
	Evidence	Evidence

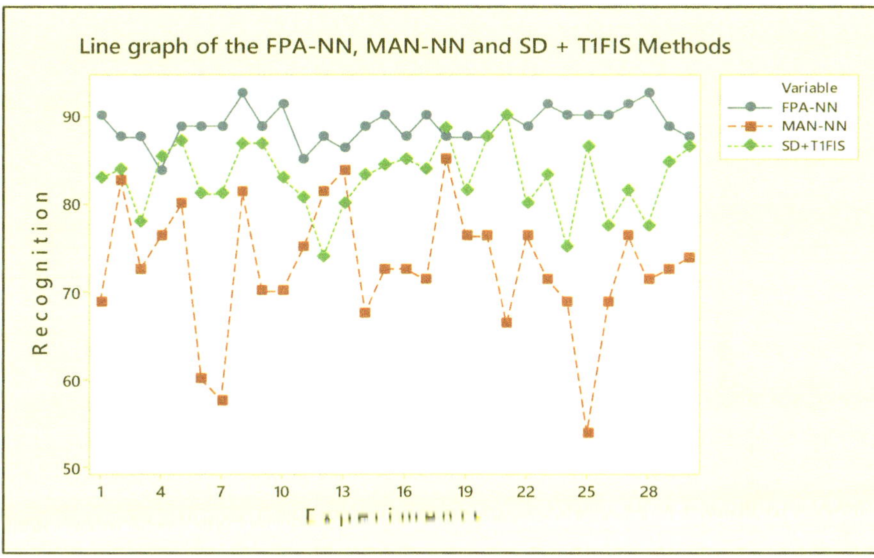

Fig. 14 Line graph of the methods FPA-NN, MAN-NN and SD + T1FIS

algorithm, that is, a neural network with a more complex architecture (SD + T1FIS) versus a less complex (monolithic) neural network architecture, better results were obtained, which leads us to conclude that the number of experiments performed (1000) and the dynamic optimization of the neural network with the FPA algorithm were determinant to obtain better results than with the other methods.

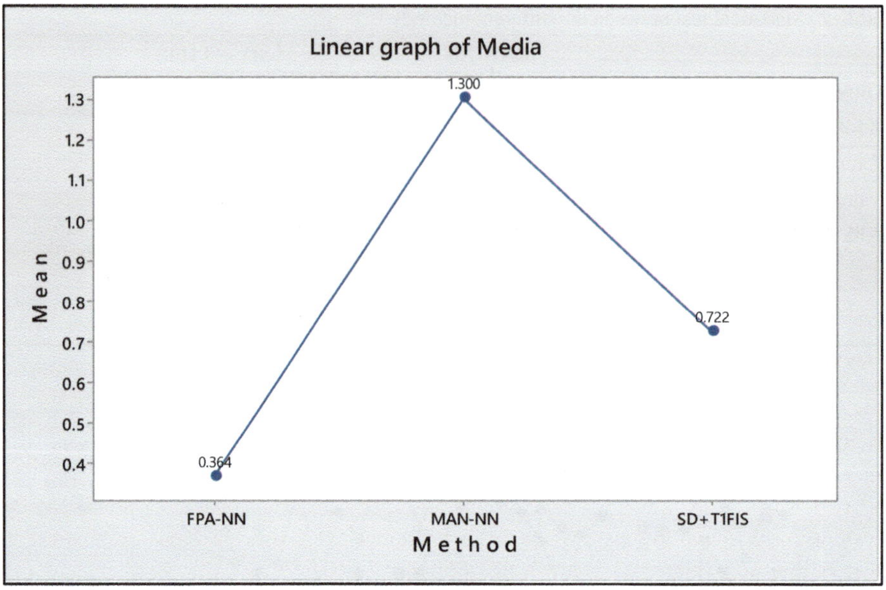

Fig. 15 Linear graph of the mean of the methods FPA-NN, MAN-NN and SD + T1FIS

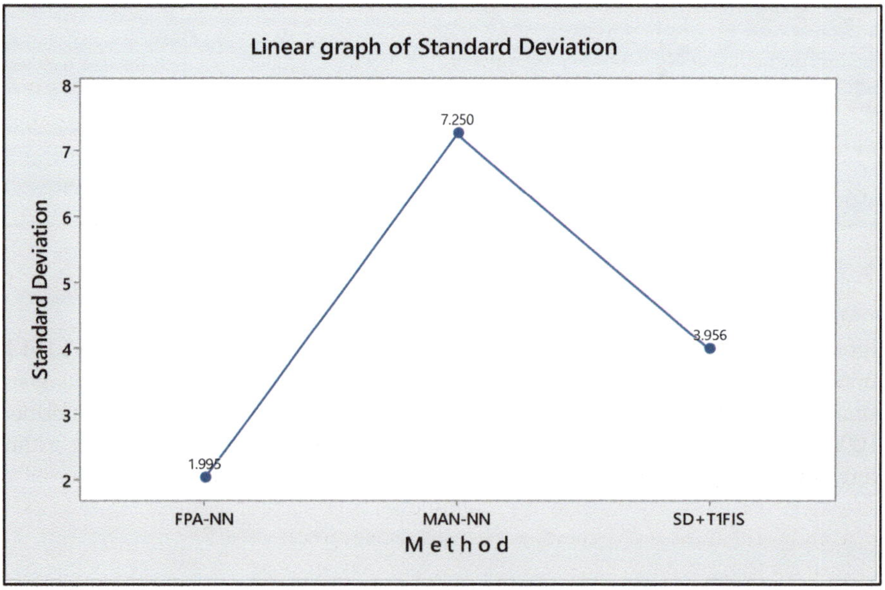

Fig. 16 Linear graph of the standard deviation of the methods FPA-NN, MAN-NN and SD + T1FIS

References

1. F. Berzal, Redes neuronales & deep learning (2018)
2. W. Cao, X. Wang, Z. Ming, J. Gao, A review on neural networks with random weights. Neurocomputing **275**, 278–287 (2018)
3. E. Cuevas, V. Osuna, D. Oliva, *Evolutionary Computation Techniques: A Comparative Perspective*, vol. 686 (2017)
4. X.-S. Yang, Flower pollination algorithm for global optimization. *Unconventional Computation and Natural Computation*. Lecture Notes in Computer Science, vol. 7445 (2012), pp. 240–249
5. C.C. Aggarwal, *Neural Networks and Deep Learning: A Textbook* (2018)
6. A. Krenker, J. Bester, A. Kos, Introduction to the artificial neural networks. *Artificial Neural Networks: Methodological Advances and Biomedical Applications* (2011)
7. D. Kriesel, A Brief Introduction to Neural Networks (2005), p. 244
8. K. Gurney, *An Introduction to Neural Networks* (UCL Press Limited, 1997)
9. I.A. Basheer, M. Hajmeer, Artificial neural networks: fundamentals, computing, design, and application. J. Microbiol. Methods **43**(1), 3–31 (2000)
10. P. Tino, L. Benuskova, A. Sperduti, *Springer Handbook of Computational Intelligence*, vol. 8, no. 3 (2015)
11. K. Halawa, *A New Multilayer Perceptron Initialisation Method with Selection of Weights on the Basis*, vol. 1 (2015), pp. 47–58
12. R. Cassani, Multilayer perceptron example (2018) [Online]. Available at: https://github.com/rcassani/mlp-example
13. A.N. Networks, H. Well, D. They, F.T. Series, R.N. Artificiales, A Review of Artificial Neural Networks (2013)
14. F. Amato, A. López, E.M. Peña-Méndez, P. Vaňhara, A. Hampl, J. Havel, Artificial neural networks in medical diagnosis. J. Appl. Biomed. **11**(2), 47–58 (2013)
15. S. Pant, A. Kumar, M. Ram, Flower pollination algorithm development: a state of art review. Int. J. Syst. Assur. Eng. Manag. (2017)
16. X. Yang, M. Karamanoglu, X. He, X. Yang, M. Karamanoglu, X. He, Flower pollination algorithm: a novel approach for multiobjective optimization. Eng. Optim. **46**, 1222–1237 (2014)
17. X.S. Yang, M. Karamanoglu, X. He, Multi-objective flower algorithm for optimization. Procedia Comput. Sci. **18**, 861–868 (2013)
18. S.D. Madasu, M.L.S. Sai Kumar, A.K. Singh, A flower pollination algorithm based automatic generation control of interconnected power system. Ain Shams Eng. J. (2015)
19. N. Sakib, A comparative study of flower pollination algorithm and bat algorithm on continuous optimization problems. Int. J. Appl. Inf. Syst. IJAIS **7**(9), 1–7 (2014)
20. D.P. Abrol, *Pollination Biology: Biodiversity Conservation and Agricultural Production* (2012)
21. A.F. Kamaruzaman, A.M. Zain, S.M. Yusuf, A. Udin, Levy flight algorithm for optimization problems—a literature review. Appl. Mech. Mater. **421**, 496–501 (2013)
22. X.-S. Yang, Random walks and optimization, *Nature-Inspired Optimization Algorithms* (2014), pp. 45–65
23. S.M. Nigdeli, G. Bekda, X. Yang, *Metaheuristics and Optimization in Civil Engineering*, vol. 7 (2016), pp. 25–42
24. X.S. Yang, G. Bekdas, S.M. Nigdeli, Application of the flower pollination algorithm in structural engineering. Model. Optim. Sci. Technol. **7**, v–vi (2016)
25. H. Chiroma, N.L.M. Shuib, S.A. Muaz, A.I. Abubakar, L.B. Ila, J.Z. Maitama, A review of the applications of bio-inspired Flower Pollination Algorithm. Procedia Comput. Sci. **62**(Scse), 435–441 (2015)
26. C. Gershenson, Artificial neural networks for beginners. ICES J. Mar. Sci. **66**(6), 1119–1129 (2009)
27. N. Comp, N. Comp, 4 Feedforward Multilayer Neural Networks—part I (2005)
28. I. Coding, M. Perceptrons, 5 feedforward multilayer neural networks—Part II. Training **1**, 1–44 (2005)

29. J. Bilski, J. Smoląg, J.M. Zurada, The parallel approach to the conjugate gradient learning algorithm for the feedforward neural networks, *Lecture Notes in Artificial Intelligence (Subseries Lecture Notes in Computer Science*, vol. 9119, no. 1 (2015), pp. 3–14
30. J. Bilski, J. Smol, Parallel approach to the Levenberg-Marquardt learning algorithm for feedforward neural networks, in *International Conference on Artificial Intelligence and Soft Computing (ICAISC 2015), Zakopane, Poland, June 14–18, 2015 Proceedings, Part I*, vol. 1, no. 1 (2015), pp. 1–1
31. A. Laboratories and Cambridge, The Database of Faces (1994) [Online]. Available at: https://www.cl.cam.ac.uk/research/dtg/attarchive/facedatabase.html
32. The BMJ, Mean and Standard Deviation (2019)
33. D.K. Lee, J. In, S. Lee, Standard deviation and standard error of the mean (2015)
34. Addinsoft, "XLSTAT" (2019) [Online]. Available at: https://www.xlstat.com/es/
35. Minitab Statistical Software, "Minitab" (2018) [Online]. Available at: http://www.minitab.com/es-mx/
36. I. Espinoza, Desarrollo de un detector de borde difuso optimizado para el procesamiento de imagenes en el reconocimiento de patrones utilizando redes neuronales (Instituto Tecnologico de Tijuana, 2018)
37. Y. Becerikli, T.M. Karan, A new fuzzy approach for edge detection 2 detection of image edges, in *Computational Intelligence and Bioinspired Systems* (LNCS, Springer, Berlin, 2005), pp. 943–951
38. C.I. Gonzalez, P. Melin, J.R. Castro, O. Mendoza, O. Castillo, An improved Sobel edge detection method based on generalized type-2 fuzzy logic. Soft Comput. **20**(2), 773–784 (2016)
39. P. Melin, C. Felix, O. Castillo, Face recognition using modular neural networks and the fuzzy Sugeno integral for response integration. Intell. Syst. **20**, 275–291 (2005)
40. P. Melin, F. Olivas, O. Castillo, F. Valdez, J. Soria, M. Valdez, Optimal design of fuzzy classification systems using PSO with dynamic parameter adaptation through fuzzy logic. Expert Syst. Appl. **40**(8), 3196–3206 (2013)
41. B. Bhattacharyya, S. Raj, PSO based bio inspired algorithms for reactive power planning. Int. J. Electr. Power Energy Syst. **74**, 396–402 (2016)
42. B.A. Garro, R.A. Vázquez, Designing artificial neural networks using particle swarm optimization algorithms. Comput. Intell. Neurosci. **2015** (2015)
43. M. Couceiro, P. Ghamisi, *Fractional Order Darwinian Particle Swarm Optimization Applications and Evaluation of an Evolutionary Algorithm* (2016)
44. Y. Maldonado, O. Castillo, Comparison between multiobjective GA and PSO for parameter optimization of AT2-FLC for a real application in FPGA, in *Comparison between multiobjective GA and PSO for parameter optimization of AT2-FLC for a real application in FPGA*, vol. 2 (2012)

Rendezvous and Docking Control of Satellites Using Chaos Synchronization Method with Intuitionistic Fuzzy Sliding Mode Control

Onur Silahtar, Fatih Kutlu, Özkan Atan, and Oscar Castillo

Abstract In this study, two different controllers have been designed to perform the rendezvous and docking tasks of two nonidentical and noncooperative cubic satellites. Firstly, the motion of cubic satellites was modeled with chaotic equations. After selecting suitable chaotic models, fuzzy sliding mode controller (FSMC) and a new intuitionistic fuzzy sliding mode controller (IFSMC), which are applied to synchronization systems under the same initial conditions, have been designed. It has been observed that both synchronizations reach stability by applying the controllers designed by considering the Lyapunov stability criteria. After a while, a short-term and random disturbance was applied to the synchronization systems and the response of both controllers was observed. The numerical results showed that the synchronization system with both controllers was stable, robust, efficient, fast and chattering-free. However, synchronization system with IFSMC was found to be more robust, faster and more efficient than synchronization system with FSMC.

Keywords Chaos synchronization · Fuzzy sliding mode control · Intuitionistic fuzzy sliding mode control · Rendezvous and docking control

1 Introduction

Researchers have recently focused on rendezvous and proximity operation due to increasing of space missions, such as refueling, the building of large space station [12, 30, 32, 33, 56]. In classical rendezvous and docking operation, knowledges of altitude, angular velocity and the mass of target are needed. However, in studies

O. Silahtar (✉) · Ö. Atan
Department of Electrical and Electronics Engineering, Faculty of Engineering, Van Yuzuncu Yil University, 65080 Van, Turkey
e-mail: onursilahtar@yyu.edu.tr

F. Kutlu
Department of Mathematics, Faculty of Science, Van Yuzuncu Yil University, 65080 Van, Turkey

O. Castillo
Division of Graduate Studies, Tijuana Institute of Technology, TecNM, Tijuana, Mexico

O. Castillo and P. Melin (eds.), *Fuzzy Logic and Neural Networks for Hybrid Intelligent System Design*, Studies in Computational Intelligence 1061,
https://doi.org/10.1007/978-3-031-22042-5_10

for approximation operation to unknown objects and meteorites, the satellite has been needing noncooperative target scenarios. In literature, there are various studies about noncooperative space target such as 6-DOF multi-constrained adaptive tracking controller [32], angels-only navigation control scheme [55], the pose estimation algorithm of rectangular structure [53], image-based visual controller for non-cooperative rendezvous maneuvers [38], adaptive fuzzy control method for rendezvous and proximity operation [43], nonlinear controller for tracking and rendezvous with a non-cooperative target [14]. The aim of all these studies was to perform the rendezvous and docking operation with minimum energy. Also, in all these papers, dynamic behavior of the satellites was modeled by using complex dynamical models.

In addition, according to above-mentioned studies, chaos was used for satellite dynamical model, such as chaos was analyzed in attitude dynamics of a flexible satellite [9], qualitative behavior of satellite systems was studied using equilibrium points, Poincaré section, bifurcation diagrams, Lyapunov exponents [20]. Moreover a sliding mode controller (SMC) was designed for synchronization of chaotic satellite systems [21], projective synchronization and stability analyzes of fractional-order satellite system were realized [50], chaos in spatial attitude dynamics of a satellite in an elliptic orbit and its analytic and numerical solution method were studied [10]. Additionally it was aimed to analyze of time-delay synchronization of chaotic satellite systems [22]. These studies were focused on chaotic behaviors of altitude dynamic of satellite and chaos synchronization. However, identical chaos synchronization and identical satellite were used in these studies. But in this study, unlike previous studies, it is focused on synchronization of non-identical chaotic satellite systems.

Non-identical chaotic systems have different parameters and sensitivity to the initial condition, so synchronizations of the chaotic systems are quite complex. For these synchronization of non-identical chaotic systems, many novel control methods were designed [13, 26, 35, 37, 41, 44, 49], and some of the studies were to compare the performance of controllers [13, 26, 35]. There are effectual applications of synchronization and control of non-identical chaotic systems in the literature. In [44], a SMC was designed for synchronization of fractional-order chaotic systems with non-identical orders, unknown parameters and disturbances, and although simulation results were shown that system had high performance against disturbance and uncertain parameters, the system had chattering problem. A backstepping controller for synchronization of identical and nonidentical chaotic and hyperchaotic systems of different orders was investigated [13]. A controller for synchronization of fractional-order chaotic systems with non-identical dimensions and different orders was designed using generalized synchronization method [37]. In [41], multi-switching synchronization of non-identical chaotic systems was investigated. For phase synchronization among non-identical fractional-order complex chaotic systems, a nonlinear controller was designed [49]. SMC and projective synchronization method were used for synchronization of multiple non-identical coupled chaotic systems [35]. Hybrid chaotic synchronization between identical and non-identical fractional-order systems was investigated [26].

In order to realize the synchronization of chaotic systems, the controller must have a high performance in control of the nonlinear systems. Among these models, SMC is considered a suitable method since it has robust performance against disturbance and parameter uncertainties [35, 44]. However, main obstacle in the performance of the SMC is chattering problem. Another control method, which has high performance in nonlinear systems is the fuzzy logic control method, and when this control method is combined with SMC, it is shown that it has the chattering free performance [11, 23, 36, 46]. An adaptive fuzzy SMC for synchronization of uncertain fractional-order chaotic systems with time delay was designed [46]. Also a fuzzy terminal SMC was designed for MIMO uncertain nonlinear systems [36]. In [23], an adaptive fuzzy SMC for chaos synchronization was designed. A fuzzy fractional order SMC was investigated for nonlinear systems [11]. With the definition of the intuitionistic fuzzy logic concept [5], which is a wider generalization of the fuzzy logic concept, intuitionistic fuzzy control (IFC) methods that are more applicable and functional to many systems in real world were developed. IFC method is seen as an easier and more effective way to design a controller for systems which involve uncertainties such as disturbance. This new generation control method was studied in many fields, mainly engineering and mathematics. Especially, in recent years, based on the concept of new generation fuzzy logic, new systems were designed and various controllers were produced [1, 2, 6, 7, 16–18, 31, 34]. Moreover, synchronization and applications of various chaotic systems were realized by using various control methods based on IFC [3, 4].

Cube satellites have high degree of uncertainty and dynamic behavior due to their nature, so the prediction and control of their behavior are quite complex. The main focus in this study is to determine how effective an IFC-based controller in cube satellite operations, where uncertainty is quite high.

Motivated by the discussion in literature, in this paper it is aimed to use chaos control and synchronization method for rendezvous and proximity operation of the satellite systems. This study consists of a chaotic model of target and service satellite, chaotic synchronization model and chaos control model of rendezvous and docking operation. The reasons given below explain why this study is original among all the other known studies the literature:

(1) Two cubic satellites were synchronized using an IFC-based controller.
(2) The target satellite and the service satellite can be not described with the identical chaotic model, due to having generally different structures. Therefore, the target and service satellites were modelled using non-identical chaotic systems in this study.
(3) In this study, FSMC and IFSMC were used for synchronization of the satellites, and the control methods were compared in terms of efficiency, settling time and robustness etc.

The rest of the paper is organized as follows: Sect. 2 presents the analysis of the dynamics and behavior of satellites and chaotic models of satellites. In Sect. 3, mathematical expressions of FSMC and IFSMC methods to be designed are shown. Section 4 consists of the application of the designed controllers to chaotic systems that are not identical, and the analysis and comparison of the results. Finally, conclusions and potential future works are presented in Sect. 5.

2 Dynamic Behavior of Satellites

The motion of two satellites moving in different elliptical orbits can be considered, as shown in Fig. 1, for modeling of rendezvous and docking systems. In this system, the distances of the service satellite moving around the ground and the target satellite to the ground and the change of angular velocity between them are shown. Accordingly, the dynamic behavior and relative motions of satellites have nonlinear dynamic behaviors dependent on many parameters, and each satellite can be modeled chaotically with its own dynamic model.

2.1 *Chaos Dynamics of the Satellites*

The satellites have complex nonlinear dynamics [40], and the dynamics are written as

$$M = I\omega \tag{1}$$

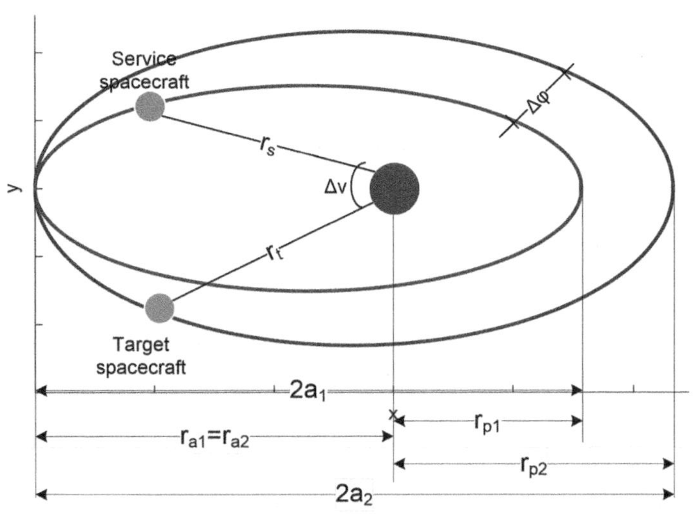

Fig. 1 Movement of two satellites in elliptical orbit [29]

where M, I, T_a, T_G, T_d, and ω are total momentum, inertia matrix, fly wheel torque, gravitational torque, disturbance and the angular velocity, respectively. The derivative of the total momentum is written as

$$\dot{M} = I\dot{\omega} + \omega \times I\omega = T_a + T_G + T_d \tag{2}$$

where T_a, T_G, T_d are, respectively, satellite control torque, gravitational torque, disturbance and $\times$ is symbolized cross product. According to the equations, angular velocities of the satellite dynamics are obtained as follows [29, 39, 40, 45, 54]:

$$\begin{aligned}
\dot{\omega}_x &= (I_x)^{-1}\omega_y\omega_z(I_y - I_z) + T_{ax} + T_{Gx} + T_{dx} + u_x \\
\dot{\omega}_y &= (I_y)^{-1}\omega_x\omega_z(I_z - I_x) + T_{ay} + T_{Gy} + T_{dy} + u_y \\
\dot{\omega}_z &= (I_z)^{-1}\omega_x\omega_{zy}(I_x - I_y) + T_{az} + T_{Gz} + T_{dz} + u_z
\end{aligned} \tag{3}$$

All perturbing disturbance torques $(T_a + T_G + T_d)$ can be written as follows:

$$\begin{aligned}
\dot{\omega}_x &= (I_x)^{-1}\big[(I_y - I_z)\omega_y\omega_z + p_1\omega_x + p_2\omega_z\big]. \\
\dot{\omega}_y &= (I_y)^{-1}\big[(I_z - I_x)\omega_x\omega_z + p_3\omega_y\big]. \\
\dot{\omega}_z &= (I_z)^{-1}\big[(I_x - I_y)\omega_x\omega_y + p_4\omega_x + p_5\omega_z\big].
\end{aligned} \tag{4}$$

Using Eq. (4), we can construct master and slave chaotic systems that are not identical, as in Eqs. (5) and (6), to express the chaotic movement of target and service satellites, respectively.

$$\begin{aligned}
\dot{\omega}_{x1} &= (I_{x1})^{-1}\big[(I_{y1} - I_{z1})\omega_{y1}\omega_{z1} + p_{11}\omega_{x1} + p_{12}\omega_{z1}\big] \\
\dot{\omega}_{y1} &= (I_{y1})^{-1}\big[(I_{z1} - I_{x1})\omega_{x1}\omega_{z1} + p_{13}\omega_{y1}\big]. \\
\dot{\omega}_{z1} &= (I_{z1})^{-1}\big[(I_{x1} - I_{y1})\omega_{x1}\omega_{y1} + p_{14}\omega_{x1} + p_{15}\omega_{z1}\big].
\end{aligned} \tag{5}$$

$$\begin{aligned}
\dot{\omega}_{x2} &= (I_{x2})^{-1}\big[(I_{y2} - I_{z2})\omega_{y2}\omega_{z2} + p_{21}\omega_{x2} + p_{22}\omega_{z2}\big]. \\
\dot{\omega}_{y2} &= (I_{y2})^{-1}\big[(I_{z2} - I_{x2})\omega_{x2}\omega_{z2} + p_{23}\omega_{y2}\big]. \\
\dot{\omega}_{z2} &= (I_{z2})^{-1}\big[(I_{x2} - I_{y2})\omega_{x2}\omega_{y2} + p_{24}\omega_{x2} + p_{25}\omega_{z2}\big].
\end{aligned} \tag{6}$$

If the constraints and initial conditions are selected as follows, chaotic behaviors in Figs. 2 and 3 are obtained as follows:

$$\big[\,\omega_{x1}(0)\;\omega_{y1}(0)\;\omega_{z1}(0)\,\big]^T = \big[\,1\;1\;1\,\big]^T$$

$$\big[\,\omega_{x2}(0)\;\omega_{y2}(0)\;\omega_{z2}(0)\,\big]^T = \big[\,1.05\;0.65\;0.87\,\big]^T. \tag{7}$$

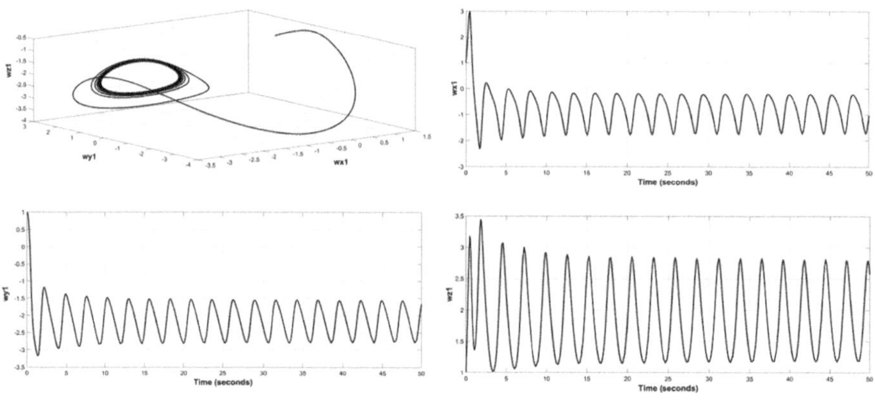

Fig. 2 Master system (target satellite) for angular velocities

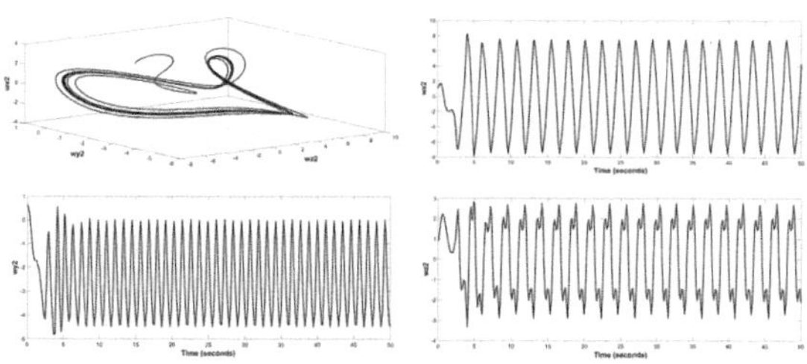

Fig. 3 Slave system (service satellite) for angular velocities

$$
\begin{aligned}
I_1 &= \left[\, I_{x1}\ I_{y1}\ I_{z1} \,\right]^T = \left[\, 10\ 20\ -5 \,\right]^T \\
I_2 &= \left[\, I_{x2}\ I_{y2}\ I_{z2} \,\right]^T = \left[\, 1\ 2\ -1 \,\right]^T
\end{aligned} \tag{8}
$$

$$
\begin{aligned}
p_1 &= \left[\, p_{11}\ p_{12}\ p_{13}\ p_{14}\ p_{15} \,\right]^T = \left[\, -30\ 40\ 12\ -12\ 4.8 \,\right]^T \\
p_2 &= \left[\, p_{21}\ p_{22}\ p_{23}\ p_{24}\ p_{25} \,\right]^T = \left[\, -1.2\ 1.22\ 2.45\ -2.45\ 0.98 \,\right]^T
\end{aligned} \tag{9}
$$

2.2 *Synchronization of Two Satellites*

Dynamic equations and relative motion of cube satellites have been discussed in many studies in the literature. In this study, for the first time, the motion of two satellites has been considered as two different chaotic systems and the angular velocities of the satellites were synchronized with synchronization control. Satellite dynamics are given for target and service satellites, respectively, as follows:

$$\dot{w}_1 = A_1 w_1 + f_1(w), \quad w_1 \in \Re^3 \tag{10}$$

$$\dot{w}_2 = A_2 w_2 + f_2(w) + \Delta(w) + u, \quad w_2 \in \Re^3 \tag{11}$$

where u is the control input, $\Delta(w)$ is parameter uncertainties and $f_1(w)$ and $f_2(w)$ are nonlinear functions. In order for the two satellites to move synchronously, the following condition must be satisfied.

$$\lim_{t \to \infty} \|w_2 - w_1\| = 0 \tag{12}$$

Synchronization errors of two satellites are defined as:

$$\begin{aligned} e_1 &= \omega_{x2} - \omega_{x1} \\ e_2 &= \omega_{y2} - \omega_{y1} \\ e_3 &= \omega_{z2} - \omega_{z1} \end{aligned} \tag{13}$$

From Eq. (13), changing of synchronization errors is obtained as follows:

$$\begin{aligned} \dot{e}_1 &= \dot{\omega}_{x_s} - \dot{\omega}_{x_m} = \alpha_1 \omega_{y_s} \omega_{z_s} - \beta_1 \omega_{x_s} + \delta_1 \omega_{z_s} + u_x - \alpha_2 \omega_{y_m} \omega_{z_m} + \beta_2 \omega_{x_m} - \delta_2 \omega_{z_m}, \\ \dot{e}_2 &= \dot{\omega}_{y_s} - \dot{\omega}_{y_m} = \sigma_1 \omega_{x_s} \omega_{z_s} + \kappa_1 \omega_{y_s} + u_y - \sigma_2 \omega_{x_m} \omega_{z_m} - \kappa_2 \omega_{y_m}, \\ \dot{e}_3 &= \dot{\omega}_{z_s} - \dot{\omega}_{z_m} = \tau_1 \omega_{x_s} \omega_{y_s} - \upsilon_1 \omega_{x_s} + \rho_1 \omega_{z_s} + u_z - \tau_2 \omega_{x_m} \omega_{y_m} + \upsilon_2 \omega_{x_m} - \rho_2 \omega_{z_m}. \end{aligned} \tag{14}$$

If it is assumed that the satellites are identical and the satellite parameters do not change:

$$\begin{aligned} &\alpha = \alpha_1 = \alpha_2, \quad \beta = \beta_1 = \beta_2 \quad \delta = \delta_1 = \delta_2 \\ &\sigma = \sigma_1 = \sigma_2, \quad \kappa = \kappa_2 = \kappa_2, \\ &\tau = \tau_1 = \tau_2, \quad \upsilon = \upsilon_1 = \upsilon_2 \quad \rho = \rho_1 = \rho_2 \\ &f_x = \omega_y \omega_z, \quad f_y = \omega_x \omega_z \quad f_z = \omega_x \omega_y \end{aligned}$$

As a result, the following change of errors equations is obtained as:

$$\begin{aligned} \dot{e}_1 &= \alpha\left(f_{x_s} - f_{x_m}\right) - \beta e_1 + \delta e_3 + u_x, \\ \dot{e}_2 &= \sigma\left(f_{y_s} - f_{y_m}\right) + \kappa e_2 + u_y, \\ \dot{e}_3 &= \tau\left(f_{z_s} - f_{z_m}\right) - \upsilon e_1 + \rho e_3 + u_z. \end{aligned}$$

Lemma 1 *For the above-defined system, Lyapunov stability criteria is satisfied.*

Proof Let Lyapunov equation be chosen as follows;

$$V(t) = \sum_{i=1}^{3} \frac{1}{2} e_i^2 \tag{15}$$

$$\dot{V}(t) = e_1\dot{e}_1 + e_2\dot{e}_2 + e_3\dot{e}_3 \le 0 \tag{16}$$

$$\begin{aligned} \dot{V}(t) &= e_1\dot{e}_1 + e_2\dot{e}_2 + e_3\dot{e}_3 \le 0 \\ &= e_1\left(\alpha\left(f_{x_s} - f_{x_m}\right) - \beta e_1 + \delta e_3 + u_x\right) + \cdots \\ &\quad + e_2\left(\sigma\left(f_{y_s} - f_{y_m}\right) + \kappa e_2 + u_y\right) + e_3\left(\tau\left(f_{z_s} - f_{z_m}\right) - \upsilon e_1 + \rho e_3 + u_z\right) \end{aligned} \tag{17}$$

If control inputs are selected as follows, Lyapunov stability condition is fulfilled.

$$\begin{aligned} u_x &= e_1(-\alpha\Delta f_x - \delta e_3) \\ u_y &= e_2\left(-\sigma\Delta f_y - \kappa e_2\right) \\ u_z &= e_3(-\tau\Delta f_z + \upsilon e_1 - \rho e_3) \end{aligned} \tag{18}$$

3 Design of Controllers

In this section, sliding mode controller (SMC), fuzzy sliding mode controller (FSMC) and intuitionistic sliding mode controller (IFSMC) are designed, respectively.

3.1 Design of SMC

SMC is a robust control method often used in the control of nonlinear systems. The overall sliding surface [48], which is important for the sliding mode controller is generally defined as;

$$s = Ce(t) \tag{19}$$

where $C = [c_1, c_2, c_3, c_4...]$ is 1xn matrix. In classical SMC, control input is defined as:

$$u = u_{eq} + u_r \tag{20}$$

where u_{eq} is equivalent control input and u_r is the reaching control input defined as $u_r = k_c u_c$ with switching gain k_c and $u_c = \text{sgn}(s)$. However as stated in [51] the sign function causes chattering problem in the system. In order to eliminate this problem, fuzzy or intuitionistic fuzzy systems can be used instead of sign function.

$$u = u_{eq} + k_{fs}u_{fs} \tag{21}$$

or

$$u = u_{eq} + k_{ifs}u_{ifs} \tag{22}$$

where $u_{eq} = \left[u_{eq1}, u_{eq2}, u_{eq3}, u_{eq4}\ldots\right]^T$ and k_{fs} or k_{ifs} is normalization parameter of output variable and u_{fs} or u_{ifs} is control output obtained from FSMC or IFSMC that is described, respectively, as follows:

$$u_{fs} = FSMC(s, \dot{s}) = [FSMC(s_1, \dot{s}_1), \ldots, FSMC(s_4, \dot{s}_4)\ldots] \tag{23}$$

or

$$u_{ifs} = IFSMC(s, \dot{s}) = [IFSMC(s_1, \dot{s}_1), \ldots, IFSMC(s_4, \dot{s}_4)\ldots] \tag{24}$$

In this study, a Lyapunov function [15] is selected as follows:

$$V = \frac{1}{2}s^2 \tag{25}$$

In order to obtain u_{eq}, the synchronization error is defined as:

$$e = w_2 - w_1 \tag{26}$$

and the error dynamics is obtained from Eq. (10) and Eq. (11) as:

$$\dot{e} = \dot{w}_2 - \dot{w}_1 = A_2 e + F(w_1, w_2) + u \tag{27}$$

where $F(w_1, w_2) = f_2(w) - f_1(w) + (A_2 - A_1)w_1$ and $\Delta(w) = 0$. Then control law is defined as follows [47]:

$$u = -F(w_1, w_2) + Kv \tag{28}$$

where K is constant gain vector. The error dynamics is obtained from Eqs. (27) and (28) as:

$$\dot{e} = A_2 e + Kv \tag{29}$$

Thus, linear time-variant control system with single input v has been designed. The following equation is obtained from Eqs. (19) and (29) as follows:

$$\dot{s} = C\dot{e}(t) = C[A_2 e + Kv] \tag{30}$$

For the synchronization system to be stable, the sliding surface must provide $s = 0$ and $\dot{s} = 0$. Thus the equivalent control law is obtained from solution of Eq. (30) for "v" as follows:

$$v_{eq} = -(CK)^{-1}(CA_2)e \tag{31}$$

where C is chosen to satisfy the condition that CK is invertible matrix. The closed-loop dynamics is obtained from Eqs. (29) and (31) as follows:

$$\dot{e} = \left[I - K(CK)^{-1}C\right]A_2 e \tag{32}$$

The matrix $\left[I - K(CK)^{-1}C\right]A_2$ has been selected as to fulfil Hurwitz condition to obtain controllable system. Therefore, FSMC or IFSMC law is obtained as

$$v = v_{eq} + k_f v_{fs} \tag{33}$$

or

$$v = v_{eq} + k_f v_{ifs} \tag{34}$$

3.2 Designing Fundamental Concepts of FSMC and IFSMC

In this section, the fundamental concepts of fuzzy and intuitionistic fuzzy systems are presented, after that common and different structures of fuzzy and intuitionistic fuzzy controller are introduced. A fuzzy set in a non-empty set X given by;

$$A = \{(x, \mu_A(x)) : x \in X\} \tag{35}$$

where $\mu_A(x) : X \to [0, 1]$ is the membership function of the fuzzy set A [52]. An intuitionistic fuzzy set in a non-empty set X given by a set of ordered triples

$$A = \{(x, \mu_A(x), \eta_A(x)) : x \in X\} \tag{36}$$

where $\mu_A(x) : X \to I$, and $\eta_A(x) : X \to I$ are functions defined such that $0 \leq \mu(x)+\eta(x) \leq 1$ for all $x \in X$, $\mu(x)$ and $\eta(x)$ represent the degree of membership and degree of non-membership of x in A respectively. For each $x \in X$, the intuitionistic fuzzy index of x in A can be defined as follows: $\pi_A(x) = 1 - \mu_A(x) - \eta_A(x)$. π_A is called the degree of hesitation or indeterminacy [5].

Fuzzy and intuitionistic fuzzy sets are used to linguistically model the structure of the control system based on expert knowledge. These modellings consist of three principal parts called fuzzification, fuzzy inference and defuzzification. Although these three parts are the same in the main idea, particularly significant performance increases are observed in control applications due to differences in intuitionistic fuzzy sets.

First of all, for both FSMC and IFSMC, 7 different triangular fuzzy sets (negative big (NB), negative medium (NM), negative small (NS), zero (ZE), positive small (PS), positive medium (PM) and positive big (PB)) for "s" and "$\dot{s}$" inputs are chosen as shown in Fig. 4. Additionally 7 different fuzzy sets (NB, NM, NS, PS, PM and PB) are chosen for "v_{fs}" and "v_{ifs}" as outputs as shown in Fig. 5.

After selecting input and output fuzzy sets, the common "if–then" rule base based on expert knowledge [24] required for FSMC and IFSMC are chosen as follows;

$$
\begin{aligned}
&R1: If\ s\ is\ PB\ and\ \dot{s}\ is\ PB,\ then\ (v_{fs}/v_{ifs})\ is\ NB\\
&R2: If\ s\ is\ PM\ and\ \dot{s}\ is\ PB,\ then\ (v_{fs}/v_{ifs})\ is\ NB\\
&\vdots\\
&R48: If\ s\ is\ NM\ and\ \dot{s}\ is\ NB,\ then\ (v_{fs}/v_{ifs})\ is\ PB\\
&R49: If\ s\ is\ NB\ and\ \dot{s}\ is\ NB,\ then\ (v_{fs}/v_{ifs})\ is\ PB
\end{aligned}
$$

The rule base consisting of 49 rules can be shown as in Table 1.

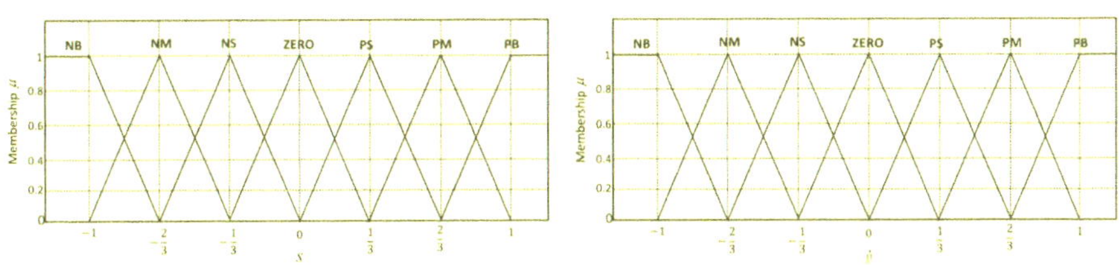

Fig. 4 Fuzzy sets for inputs of FSMC and IFSMC

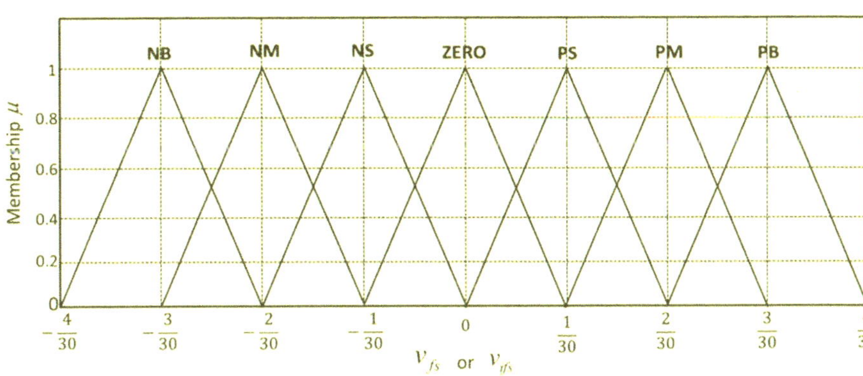

Fig. 5 Fuzzy sets for outputs of FSMC and IFSMC

Table 1 Fuzzy sets for outputs of FSMC and IFSMC

v_{fs}/v_{ifs}		s						
		PB	PM	PS	ZERO	NS	NM	NB
$\dot{s}$	PB	NB	NB	NB	NB	NM	NS	ZERO
	PM	NB	NB	NB	NM	NS	ZERO	PS
	PS	NB	NB	NM	NS	ZERO	PS	PM
	ZERO	NB	NM	NS	ZERO	PS	PM	PB
	NS	NM	NS	ZERO	PS	PM	PB	PB
	NM	NS	ZERO	PS	PM	PB	PB	PB
	NB	ZERO	PS	PM	PB	PB	PB	PB

3.2.1 Design of FSMC

A FSMC consists of 3 parts called fuzzification, inference and defuzzification. Firstly, in fuzzification block, the crisp input values are converted into membership degrees by using the triangular functions in Fig. 4. The second part of the system is fuzzy inference process which is defined with an if–then rules base, which is based on the fuzzy modal operator, fuzzy implication and fuzzy aggregation function. Here, the fuzzy values are subjected to intersection, implication and aggregation, respectively, and the resulting fuzzy numbers are sent to the defuzzification block. In this study, intersection and implication methods are chosen as "min T-norm" (Lee., 2004). Let's rewrite any rule base as follows.

R$_j$: *If s is A_j and $\dot{s}$ is B_j,, then ($\overline{v}_{fs}$) is C_j* where $j = 1, 2, \ldots, 49$.

Then intersection and implication process can be performed respectively for all rule base:

$$\alpha_j = \mu_{A_j \cap B_j} = \min(\mu_{A_j}(s), \mu_{B_j}(\dot{s})) \tag{37}$$

$$\mu_{implied} = \mu_{C'_j}(\overline{v}_{fs}) = \min(\alpha_j, \mu_{C_j}(\overline{v}_{fs})) \tag{38}$$

The 49 fuzzy numbers obtained as a result of the implication process are aggregated [27] as follows.

$$\begin{aligned} \mu_{C'}(\overline{v}_{fs}) &= \mu_{C'_1}(\overline{v}_{fs}) \cup \mu_{C'_2}(\overline{v}_{fs}) \ldots \cup \mu_{C'_{49}}(\overline{v}_{fs}) \\ &= \max(\mu_{C'_1}(\overline{v}_{fs}), \mu_{C'_2}(\overline{v}_{fs}) \ldots \mu_{C'_{49}}(\overline{v}_{fs})) \end{aligned} \tag{39}$$

Finally, defuzzification process was performed using the "center of area (COA)" defuzzification method [42] and the crisp v_{fs} is obtained as follows.

$$v_{fs} = \frac{\int \mu_{C'}(\overline{v}_{fs})(\overline{v}_{fs}).d\overline{v}_{fs}}{\int (\overline{v}_{fs}).d\overline{v}_{fs}} \tag{40}$$

3.2.2 Design of IFSMC

As in FSMC design, although fuzzification, inference and defuzzification processes are performed respectively in IFSMC design, the used methods in some processes can differ. Firstly, intuitionistic fuzzy sets are chosen for inputs (s and $\dot{s}$) and outputs (v_{fs}) as Figs. 4 and 5 and degrees of membership are obtained by performing i-fuzzification. Then intuitionistic fuzzy pairs (IFP), which is defined in Eq. (36), are produced by obtaining degrees of non-membership of each element using "Sugeno Generator" [8] as shown.

$$\eta_A(x) = \frac{1 - \mu_A(x)}{1 + \lambda\mu_A(x)} \tag{41}$$

where $\lambda \in (-1, \infty)$ is Sugeno constant.

In IFSMC, although the same rule base is defined in the inference part of the system as in the FSMC design, the inference process is performed by using ordered pairs, not just degrees of membership. In this study "Mamdani intersection and implication" methods are chosen as shown respectively [28]. Let's rewrite any rule base as follows.

R_j: *If s is A_j and $\dot{s}$ is B_j, then $(\overline{v}_{ifs})$ is C_j* where $j = 1, 2, \ldots, 49$.

Intersection for a rule base;

$$(\alpha_j, \beta_j) = (\mu_{A_j \cap B_j}, \eta_{A_j \cap B_j}) = \min(\mu_{A_j}(s), \mu_{B_j}(\dot{s})), \max(\eta_{A_j}(s), \eta_{B_j}(\dot{s}))) \tag{42}$$

Implication for a rule base;

$$R_{M_j}(s, \dot{s}, \overline{v}_{ifs}) = (\min\{\alpha_j, \mu_{C_j}(\overline{v}_{ifs})\}, max\{\beta_j, \eta_{C_j}(\overline{v}_{ifs})\}) = \{D_{\mu_j}, D_{\eta_j}\} \tag{43}$$

After the implication process is performed for the all rule base, obtained fuzzy numbers are aggregated [19] separately as follows.

$$\mu_{C'}(\overline{v}_{ifs}) = \max(D_{\mu_1}, D_{\mu_2} \ldots D_{\mu_{49}}) \tag{44}$$

$$\eta_{C'}(\overline{v}_{ifs}) = \min(D_{\eta_1}, D_{\eta_2} \ldots D_{\eta_{49}}) \tag{45}$$

Additionally, using the fuzzy numbers $\mu_{C'}(\overline{v}_{ifs})$ and $\eta_{C'}(\overline{v}_{ifs})$, the fuzzy number $\pi_{C'}(\overline{v}_{ifs})$ showing the indeterminacy is found as follows.

$$\pi_{C'}(\overline{v}_{ifs}) = 1 - \mu_{C'}(\overline{v}_{ifs}) - \eta_{C'}(\overline{v}_{ifs}) \tag{46}$$

After the inference process is completed, the IFSMC controller design is completed by converting the intuitionistic fuzzy numbers obtained to the v_{ifs} crisp output using a novel intuitionistic defuzzification method [25] as shown.

$$v_{ifs} = \frac{\int \overline{v}_{ifs}.((1 - \pi_{C'}(\overline{v}_{ifs}))\mu_{C'}(\overline{v}_{ifs}) + \mu_{C'}(\overline{v}_{ifs})(\overline{v}_{ifs})).d\overline{v}_{ifs}}{\int ((1 - \pi_{C'}(\overline{v}_{ifs}))\mu_{C'}(\overline{v}_{ifs}) + \mu_{C'}(\overline{v}_{ifs})(\overline{v}_{ifs})).d\overline{v}_{ifs}} \tag{47}$$

4 Numerical Simulation for Satellite Synchronization

In this section, FSMC and IFSMCs have been designed separately and applied to the satellite synchronization system with the same initial conditions and system parameters. The performance and efficiency of the satellite synchronization system have been compared with each other under the proposed FSMC and IFSMC methods in terms of steady state error, setting time, maximum overshoot, chattering and robustness.

The parameters used for investigation of satellite system as in Eq. (7) with initial condition are given as follows:

$$\begin{aligned} \left[\, \omega_{x1}(0)\ \omega_{y1}(0)\ \omega_{z1}(0) \,\right]^T &= \left[\, 1\ 1\ 1 \,\right]^T \\ \left[\, \omega_{x2}(0)\ \omega_{y2}(0)\ \omega_{z2}(0) \,\right]^T &= \left[\, 1.05\ 0.65\ 0.87 \,\right]^T \end{aligned} \tag{48}$$

The block diagram of the satellite synchronization system is shown as Fig. 6, and all the numerical results for proposed controllers have been simulated in Matlab/Simulink. In the simulation, 4th Runge–Kutta method is used with the step size of 0.00001 s.

The detailed block diagrams of satellite synchronization systems are shown in Figs. 7 and 8.

4.1 Numerical Simulation Results for FSMC

The designed FSMC block diagram is shown in Fig. 7, where $k_f = -30$.

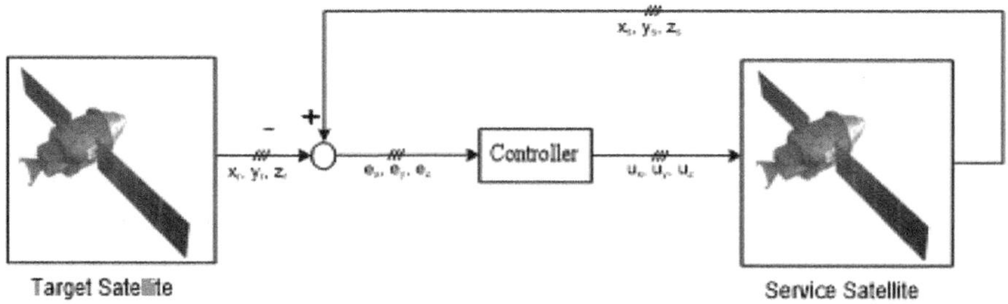

Fig. 6 Block diagram of satellite synchronization system

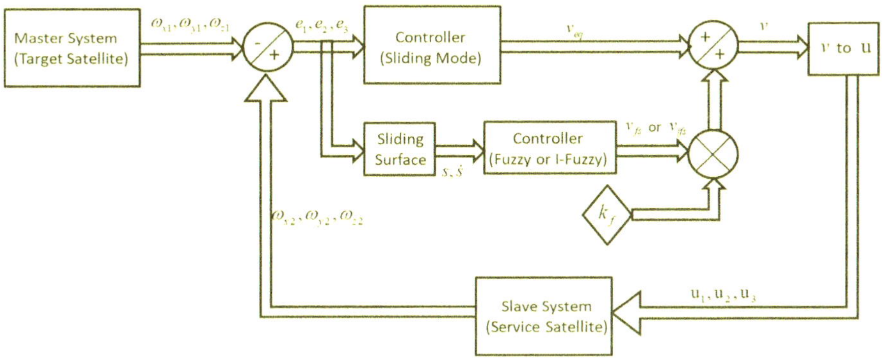

Fig. 7 Detailed block diagram of satellite synchronization system

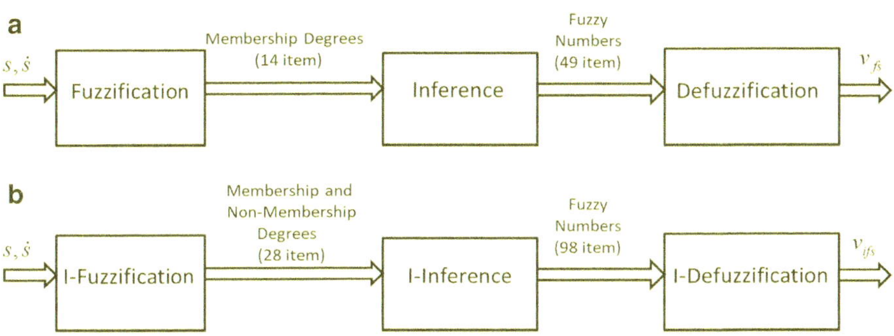

Fig. 8 **a** Fuzzy controller. **b** Intuitionistic fuzzy controller

As shown Fig. 9, it is observed that FSMC reaches a stable performance in about 40th seconds for e_1, e_2, e_3 and maximum overshoot is -4.23 for e_3, without noise at beginning. However, when a random noise signal with duration of 1 s, as shown Fig. 10, is applied to the system in the 50th seconds, it is seen that system reacts hardly and loses its stability for a short time. When the effect of the noise disappears, it is seen that the system is stable and the maximum steady state error is -1.69×10^{-4} for e_3 at 90th seconds and chattering does not occur.

4.2 *Numerical Simulation Results for IFSMC Method*

The designed IFSMC block diagram is given in Fig. 7 with $k_f = -30$ and $\lambda = 0.07$. The main difference of this design according to FSMC is that the intuitionistic controller is added to the sliding mode controller instead of fuzzy controller. All system parameters and initial conditions are selected the same with FSMC for objective comparison.

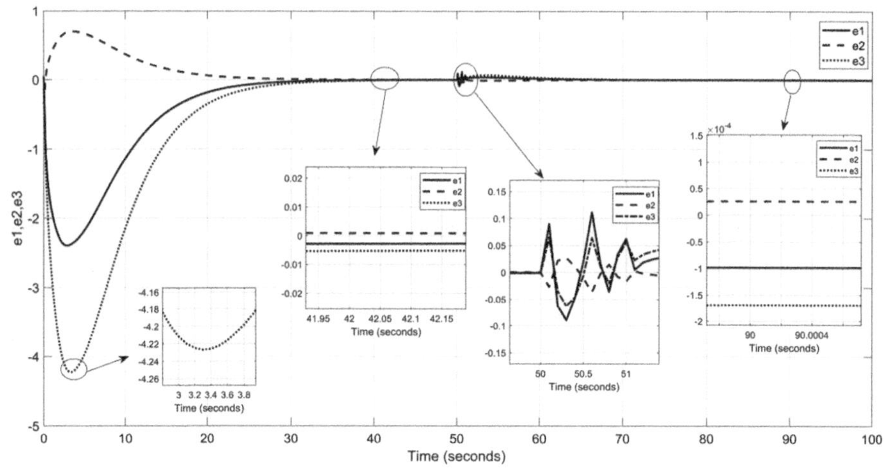

Fig. 9 e_1, e_2, e_3 for FSMC

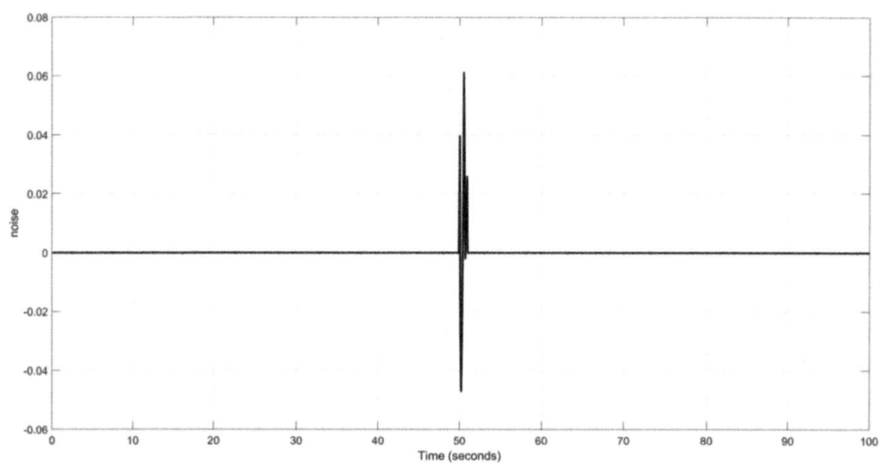

Fig. 10 Noise signal

As shown Fig. 11, it is observed that IFSMC reaches stable performance about in 34th seconds for e_1, e_2, e_3 and maximum overshoot is -3.56 for e_3, without noise at beginning. However, when a random noise signal with duration of 1 s, as shown Fig. 10, is applied to the system in the 50th seconds, it is seen that system reacts softly and loses its stability for a short time. When the effect of the noise disappears, it is seen that the system is stable and the maximum steady state error is -4.72×10^{-5} for e_3 at 85th seconds and chattering does not occur.

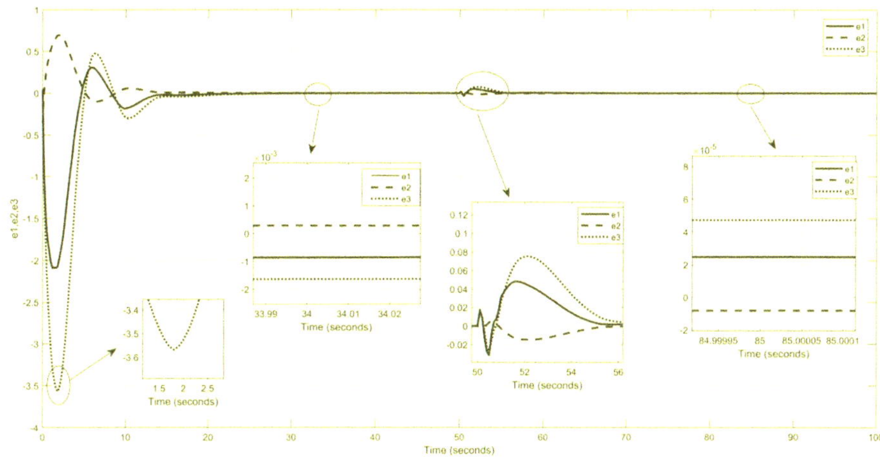

Fig. 11 e_1, e_2, e_3 for IFSMC

4.3 Numerical Simulation Results According to k_f

The simulation results are shown in Sects. 4.1 and 4.2 according to the fixed k_f constant. However, another issue to consider is how the simulation results change according to the change in the k_f constant. These results are listed according to the some k_f values as shown.

Simulation results are compared with $-60 \leq k_f \leq -20$. Because outside of this range, both FSMC and IFSMC designs are observed to be unstable. According to Table 2, as k_f approaches − 20, it is seen that for both FSMC and IFSMC designs, settling time and steady-state error decrease and maximum overshoot increases. However, it is observed that as k_f approaches − 20, the steady-state error decreases for FSMC design and increases for IFSMC design. It is noticed that for all $-60 \leq k_f \leq -20$ both designs reach steady state stability and no chattering occurs.

5 Conclusions and Future Works

When the simulation results are examined, basic results are obtained as follows.

1. At the same k_f values, it has been observed that IFSMC gives better results than FSMC in terms of maximum overshoot, settling time, steady state error and response to noise.
2. It was observed that the response to noise is hard (triangular) in FSMC and soft (parabolic) in IFSMC.
3. It has been observed that the system with IFSMC is faster and more robust than FSMC.

Table 2 Simulation results according to k_f

k_f	FSMC				IFSMC			
	Max. overshoot (for e_3)	Settling time (without noise) (s)	State steady error at 90th sec. (for e_3)	Max. response to noise for e_1 (amplitude)	Max. overshoot (for e_3)	Settling time (without noise) (s)	State steady error at 85th sec. (for e_3)	Max. response to noise for e_3 (amplitude)
− 60	− 3.73	~ 48	− 15.47 × 10^{-4}	0.379	− 3.03	~ 39	6.81 × 10^{-5}	0.069
− 50	− 3.85	~ 44	− 2.58 × 10^{-4}	0.250	− 3.19	~ 36	5.43 × 10^{-5}	0.070
− 30	− 4.23	~ 42	− 1.69 × 10^{-4}	0.116	− 3.56	~ 34	4.72 × 10^{-5}	0.075
− 20	− 4.67	~ 41	− 1.10 × 10^{-4}	0.094	− 3.94	~ 33	3.39 × 10^{-5}	0.081

It is thought that the main reason why IFSMC design gives better results compared to FSMC design is that system uncertainties are taken into account while designing IFSMC, unlike FSMC. This can be seen as the main reason for system acceleration, decrease of steady-state error and maximum overshoot. In addition, the smoother response of the design and the quicker damping of the distorting effect of the noise when noise is applied to the system can be seen as a great advantage. In this way, since soft switching is achieved, less wear of the switching elements is provided, especially in mechanical and electrical systems, and a more efficient control method is obtained in general.

In this study, controllers were designed according to predetermined k_f and Sugeno coefficients. In future studies, it is aimed to use adaptive parameters instead of fixed parameters such as k_f and Sugeno coefficients in both FSMC and IFSMC designs in order to make the system work more efficiently. In addition, introducing novel AFSMC and AIFSMC designs by using adaptive fuzzy sets instead of fixed and triangular fuzzy sets during the fuzzification process is another goal.

References

1. M. Akram, S. Habib, I. Javed, Intuitionistic fuzzy logic control for washing machines. Indian J. Sci. Technol. **7**, 654–661 (2014). https://doi.org/10.17485/ijst/2014/v7i5.20
2. M. Akram, S. Shahzad, A. Butt, A. Khaliq, Intuitionistic fuzzy logic control for heater fans. Math. Comput. Sci. **7**, 367–378 (2013). https://doi.org/10.1007/s11786-013-0161-x
3. Ö. Atan, F. Kutlu, Synchronization control of two chaotic systems via a novel fuzzy control method, in *2nd International Conference on Pure and Applied Mathematics* (2018), p. 51

4. Ö. Atan, F. Kutlu, O. Castillo, Intuitionistic fuzzy sliding controller for uncertain hyperchaotic synchronization. Int. J. Fuzzy Syst. **22**, 1430–1443 (2020). https://doi.org/10.1007/s40815-020-00878-x
5. K. Atanassov, *Intuitionistic Fuzzy Sets* (Physica, Heidelberg, 1999)
6. O. Castillo, Framework for optimization of intuitionistic and type-2 fuzzy systems in control applications, in *Recent Advances in Intuitionistic Fuzzy Logic Systems* (Springer, 2019), pp. 79–86. https://doi.org/10.1007/978-3-030-02155-9_7
7. O. Castillo, F. Kutlu, Ö. Atan, Intuitionistic fuzzy control of twin rotor multiple input multiple output systems. J. Intell. Fuzzy Syst. **38**, 821–833 (2020). https://doi.org/10.3233/JIFS-179451
8. T. Chaira, *Fuzzy Set and Its Extension: The Intuitionistic Fuzzy Set* (Wiley, 2019). https://doi.org/10.1002/9781119544203
9. M. Chegini, H. Sadati, H. Salarieh, Chaos analysis in attitude dynamics of a flexible satellite. Nonlinear Dyn. **93**, 1421–1438 (2018). https://doi.org/10.1007/s11071-018-4269-z
10. M. Chegini, H. Sadati, H. Salarieh, Analytical and numerical study of chaos in spatial attitude dynamics of a satellite in an elliptic orbit. Proc. Inst. Mech. Eng. Part C J. Mech. Eng. Sci. **233**, 561–577 (2018). https://doi.org/10.1177/0954406218762019
11. H. Delavari, R. Ghaderi, A. Ranjbar, S. Momani, Fuzzy fractional order sliding mode controller for nonlinear systems. Commun. Nonlinear Sci. Numer. Simul. **15**, 963–978 (2010). https://doi.org/10.1016/j.cnsns.2009.05.025
12. K. Dong, J. Luo, Z. Dang, L. Wei, Tube-based robust output feedback model predictive control for autonomous rendezvous and docking with a tumbling target. Adv. Space Res. **65**, 1158–1181 (2020). https://doi.org/10.1016/j.asr.2019.11.014
13. E.D. Dongmo, K.S. Ojo, P. Woafo, A.N. Njah, Difference synchronization of identical and nonidentical chaotic and hyperchaotic systems of different orders using active backstepping design. J. Comput. Nonlinear Dyn. **13** (2018). https://doi.org/10.1115/1.4039626
14. D. Gao, J. Luo, W. Ma, B. Englot, Parameterized nonlinear suboptimal control for tracking and rendezvous with a non-cooperative target. Aerosp. Sci. Technol. **87**, 15–24 (2019). https://doi.org/10.1016/j.ast.2019.01.044
15. W. Hahn, *Stability of Motion* (Springer, Berlin, 1967)
16. P. Hajek, V. Olej, Defuzzification methods in intuitionistic fuzzy inference systems of Takagi-Sugeno type: the case of corporate bankruptcy prediction, in *2014 11th International Conference on Fuzzy Systems and Knowledge Discovery (FKSD)* (2014), pp. 232–236. https://doi.org/10.1109/FSKD.2014.6980838
17. P. Hájek, V. Olej, Adaptive intuitionistic fuzzy inference systems of Takagi-Sugeno type for regression problems, in *Artificial Intelligence Applications and Innovations* (2012), pp. 206–216. https://doi.org/10.1007/978-3-642-33409-2_22
18. I. Iancu, M. Gabroveanu, M. Cosulschi, Intuitionistic fuzzy control based on association rules, in *Computational Collective Intelligence. Technologies and Applications* (2013), pp. 235–244. https://doi.org/10.1007/978-3-642-40495-5_24
19. W. Jiang, B. Wei, X. Liu, X. Li, H. Zheng, Intuitionistic fuzzy power aggregation operator based on entropy and its application in decision making. Int. J. Intell. Syst. **33**, 49–67 (2018). https://doi.org/10.1002/int.21939
20. A. Khan, S. Kumar, Study of chaos in chaotic satellite systems. Pramana J. Phys. **90**, 1–9 (2018). https://doi.org/10.1007/s12043-017-1502-0
21. A. Khan, S. Kumar, Measuring chaos and synchronization of chaotic satellite systems using sliding mode control. Optim. Control Appl. Methods **39**, 1597–1609 (2018). https://doi.org/10.1002/oca.2428
22. A. Khan, S. Kumar, Analysis and time-delay synchronisation of chaotic satellite systems. Pramana J. Phys. **91**, 1–13 (2018). https://doi.org/10.1007/s12043-018-1610-5
23. C.-L. Kuo, Design of an adaptive fuzzy sliding-mode controller for chaos synchronization. Int. J. Nonlinear Sci. Numer. Simul. **8**, 631–636 (2007). https://doi.org/10.1515/IJNSNS.2007.8.4.631
24. C.L. Kuo, T.H.S. Li, N.R. Guo, Design of a novel fuzzy sliding-mode control for magnetic ball levitation system. J. Intell. Robot. Syst. **42**, 295–316 (2005). https://doi.org/10.1007/s10846-004-3026-3

25. F. Kutlu, Ö. Atan, O. Silahtar, Intuitionistic fuzzy adaptive sliding mode control of nonlinear systems. Soft Comput. **24**, 53–64 (2020). https://doi.org/10.1007/s00500-019-04286-8
26. A. Lassoued, O. Boubaker, Hybrid chaotic synchronisation between identical and non-identical fractional-order systems. Int. J. Comput. Appl. Technol. **60**, 134 (2019). https://doi.org/10.1504/IJCAT.2019.100134
27. K.H. Lee, *First Course on Fuzzy Theory and Applications* (Springer, 2004). https://doi.org/10.1007/3-540-32366-x
28. J. Li, Z. Gong, SISO intuitionistic fuzzy systems: IF-t-norm, IF-R-implication, and universal approximators. IEEE Access **7**, 70265–70278 (2019). https://doi.org/10.1109/ACCESS.2019.2918169
29. P. Li, Z.H. Zhu, Model predictive control for spacecraft rendezvous in elliptical orbit. Acta Astron. **146**, 339–348 (2018). https://doi.org/10.1016/j.actaastro.2018.03.025
30. Q. Li, B. Zhang, J. Yuan, H. Wang, Potential function based robust safety control for spacecraft rendezvous and proximity operations under path constraint. Adv. Space Res. **62**, 2586–2598 (2018). https://doi.org/10.1016/j.asr.2018.08.003
31. Y. Lin, X. Zhou, S. Gu, S. Wang, The Takagi-Sugeno ıntuitionistic fuzzy systems are universal approximators, in *2012 2nd International Conference on Consumer Electronics, Communications and Networks (CECNet)* (IEEE, 2012), pp. 2214–2217. https://doi.org/10.1109/CECNet.2012.6202025
32. Y. Liu, Y. Lyu, G. Ma, 6-DOF multi-constrained adaptive tracking control for noncooperative space target. IEEE Access **7**, 48739–48752 (2019). https://doi.org/10.1109/ACCESS.2019.2910304
33. A.M. Long, M.G. Richards, D.E. Hastings, On-orbit servicing: a new value proposition for satellite design and operation. J. Spacecraft Rockets **44**, 964–976 (2007). https://doi.org/10.2514/1.27117
34. M. Marinov, V. Lazarov, Intuitionistic fuzzy robot motion control. Probl. Eng. Cybern. Robot. **69**, 40–51 (2018)
35. M.R. Mufti, H. Afzal, F. Ur-Rehman, W. Aslam, M.I. Qureshi, Transmission projective synchronization of multiple non-identical coupled chaotic systems using sliding mode control. IEEE Access **7**, 17847–17861 (2019). https://doi.org/10.1109/ACCESS.2019.2895067
36. V. Nekoukar, A. Erfanian, Adaptive fuzzy terminal sliding mode control for a class of MIMO uncertain nonlinear systems. Fuzzy Sets Syst. **179**, 34–49 (2011). https://doi.org/10.1016/j.fss.2011.05.009
37. A. Ouannas, G. Grassi, A.T. Azar, A new generalized synchronization scheme to control fractional chaotic systems with non-identical dimensions and different orders. *Advances in Intelligent Systems and Computing* (Springer, 2020). https://doi.org/10.1007/978-3-030-14118-9_42
38. J. Pomares, L. Felicetti, J. Pérez, M.R. Emami, Concurrent image-based visual servoing with adaptive zooming for non-cooperative rendezvous maneuvers. Adv. Space Res. **61**, 862–878 (2018). https://doi.org/10.1016/j.asr.2017.10.054
39. L.L. Show, J.C. Juang, Y.W. Jan, An LMI-based nonlinear attitude control approach. IEEE Trans. Control Syst. Technol. **11**, 73–83 (2003). https://doi.org/10.1109/TCST.2002.806450
40. M.J. Sidi, *Spacecraft Dynamics and Control: A Practical Engineering Approach* (Cambridge University Press, 1997)
41. S. Singh, A.T. Azar, Q. Zhu, Multi-switching master–slave synchronization of non-identical chaotic systems, in *Innovative Techniques and Applications of Modelling, Identification and Control*. Lecture Notes in Electrical Engineering, vol. 467 (2018), pp. 321–330. https://doi.org/10.1007/978-981-10-7212-3
42. S.N. Sivanandam, S. Sumathi, S.N. Deepa, *Introduction to Fuzzy Logic Using MATLAB* (Springer, Berlin, 2007). https://doi.org/10.1007/978-3-540-35781-0
43. L. Sun, W. He, C. Sun, Adaptive fuzzy relative pose control of spacecraft during rendezvous and proximity maneuvers. IEEE Trans. Fuzzy Syst. **26**, 3440–3451 (2018). https://doi.org/10.1109/TFUZZ.2018.2833028

44. Z. Sun, Synchronization of fractional-order chaotic systems with non-identical orders, unknown parameters and disturbances via sliding mode control. Chin. J. Phys. **56**, 2553–2559 (2018). https://doi.org/10.1016/j.cjph.2018.08.007
45. A.P.M. Tsui, A.J. Jones, The control of higher dimensional chaos: comparative results for the chaotic satellite attitude control problem. Phys. D Nonlinear Phenom. **135**, 41–62 (2000). https://doi.org/10.1016/S0167-2789(99)00114-1
46. T.-C. Lin, T.-Y. Lee, Chaos synchronization of uncertain fractional-order chaotic systems with time delay based on adaptive fuzzy sliding mode control. IEEE Trans. Fuzzy Syst. **19**, 623–635 (2011). https://doi.org/10.1109/TFUZZ.2011.2127482
47. S. Vaidyanathan, Analysis and synchronization of the hyperchaotic Yujun systems via sliding mode control. Adv. Intell. Syst. Comput. (AISC) **176**, 329–337 (2012). https://doi.org/10.1007/978-3-642-31513-8_34
48. S. Vaidyanathan, S. Sampath, Global chaos synchronization of hyperchaotic Lorenz systems by sliding mode control, in *Advances in Digital Image Processing and Information Technology* (2011), pp. 156–164. https://doi.org/10.1007/978-3-642-24055-3_16
49. V.K. Yadav, G. Prasad, M. Srivastava, S. Das, Combination–combination phase synchronization among non-identical fractional order complex chaotic systems via nonlinear control. Int. J. Dyn. Control **7**, 330–340 (2019). https://doi.org/10.1007/s40435-018-0432-0
50. V.K. Yadav, V.K. Shukla, S. Das, A.Y.T. Leung, M. Srivastava, Function projective synchronization of fractional order satellite system and its stability analysis for incommensurate case. Chin. J. Phys. **56**, 696–707 (2018). https://doi.org/10.1016/j.cjph.2018.01.008
51. H.T. Yau, C.L. Chen, Chattering-free fuzzy sliding-mode control strategy for uncertain chaotic systems. Chaos Solitons Fractals **30**, 709–718 (2006). https://doi.org/10.1016/j.chaos.2006.03.077
52. L.A. Zadeh, Fuzzy Sets, Information and Control (1965)
53. L. Zhang, F. Zhu, Y. Hao, W. Pan, Rectangular-structure-based pose estimation method for non-cooperative rendezvous. Appl. Opt. **57**, 6164–6173 (2018). https://doi.org/10.1364/ao.57.006164
54. R. Zhang, *Satellite Orbit Attitude Dynamics and Control* (Univ. Aeronaut. Astronaut. Press, Beijing, 1998), p. 115
55. Y. Zhang, P. Huang, K. Song, Z. Meng, An angles-only navigation and control scheme for noncooperative rendezvous operations. IEEE Trans. Ind. Electron. **66**, 8618–8627 (2019). https://doi.org/10.1109/TIE.2018.2884213
56. B.Z. Zhou, X.F. Liu, G.P. Cai, Motion-planning and pose-tracking based rendezvous and docking with a tumbling target. Adv. Space Res. **65**, 1139–1157 (2020). https://doi.org/10.1016/j.asr.2019.11.013

Optimizing a Convolutional Neural Network with a Hierarchical Genetic Algorithm for Diabetic Retinopathy Detection

Rodrigo Cordero-Martínez, Daniela Sánchez, and Patricia Melin

Abstract One of the worse conditions caused by diabetes mellitus (DM) is diabetic retinopathy (DR) and it can be irreversible if it is not treated in time. The patient with this condition can be completely blind because DR does not have symptoms until advanced stages. Because of this, some authors have been searching for a solution to an early detection of DR. One of the most used technologies for the detection of DR is the neural networks called: Convolutional neural networks (CNN). But design a CNN model from beginning could be slow. Along this work, we proposed the design of a hierarchical genetic algorithm (HGA) to find the best hyperparameters for a CNN model for the detection of DR. Before designing the hierarchical genetic algorithm, we applied pre-processing to the APTOS 2019 database. Then we executed 30 times the hierarchical genetic algorithm and achieved 0.9650 of accuracy mean and 0.007665 of standard deviation. The best CNN model got an accuracy of 0.9781 for DR detection.

Keywords Genetic algorithm · Diabetic retinopathy · Neural networks

1 Introduction

Diabetic retinopathy (DR) causes a progressive decrease in visual acuity and can lead to blindness [1]. This loss of vision decreases the quality of life of the patient and his relatives and generates high economic costs for the family and the country [2].

The probability that a person in the world in 2002 would go blind, because of DR, was only 0.75% [3]. But due to the growth in the number of patients with Diabetes Mellitus (DM), worldwide in 2013 there were 382 million people with approximately 35% of these patients had some level of DR, and by the year 2035 it is calculated that there will be 592 million people diagnosed with DM, thus having more than 4 million people blind because of DR [4].

R. Cordero-Martínez · D. Sánchez · P. Melin (✉)
Tijuana Institute of Technology/TecNM, Tijuana, Mexico
e-mail: pmelin@tectijuana.mx

O. Castillo and P. Melin (eds.), *Fuzzy Logic and Neural Networks for Hybrid Intelligent System Design*, Studies in Computational Intelligence 1061,
https://doi.org/10.1007/978-3-031-22042-5_11

One of the most used technologies in recent years for the detection of DR is the use of neural networks [5]. Promising results have been obtained for the detection of DR using this technology.

Our main objective for this work was the design of a hierarchical genetic algorithm, with the goal to increase the precision of a convolutional neural network that detects in photos of the retina if the patient has diabetic retinopathy. The chromosomes and their genes of the hierarchical genetic algorithm modify the hyperparameters of the layers of the convolutional neural network until the optimal results are found.

In this work is implemented the use of convolutional neural networks for the detection of diabetic retinopathy and the network is optimized with the implementation of a hierarchical genetic algorithm. Works by other authors have already implemented this algorithm to optimize neural networks, but no work has focused on the detection or classification of diabetic retinopathy.

In Sect. 2, are described the basic concepts to fully understand this work. In Sect. 3, are explained and described in detail the methods applied. In Sect. 4 the results obtained by the experiments are presented, and finally, in Sect. 5 the conclusions are presented.

2 Basic Concepts

To fully understand this work, it is important to have knowledge of the concepts presented below.

2.1 Neural Networks

The use of artificial intelligence in the medical area is a widely used resource, being neural networks a resource that is still used today for the detection of diseases, pattern recognition and prediction [6]. To create a neural network model that can be used to detect diseases or anomalies in photographs, it is best if the network has supervised learning [7]. This means that the images in the database need to be tagged for training, validation, and testing.

For a neural network to learn as much as possible, it often involves skilled technicians extracting important features from images. However, in recent years a kind of neural network has been used that performs this feature extraction without the need for an expert technician. These are called: convolutional neural networks (CNN) [8].

The CNNs have different layers connected that are in charge of extracting these characteristics from the images [9]. CNNs start with the input of the images and a convolutional layer with an activation function. The recommended activation function is the Rectified Linear Unit (better known as ReLU) but it is not the only one. The first convolutional layers are straightforward and detect lines and curves, but the more layers you use, the more complex the shapes it recognizes. Then a layer called

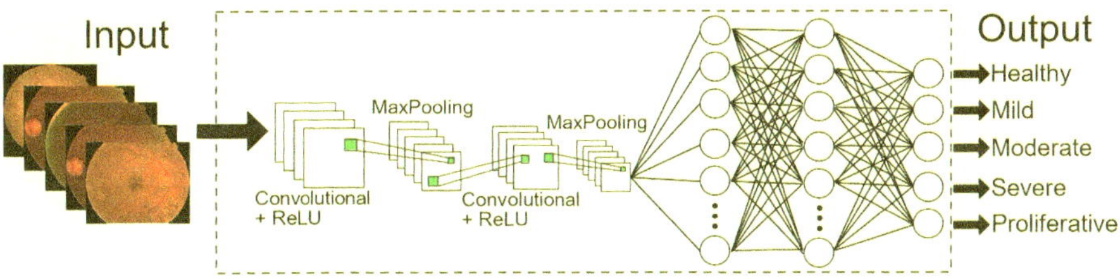

Fig. 1 Graphic of how a CNN model works

MaxPooling is used to minimize the size of the images, keeping the maximum values according to a kernel (2 pixels high and wide is the most used) that will be traversing the images. Finally, there are layers of fully connected neurons and an output layer that classifies the images. In this work, there are just 2 classifications: the patience has DR or does not. A graphic example of a CNN model can be seen in Fig. 1.

2.2 *Hierarchical Genetic Algorithms*

Evolution is a topic in biology that has been widely discussed for many years [10], and this process served as inspiration for the development of an optimization algorithm called the genetic algorithm (GA). Simple genetic algorithms are based on Darwin's theory of evolution [11], in which he comments that a child will have characteristics of his parents and will improve over the generations.

One of the extensions of simple genetic algorithms is hierarchical genetic algorithms (HGA). The main difference is that in hierarchical genetic algorithms there are two types of genes: control and parametric [12]. They were developed to solve a particular class of hierarchical problems. It can be represented as a tree where genes in higher levels (control genes) are in charge of controlling genes in lower levels (parametric genes).

3 Proposed Method

For the development of this work, the pre-processing applied in the database and the properties of the genetic algorithm are explained below.

3.1 *Pre-processing Applied to APTOS 2019*

Apply a pre-processing to the database images is a very important step that help the CNN models to extract the important features. For this work, we used the APTOS

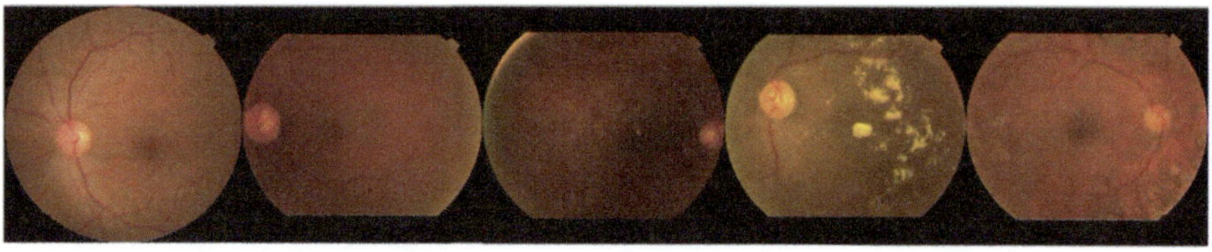

Fig. 2 Images taken from APTOS 2019 after applying the pre-processing

2019 database. This database has 3662 tagged images with five different stages of DR [13, 14]. There are 1805 images of healthy retinas and 1857 images that contain some level of DR. Database images have different sizes and lot of noise that can affect the learning of the CNN [15]. To solve this, we applied pre-processing to all images.

First, we extracted the retina from the image. With this, we avoided the noise in the black zones. Then, we added black pixels at the top and bottom (or on the left and right edges) of the image and create an image completely square. With this shape, the images are not deformed while they are inserted in the CNN. Examples taken from database after apply pre-processing can be observed in Fig. 2.

3.2 Genes of the Chromosomes

As we mentioned before, the chromosomes used in hierarchical genetic algorithm have control genes and parametric genes. The data inside each gen has a different purpose. First, is necessary to about the genes for control to understand what parametric genes do.

Control Genes. The first control gen has an integer value that decides how many convolutional layers will be used in the CNN model, with a minimum value of 3 and a maximum of 6. The second gen also has an integer value and specify how many fully connected layers will be after the convolutional layers and before the classification layer. Finally, the third gen can have values of 10, 20, 30, 40, 50, 60, 70, 80, 90 or 100, and represents the number of epochs. A graphic representation of the chromosome with control genes can be observed on Fig. 3.

Parametric genes for convolutional layers. As we mentioned before, the first gen of the control genes decides how many convolutional layers will have the CNN model. The first gen of the parametric genes, for convolutional layers, has an integer value of 1, 2 or 3. The value 1 means that the layer will be just a convolutional layer using ReLU activation function. Value 2 means all of value 1 and MaxPooling layer with a kernel with 2×2 size and stride of 2. Finally, if the value is 3 means the convolutional layer with ReLU activation function, MaxPooling layer as same as detailed before, and a Dropout probability.

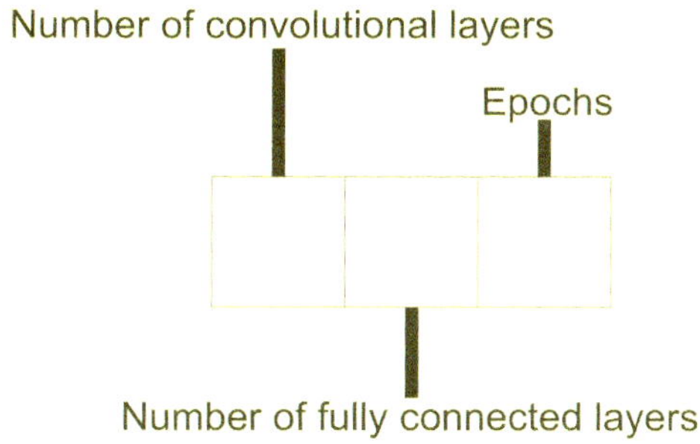

Fig. 3 Control genes

The second parametric gen has the size of the filter that uses the convolutional layer. The values can be one of the integers 3, 4 or 5. The third parametric gen has the number of filters for the convolution layer. The minimum and maximum number of filters depends on the number of the convolutional layer. If the convolutional layer is the number 1 or 2, the number of filters will be between 16 and 32, inclusive. If the number is 3 or 4, the number of filters will be between 64 and 128, inclusive.

Finally, if the number is 5 or 6, the number of filters will be between 256 and 512, inclusive. The fourth and last parametric gen has the value used for the Dropout in case the value of the first gen is 3. Dropout values are integer and can be 10, 20, 30, 40 or 50. Before entering the value in the CNN model, the Dropout value is divided by 100. A graphic representation of the parametric genes for convolutional layers can be observed in Fig. 4.

Parametric genes for fully connected layers. As we mentioned before, the second gen of the control genes decides how many fully connected layers will have the model. The first gen of the parametric genes, for fully connected layers, has an integer value of 1 or 2. The value 1 means that it will adds a fully connected layer with its ReLU activation function. Value 2 means all of value 1 and a Dropout probability.

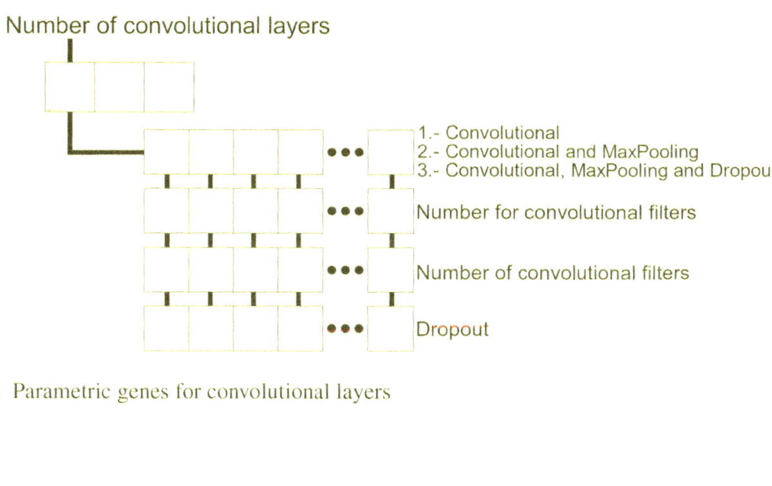

Fig. 4 Parametric genes for convolutional layers

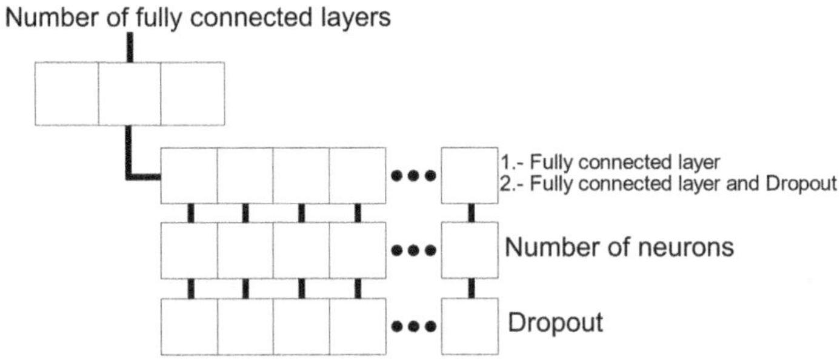

Fig. 5 Parametric genes for fully connected layers

The second parametric gen has how many neurons will have the fully connected layer. Depending on the position of the fully connected layer, the minimum and maximum number of neurons will be different. If the position is the number 1 or 2, the number of neurons will be between 64 and 128, inclusive. If the number is 3 or 4, the number of neurons will be between 128 and 256, inclusive. Finally, if the number is 5 or 6, the number of neurons will be between 256 and 512, inclusive. The third and last parametric gen has the value used for the Dropout in case the value of the first gen is 2. Dropout values are integer and can be 10, 20, 30, 40 or 50. Before entering the value in the CNN model, the Dropout value is divided by 100. A graphic representation of the parametric genes for fully connected layers can be observed in Fig. 5.

4 Experimental Results

One experiment was carried out using the hierarchical genetic algorithm using the chromosomes detailed before. The experiment was executed 30 times. All the executions used the same image pre-processing for the database, and it was distributed as follows: 72% of training images, 8% for validation and 20% for testing. For the CNN properties, each execute used the Adaptive Moment Estimation Optimizer, better known as Adam Optimizer, and in the last layer of the CNN the use of sigmoid activation function.

For the hierarchical genetic algorithm properties, each execute begins with 10 individuals for the population. Also 10 generations, 80% for selection probability, 50% for crossover probability and 80% for mutation probability. The best children replace the worst fathers. The binary crossover was used for this experiment and for mutation, it was implemented by taking a random row of each parametric gen and changing its value in the same way as detailed in previous section. The fitness for each chromosome was the accuracy obtained for the CNN model using the testing images. In Table 1 can be seen the best fitness obtained for the hierarchical genetic algorithm in each run.

Table 1 Results obtained by the hierarchical genetic algorithm

Number of the execution	Best fitness obtained	Number of the execution	Best fitness obtained	Number of the execution	Best fitness obtained
1	0.9658	11	0.9713	21	0.9658
2	0.9399	12	0.9563	22	0.9686
3	0.9727	13	0.9590	23	0.9658
4	0.9563	14	0.9693	24	0.9549
5	0.9720	15	0.9618	25	0.9563
6	0.9597	16	0.9699	26	0.9631
7	0.9645	17	0.9713	27	0.9706
8	0.9781	18	0.9727	28	0.9686
9	0.9652	19	0.9631	29	0.9699
10	0.9604	20	0.9631	30	0.9747

As it can be seen in the execution #8, the best CNN model obtained by the hierarchical genetic algorithm achieved an accuracy of 0.9781 using 100 epochs. We achieved an accuracy mean of 0.9650. With the information of Table 1 we could also achieved a standard deviation of 0.007665. In Fig. 6 can be observed the layers and hyperparameters of the CNN model.

One of the works that has used APTOS 2019 [16] implemented a modification of InceptionV3, CNN that was presented in 2015 [17]. It is one of the most recent works for detection of DR and they got an accuracy of 0.9446 in their best experiment.

5 Conclusions

In this work, we designed a HGA for the creation of CNN models. Before that, we applied pre-processing to the APTOS 2019 database to help in the extraction of the features of DR. In the experiment we got a CNN model with good results for the detection of DR. Recent works preferred the use of existing CNN models, but we think that using algorithms for optimization could create better CNN models.

As future work, we could use the hierarchical genetic algorithm for classification of DR and not just the detection. APTOS 2019 have 5 stages for classification of DR, so we can use the same pre-processed images for that work. Other thing is adding more hyperparameters to optimize, like changing the size of the kernel of the MaxPooling layer or replacing the ReLU activation function with other function. Our results improve the accuracy for the CNN models, but there is room for improvement, so we could change the pre-processing applied in the database in addition of the mentioned before.

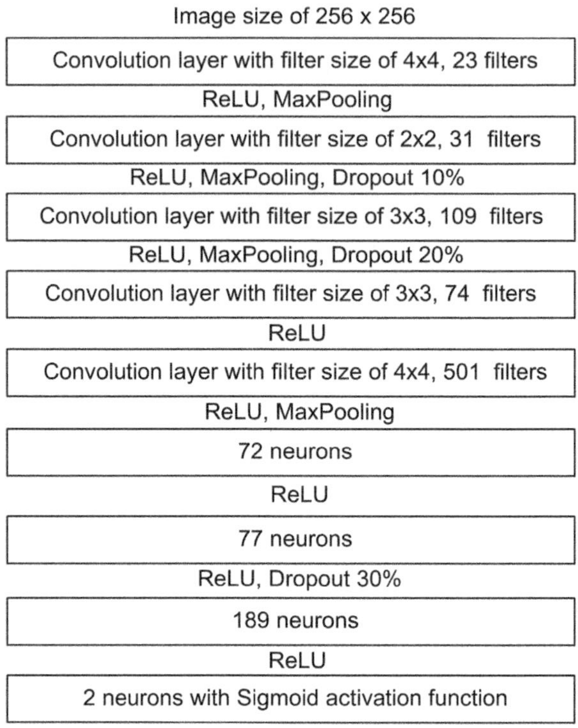

Fig. 6 CNN model with the best accuracy

Also, we could implement other optimization algorithm like Particle Swarm Optimization to improve the accuracy of this work. There are many databases for DR that can been used with the hierarchical genetic algorithm explained in this work, but also it can be used to find a CNN model for any image database with just a few modifications (like the size and shape of the input images).

One thing that could help the HGA, is an addition of different types of pre-processing for the database. Some authors use more than one pre-processing searching which one improve their accuracies. Also, combining different database for DR could improve the training of the CNN models. APTOS 2019 is not the larger database for DR, so using two or more databases at same time could offer better results in less time with an appropriate CNN model.

As future work we plan to consider the proposed model in other different applications, like in optimization problems [18, 19], time series prediction [20–22] and in medical problems [23–25].

References

1. L. Qiao, Y. Zhu, H. Zhou, Diabetic retinopathy detection using prognosis of microaneurysm and early diagnosis system for non-proliferative diabetic retinopathy based on deep learning algorithms. IEEE Access **8**, 104292–104302 (2020). https://doi.org/10.1109/access.2020.2993937
2. E.K. Fenwick, K. Pesudovs, J. Khadka et al., The impact of diabetic retinopathy on quality of life: qualitative findings from an item bank development project. Qual. Life Res. **21**, 1771–1782 (2012). https://doi.org/10.1007/s11136-012-0110-1
3. A. Das, S. Stroud, A. Mehta, S. Rangasamy, New treatments for diabetic retinopathy. Diabetes Obes. Metab. **17**, 219–230 (2014). https://doi.org/10.1111/dom.12384
4. L. Guariguata, D.R. Whiting, I. Hambleton et al., Global estimates of diabetes prevalence for 2013 and projections for 2035. Diabetes Res. Clin. Pract. **103**, 137–149 (2014). https://doi.org/10.1016/j.diabres.2013.11.002
5. G.M. Somfai, E. Tátrai, L. Laurik et al., Automated classifiers for early detection and diagnosis of retinopathy in diabetic eyes. BMC Bioinform. (2014). https://doi.org/10.1186/1471-2105-15-106
6. D.A. Gomez-Cravioto, R.E. Diaz-Ramos, F.J. Cantu-Ortiz, H.G. Ceballos, Data analysis and forecasting of the COVID-19 spread: a comparison of recurrent neural networks and time series models. Cogn. Comput. (2021). https://doi.org/10.1007/s12559-021-09885-y
7. G. Yang, C. Wang, J. Yang et al., Weakly-supervised convolutional neural networks of renal tumor segmentation in abdominal CTA images. BMC Med. Imaging **20**, 37 (2020). https://doi.org/10.1186/s12880-020-00435-w
8. Y. Xu, Y. Chi, Y. Tian, Deep convolutional neural networks for feature extraction of images generated from complex networks topologies. Wirel. Pers. Commun. **103**, 327–338 (2018). https://doi.org/10.1007/s11277-018-5445-7
9. A. Bhandari, J. Koppen, M. Agzarian, Convolutional neural networks for brain tumor segmentation. Insights Imaging **11**, 77 (2020). https://doi.org/10.1186/s13244-020-00869-4
10. J. Rosenau, Evolution and biogeography: leading students in Darwin's and Wallace's footsteps. Evol.: Educ. Outreach **5**, 582–584 (2012). https://doi.org/10.1007/s12052-012-0459-1
11. J.I. Serrano, M.D. del Castillo, On the origin of the evolutionary computation species influences of Darwin's theories on computer science. Artif. Intell. Rev. **38**, 41–54 (2012). https://doi.org/10.1007/s10[illegible]-011-[illegible]-6
12. Z.-y. Xing, X.-m. Pang, H.-y. Ji et al., Hierarchical genetic algorithm based RBF neural networks and application for modelling of the automatic depth control electrohydraulic system. Int. J. Control Autom. Syst. **9**, 759–767 (2011). https://doi.org/10.1007/s12555-011-0418-6
13. J.D. Bodapati, N.S. Shaik, V. Naralasetti, Composite deep neural network with gated-attention mechanism for diabetic retinopathy severity classification. J. Ambient. Intell. Hum. Comput. **12**, 9825–9839 (2021). https://doi.org/10.1007/s12652-020-02727-z
14. Y. Li, Z. Song, S. Kang et al., Semi-supervised auto-encoder graph network for diabetic retinopathy grading. IEEE Access **9**, 140759–140767 (2021). https://doi.org/10.1109/access.2021.3119434
15. M. Nahiduzzaman, M.R. Islam, S.M. Islam et al., Hybrid CNN-SVD based prominent feature extraction and selection for grading diabetic retinopathy using extreme learning machine algorithm. IEEE Access **9**, 152261–152274 (2021). https://doi.org/10.1109/access.2021.3125791
16. V. Vives-Boix, D. Ruiz-Fernández, Diabetic retinopathy detection through convolutional neural networks with synaptic metaplasticity. Comput. Methods Programs Biomed. **206**, 106094–106094 (2021). https://doi.org/10.1016/j.cmpb.2021.106094
17. C. Szegedy, W. Liu, Y. Jia et al., Going deeper with convolutions, in *2015 IEEE Conference on Computer Vision and Pattern Recognition (CVPR)* (2015), pp. 1–9. https://doi.org/10.1109/cvpr.2015.7298594
18. F. Olivas, F. Valdez, P. Melin, A. Sombra, O. Castillo, Interval type-2 fuzzy logic for dynamic parameter adaptation in a modified gravitational search algorithm. Inf. Sci. **476**, 159–175 (2019)

19. P. Melin, D. Sanchez, Multi-objective optimization for modular granular neural networks applied to pattern recognition. Inf. Sci. **460–461**, 594–610 (2018)
20. O. Castillo, J.R. Castro, P. Melin, A. Rodriguez Dias, Application of interval type-2 fuzzy neural networks in non-linear identification and time series prediction. Soft Comput. **18**(6), 1213–1224 (2014)
21. P. Melin, J.C. Monica, D. Sanchez, O. Castillo, Multiple ensemble neural network models with fuzzy response aggregation for predicting COVID-19 time series: the case of Mexico. Healthcare **8**, 181 (2020)
22. P. Melin, J.C. Monica, D. Sanchez, O. Castillo, Analysis of spatial spread relationships of coronavirus (COVID-19) pandemic in the world using self organizing maps. Chaos Solitons Fractals **138**(109917), 1–7 (2020)
23. B.R. Beck, B. Shin, Y. Choi, S. Park, K. Kang, Predicting commercially available antiviral drugs that may act on the novel coronavirus (SARS-CoV-2) through a drug-target interaction deep learning model. Comput. Struct. Biotechnol. J. **18**, 784–790 (2020)
24. B. González, F. Valdez, P. Melin, G. Prado-Arechiga, Fuzzy logic in the gravitational search algorithm enhanced using fuzzy logic with dynamic alpha parameter value adaptation for the optimization of modular neural networks in echocardiogram recognition. Appl. Soft Comput. **37**, 245–254 (2015)
25. F. Gaxiola, P. Melin, F. Valdez, J.R. Castro, O. Castillo, Optimization of type-2 fuzzy weights in backpropagation learning for neural networks using GAs and PSO. Appl. Soft Comput. **38**, 860–871 (2016)

Interval Type-3 Fuzzy Systems: A Natural Evolution from Type-1 and Type-2 Fuzzy Systems

Oscar Castillo, Juan R. Castro, and Patricia Melin

Abstract This article discusses the natural evolution that took place from type-1 to type-2, and now recently is occurring from type-2 to type-3 fuzzy systems. Prof. Zadeh originally proposed the idea of type-1 fuzzy sets in 1965 and later of type-2 fuzzy in 1975. The goal was to model the uncertainty existing in the real world, and it is well known now that type-2 fuzzy models are better to handle the levels of uncertainty in the real-world, coming from noisy, dynamic, non-linear environments and systems. In addition, subjectivity that is handled by humans is also better represented by type-2 fuzzy sets. As a consequence, type-2 systems have been able to overcome type-1 fuzzy systems in many application areas, such as, intelligent control, pattern recognition, and diagnosis. More recently, we have witnessed the rise of the interval type-3 fuzzy sets and their utilization in control and identification of non-linear systems, showing better results than type-2 and type-1, so we expect that also this pattern will continue to other areas of application. In this article, the main differences among the concepts of type-3, type-2 and type-1 will be discussed and then an account of the existing applications of type-2 will be highlighted and finally, future areas of research will be outlined.

Keywords Interval type-3 fuzzy logic · Type-2 fuzzy · Type-1 fuzzy

O. Castillo (✉) · P. Melin
Tijuana Institute of Technology, TecNM, Tijuana, Mexico
e-mail: ocastillo@tectijuana.mx

P. Melin
e-mail: pmelin@tectijuana.mx

J. R. Castro
UABC University, Tijuana, Mexico
e-mail: jrcastror@uabc.edu.mx

O. Castillo and P. Melin (eds.), *Fuzzy Logic and Neural Networks for Hybrid Intelligent System Design*, Studies in Computational Intelligence 1061,
https://doi.org/10.1007/978-3-031-22042-5_12

1 Introduction

Within the field of artificial intelligence, it is well known that fuzzy systems can have successful utilization in a plethora of areas, such as: Intelligent control, recognition, and diagnosis. Originally, fuzzy sets (now called type-1) were proposed by Lotfi Zadeh in 1965. Later fuzzy logic and fuzzy systems were also proposed by Zadeh, and many applications follow, mainly in control [1]. Type-1 fuzzy systems evolved to type-2 fuzzy systems with the works by Mendel in 2001 [2]. Initially, interval type-2 fuzzy systems were studied and applied to several problems [3]. These systems were applied to many problems in areas such as: robotics, intelligent control and others [4, 5]. Simulation and experimental results show that interval type-2 outperform type-1 fuzzy systems in situations with higher levels of noise, dynamic environments or highly nonlinear problems [6–8]. Later, general type-2 fuzzy systems were considered to manage higher levels of uncertainty, and good results have been achieved in several areas of application [9–11]. Recently, is becoming apparent that type-3 fuzzy systems could help solve even more complex problems. For this reason, in this paper we are putting forward the basic constructs of type-3 fuzzy systems by extending the ideas of type-2 fuzzy systems [12–14].

The contribution is the proposal of mathematical definitions of interval type-3 fuzzy sets, which were obtained by using the extension principle on the type-2 fuzzy definitions. In addition, the proposal of a way to define interval type-3 fuzzy systems for the applications. A particular case of control is used to illustrate the proposed ideas, showing that interval type-3 is able to outperform type-2 and type-1 in control. We consider that these are is key contributions to fuzzy theory.

The remaining of the article is: Sect. 2 is presenting the proposed definitions of interval type-3 fuzzy, in Sect. 3 we are presenting the proposed definition of the footprint of uncertainty in type-3, Sect. 4 describes interval type-3 fuzzy systems, Sect. 5 summarizes simulation results and in Sect. 6 we are presenting the conclusions that we have reached with the research done in this work.

2 Interval Type-3 Fuzzy Sets

Interval type-3 fuzzy can be viewed as an extension of type-2 models. We offer basic terminology of interval type-3 fuzzy sets to give an idea of the difference with respect to their type-2 counterparts.

Definition 1 A type-3 fuzzy set (T3 FS) [15, 16], denoted by $A^{(3)}$, is represented by the plot of a trivariate function, called membership function (MF) of $A^{(3)}$, in the Cartesian product $X \times [0, 1] \times [0, 1]$ in $[0, 1]$, where X. is the universe of the primary variable of $A^{(3)}$, x. The MF of $\mu_{A^{(3)}}$ is denoted by $\mu_{A^{(3)}}(x, u, v)$ (or $\mu_{A^{(3)}}$ to abbreviate) and it is called a type-3 membership function (T3 MF) of the T3 FS. In other words,

$$\mu_{A^{(3)}} : X \times [0, 1] \times [0, 1] \rightarrow [0, 1]$$

$$A^{(3)} = \{(x, u(x), v(x, u), \mu_{A^{(3)}}(x, u, v)) | x \in X, u \in U \subseteq [0, 1], v \in V \subseteq [0, 1]\} \tag{1}$$

where U is the universe for the secondary variable u and V is the universe for tertiary variable v. A T3FS, $A^{(3)}$ can also be formulated in continuous notation as:

$$A^{(3)} = \int_{x \in X} \int_{u \in [0,1]} \int_{v \in [0,1]} \mu_{A^{(3)}}(x, u, v)/(x, u, v) \tag{2}$$

$$A^{(3)} = \int_{x \in X} \left[\int_{u \in [0,1]} \left[\int_{v \in [0,1]} \mu_{A^{(3)}}(x, u, v)/v \right] /u \right] /x \tag{3}$$

where $\iiint$ is notation for the union over all the admissible x, u, v values.

Equation (3) is represented as a mapping of the T3 FS MFs with the following equations:

$$A^{(3)} = \int_{x \in X} \mu_{A_x^{(3)}}(u, v)/x$$

$$\mu_{A_x^{(3)}}(u, v) = \int_{u \in [0,1]} \mu_{A_{(x,u)}^{(3)}}(v)/u$$

$$\mu_{A_{(x,u)}^{(3)}}(v) = \int_{v \in [0,1]} \mu_{A^{(3)}}(x, u, v)/v$$

where $\mu_{A_x^{(3)}}(u, v)$ is the primary MF, $\mu_{A_x^{(3)}}(u, v)$ is the secondary function and $\mu_{A_{(x,u)}^{(3)}}(v)$ is the tertiary function of the T3 FS.

If $\mu_{A^{(3)}}(x, u, v) = 1$, the T3 FS, $A^{(3)}$, is reduced to an interval type-3 fuzzy set (IT3 FS) with the notation $\mathbb{A}$, defined by Eq. (4).

$$\mathbb{A} = \int_{x \in X} \left[\int_{u \in [0,1]} \left[\int_{v \in [\underline{\mu}_{\mathbb{A}}(x,u), \overline{\mu}_{\mathbb{A}}(x,u)]} 1/v \right] /u \right] /x \tag{4}$$

where

$$\mu_{\mathbb{A}(x,u)}(v) = \int_{v \in [\underline{\mu}_{\mathbb{A}}(x,u), \overline{\mu}_{\mathbb{A}}(x,u)} 1/v$$

$$\mu_{\mathbb{A}(x)}(u, v) = \int_{u \in [0,1]} \left[\int_{v \in [\underline{\mu}_{\mathbb{A}}(x,u), \overline{\mu}_{\mathbb{A}}(x,u)} 1/v \right] / u$$

$$\mathbb{A} = \int_{x \in X} \mu_{\mathbb{A}_{(x)}}(u, v)/x$$

Assuming that $v \in \left[\underline{\mu}_{\mathbb{A}}(x, u), \overline{\mu}_{\mathbb{A}}(x, u)\right]$ and the lower and upper membership functions $\underline{\mu}_{\mathbb{A}}(x, u), \overline{\mu}_{\mathbb{A}}(x, u)$ are general type-2 membership functions (T2 MF) over the plane (x, u), Eq. (4) can be simplified as a bivariate isosurface simplifica with an interval type-3 membership function (IT3 MF), $\tilde{\mu}_{\mathbb{A}}(x, u) \in \left[\underline{\mu}_{\mathbb{A}}(x, u), \overline{\mu}_{\mathbb{A}}(x, u)\right]$, defined by Eq. (5).

$$\mathbb{A} = \int_{x \in X} \int_{u \in [0,1]} \tilde{\mu}_{\mathbb{A}}(x, u)/(x, u) \tag{5}$$

where the lower T2 MF $\underline{\mu}_{\mathbb{A}}(x, u)$, is contained in the upper T2 MF $\overline{\mu}_{\mathbb{A}}(x, u)$, this is, $\underline{\mu}_{\mathbb{A}}(x, u) \subseteq \overline{\mu}_{\mathbb{A}}(x, u)$, then $\underline{\mu}_{\mathbb{A}}(x, u) \leq \overline{\mu}_{\mathbb{A}}(x, u)$, and as a consequence, an IT3 FS is represented by two T2 FSs, one inferior $\underline{\boldsymbol{A}}$ with T2 MF $\underline{\mu}_{\mathbb{A}}(x, u)$ and another superior $\overline{\boldsymbol{A}}$, with T2 MF $\overline{\mu}_{\mathbb{A}}(x, u)$ defined by Eqs. (6) and (7) (see Fig. 1)

$$\underline{\boldsymbol{A}} = \int_{x \in X} \int_{u \in [0,1]} \underline{\mu}_{\mathbb{A}}(x, u)/(x, u) = \int_{x \in X} \left[\int_{u \in [0,1]} \underline{f}_x(u)/u \right] / x \tag{6}$$

$$\overline{\boldsymbol{A}} = \int_{x \in X} \int_{u \in [0,1]} \overline{\mu}_{\mathbb{A}}(x, u)/(x, u) = \int_{x \in X} \left[\int_{u \in [0,1]} \overline{f}_x(u)/u \right] / x \tag{7}$$

where, the secondary MFs of $\underline{\boldsymbol{A}}$ and $\overline{\boldsymbol{A}}$ are T1 MFs of T1FS given by the Eqs. (8) and (9)

$$\mu_{\underline{A}(x)}(u) = \int_{u \in J_x} \underline{f}_x(u)/u \tag{8}$$

$$\mu_{\overline{A}(x)}(u) = \int_{u \in J_x} \overline{f}_x(u)/u \tag{9}$$

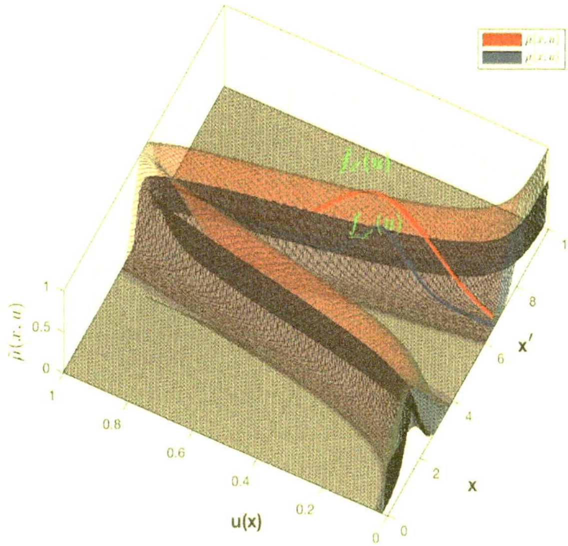

Fig. 1 IT3 FS with IT3MF $\tilde{\mu}(x,u)$ where $\underline{\mu}(x,u)$ is the LMF and $\overline{\mu}(x,u)$ is the UMF. The embedded secondary T1 MFs in x' of $\underline{A}$ and $\overline{A}$ are $\underline{f}_{x'}(u)$ and $\overline{f}_{x'}(u)$ respectively

3 Footprint of Uncertainty

Another simple way to represent the IT3 MF from IT3 FS, $\mathbb{A}$, is like a bivariate isosurface, $\tilde{\mu}_{\mathbb{A}}(x,u)$, because of the union of vertical-slices in x, where each vertical cut in $x = x'$ is an embedded secondary interval type-2 membership function (IT2 MF) [16, 17] $\tilde{f}_{x'}(u)$, in other words, $\tilde{f}_{x'}(u) \in 1/\left[\underline{\mu}_{\mathbb{A}}(x',u), \overline{\mu}_{\mathbb{A}}(x',u)\right]$ or $\tilde{f}_{x'}(u) \in \left[\underline{f}_{x'}(u), \overline{f}_{x'}(u)\right]$. As a consequence, Eq. (5) can be simplified to Eq. (10) (see Fig. 1).

$$\mathbb{A} = \int_{x\in X}\int_{u\in[0,1]} \tilde{\mu}_{\mathbb{A}}(x,u)/(x,u) = \int_{x\in X}\left[\int_{u\in[0,1]} \tilde{f}_x(u)/u\right]/x \tag{10}$$

where $\tilde{\mu}_{\mathbb{A}}(x,u) \in \left[\underline{\mu}_{\mathbb{A}}(x,u), \overline{\mu}_{\mathbb{A}}(x,u)\right]$.

Equation (10) can be expressed as Eqs. (11) and (12)

$$\mathbb{A} = \int_{x\in X} \mu_{\mathbb{A}(x)}(u)/x = \int_{x\in X}\left[\int_{u\in[0,1]} \tilde{f}_x(u)/u\right]/x \tag{11}$$

where $\mu_{\mathbb{A}(x)}(u)$ is an Interval type-2 fuzzy set (IT2 FS), as shown in Eq. (11).

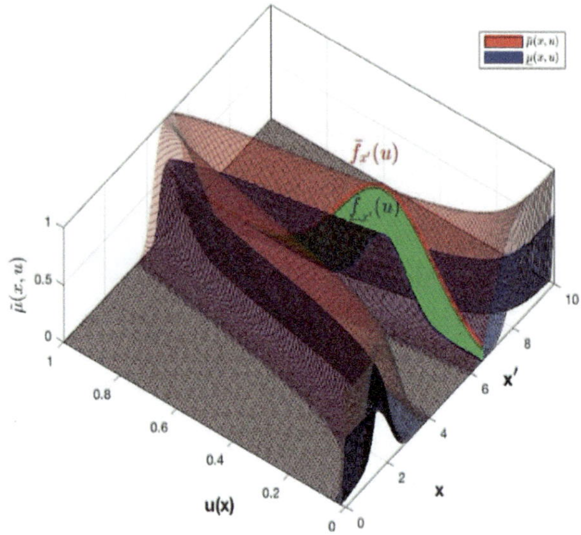

Fig. 2 IT3 FS, $\mathbb{A}$, with IT3 MF, $\tilde{\mu}_{\mathbb{A}}(x,u)$ and an embedded vertical cut, $\mu_{\mathbb{A}(x')}(u) \in \left[\underline{f}_{x'}(u), \overline{f}_{x'}(u)\right]$ with the FOU in green color

$$\mu_{\mathbb{A}(x)}(u) = \int_{u \in J_x} \tilde{f}_x(u)/u = \int_{u \in [\underline{f}_x(u), \overline{f}_x(u)} 1/u \tag{12}$$

Definition 2 An Interval type-3 fuzzy set (IT3 FS), represented by $\mathbb{A}$, is an isosurface with a bivariate function, called MF of $\mathbb{A}$ (see Fig. 4) over the Cartesian product $X \times [0, 1]$ in $[0, 1]$, where X is the universe for the primary variable of $\mathbb{A}$, x. The MF of $\mathbb{A}$ is denoted by $\tilde{\mu}_{\mathbb{A}}(x,u)$, (or $\tilde{\mu}_{\tilde{A}}$ for simplicity) and it is called Interval type-3 membership function MF (IT3 MF), in other words,

$$\mathbb{A} = \{(x, u, \tilde{\mu}_{\mathbb{A}}(x,u)) | x \in X, u \in U \equiv [0, 1]\}$$

In which $\tilde{\mu}_{\mathbb{A}}(x,u) \subseteq [0, 1]$. U is the universe for the secondary variable u, and in this work is assumed that U is $[0, 1]$.

An IT3 FS, $\mathbb{A}$ can also be formulated in notation of fuzzy sets as in Eqs. (9) and (10).

$$\mathbb{A} = \int_{x \in X} \int_{u \in [0,1]} \tilde{\mu}_{\mathbb{A}}(x,u)/(x,u) = \int_{x \in X} \mu_{\mathbb{A}(x)}(u)/x = \int_{x \in X} \left[\int_{u \in [0,1]} \tilde{f}_x(u)/u \right] /x$$

where $\mu_{\mathbb{A}(x)}(u)$ is an Interval type-2 fuzzy set (IT2 MF).

$$\mu_{\mathbb{A}(x)}(u) = \int_{u \in [0,1]} \tilde{f}_x(u)/u = \int_{u \in [\underline{f}_x(u), \overline{f}_x(u)} 1/u$$

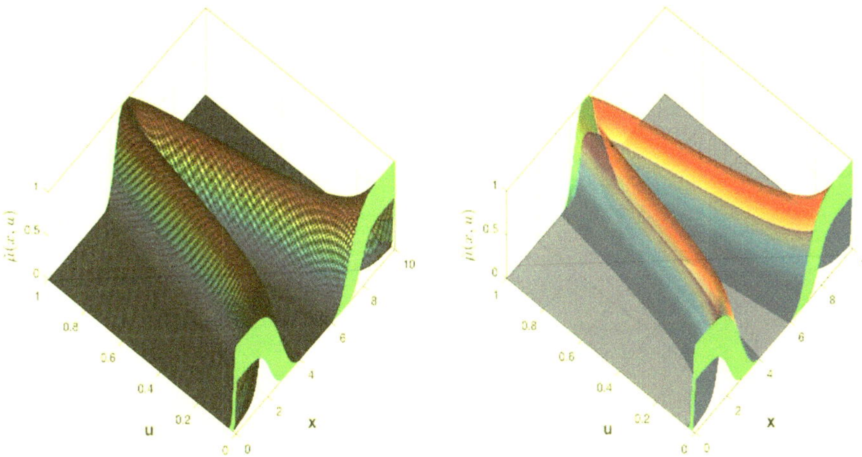

Fig. 3 Isosurface of the membership function of the IT3 FS

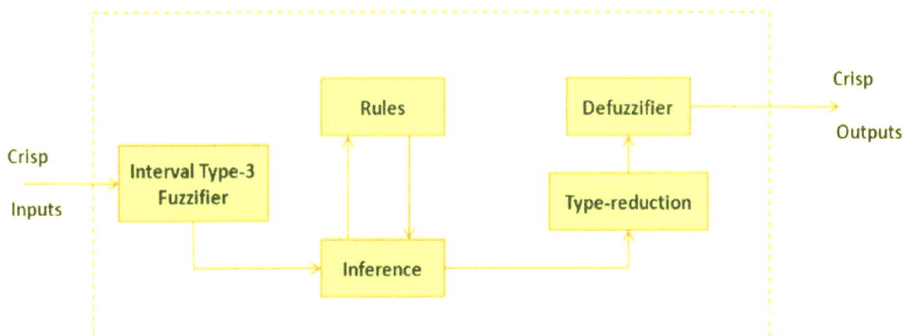

Fig. 4 General structure of an interval type-3 fuzzy system

and $\iint$ denotes the union over all the admisible x and u.

The 3D plot of the IT3MF is an isosurface with volume in between the layers of the Surface formed by all the secondary IT2MFs $\mu_{\tilde{A}(x)}(u)$ in green color in Fig. 3, which forms the domain of uncertainty (DOU) of IT3 FS.

4 Interval Type-3 Fuzzy Systems

Here we offer an overview of how interval type-3 fuzzy systems can be defined and applied in many possible areas. First, we can state that the general structure of an interval type-3 fuzzy system is similar to the structures of type-2 and type-1, consisting of fuzzifier, fuzzy rules, inference, type-reduction and defuzzifier. The

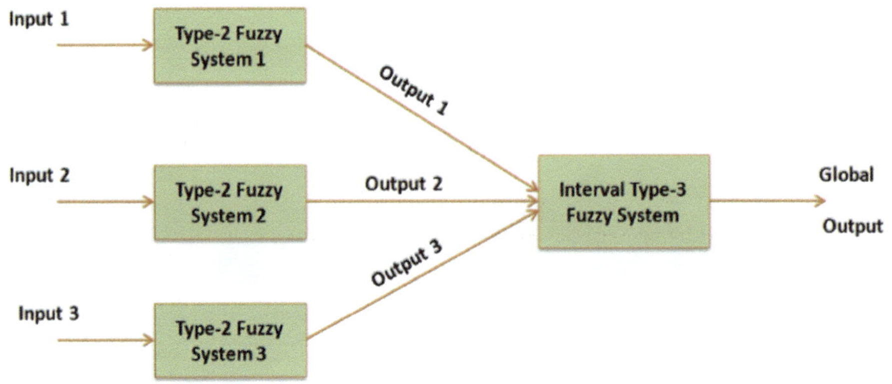

Fig. 5 Hierarchical architecture using an interval type-3 fuzzy system to combine the outputs of simpler systems

main difference with respect to type-2 and type-1 is that now interval type-3 fuzzy sets are used in the fuzzifier and also are the outputs of the inference and need to be type reduced. In Fig. 4 the structure of an interval type-3 fuzzy system is found.

The structure of Fig. 5 can be used to build interval type-3 fuzzy systems for different applications, in a similar way as it is done in type-2 and type-1, just the difference would be that we need to now define the type-3 membership functions. Of course, this would in theory enable the fuzzy system to deal with the uncertainty for the problem at hand. It is also possible that for more complex problems a modular structure could be used, meaning that the complete problem is divided into subproblems that simpler fuzzy systems can handle and then fuzzy systems (as aggregator) can combine the results to achieve a global result. If we assume that three fuzzy systems can solve the three subproblems (maybe using type-2 systems) then the aggregator could an interval type-3 system, as it is illustrated in Fig. 5. The main idea of using an interval type-3 system for combining the outputs of the type-2 systems is to manage the uncertainty of the results produced by these systems.

This is just shown as an illustration of the possible form of this fuzzy system. In general, we can say that there can be other possible architectures for the hierarchical idea. For example, we could have individual models of type-1 providing inputs to the type-3 aggregator, or the individual controllers could be neural networks or neuro-fuzzy systems. It is even possible to have classical controllers combined with fuzzy controllers. At the end, the main idea is that the type-3 aggregator should handle the uncertainty coming out of the individual controllers. In particular, the interval type-3 fuzzy systems can have a structure like the one shown in Fig. 6, where we can find the three inputs and one outputs, showing interval type-3 Gaussian membership functions.

In the following section we show simulation results to illustrate the results of an interval type-3 system, as the one illustrated in Fig. 6, in a control application.

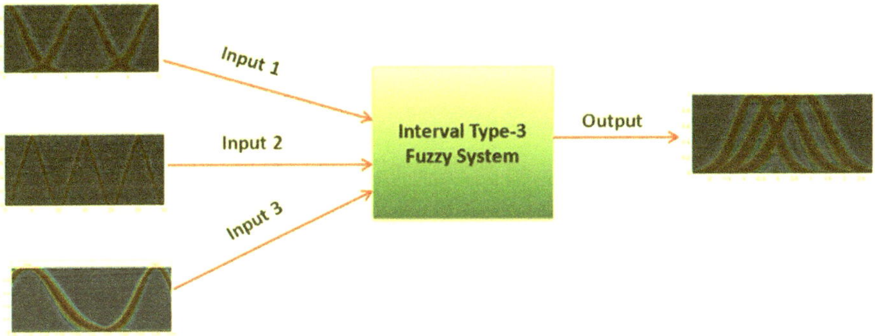

Fig. 6 Example of an interval type-3 system

5 Simulation Results

In this section we consider a control problem in which three individual fuzzy controllers are used to control simpler parts of the problem and then an interval type-3 fuzzy controller is used to combine the outputs of the individual fuzzy controllers. The particular problem is the one of controlling the flight an airplane [18].

In this problem the total flight control is achieved by combination of three individual controllers (longitudinal, lateral and direction control). For obtaining the longitudinal part, a system with one input and one output was put forward, the output is the elevator action. In this case the linguistic values are *push*, *pull* or *not move*, and the stick depending on the movement are, then the action in the elevator is obtained. In lateral control, the outputs are the movements of the aileron because when they are moved the aircraft can change the direction in the z axis and the input is the stick of the pilot and when this is moved (right or left) the ailerons have a reaction and the linguistic values are: *move right*, *move left*, *center*. In direction control the output consists of the movements of the rudder and these determine if the aircraft changes the trajectory in the x axis, in the input the pedals are consider (*left* and *right*). The previously explained expert knowledge can be expressed as fuzzy rules in the following form, for more details please check the work in [18]:

The fuzzy rules of longitudinal control are:

If wheel is push then elevator is down
If wheel is center then elevator is center
If wheel is pull then elevator is up

The fuzzy rules of lateral control are:

If wheel is left then aileron is left_up_right_down
If wheel is center then aileron is center
If wheel is right then aileron is right_up_left_down

The fuzzy rules for direction control are:

If pedals is left then rudder is left
If pedals is center then rudder is center
If pedals is right then rudder is right

The fuzzy rules (we only show 3 rules as illustration to offer an idea of the global controller) of the aggregator in Fig. 6 that combine the outputs of the three individual fuzzy controllers are as follows [18].

1. If (Elevator is *low*) and (rudder is *medium*) and (Aileron is *low*) then (Nelevator is *low*) (Nrudder is *low*) (Naileron is *medium*).
2. If (Elevator is *high*) and (rudder is *medium*) and (Aileron is *high*) then (Nelevator is *medium*) (Nrudder is *high*) (Naileron is *medium*).
3. If (Elevator is *high*) and (rudder is *high*) and (Aileron is *low*) then (Nelevator is *high*) (Nrudder is *medium*) (Naileron is *medium*).

After obtaining the rules of the aggregator, the outputs of the individual controllers are combined with an aggregator using type-2 fuzzy and the results are summarized in Tables 1, 2 and 3 based on the work in [18]. When the type-2 fuzzy aggregator was applied in [18] the behavior was improved, but in this article we decided to also use an interval type-3 fuzzy aggregator (like in Fig. 6) to test if the performance could be improved even more for the behavior of the aircraft and this is also shown in Tables 1, 2 and 3. The means are from 15 simulations of airplane control for each case. It is important to highlight that this aggregator is an interval type-3 system to achieve the global control by managing the uncertainties in this problem in a better fashion.

The simulation results presented in Tables 1, 2 and 3, show evidence that there exists a significant improvement when the interval type-3 system is used instead of type-2 in the aggregator for the proposed approach for global control combining the individual controllers. The main reason for the improvement is based on theoretical

Table 1 Comparison of results for the aileron

Case	Comparison of results		
	Controller	Mean	Deviation
1	Type-1 fuzzy aggregator [18]	0.4100	0.1070
2	Type-2 fuzzy aggregator [18]	0.2049	0.0533
3	Interval type-3 fuzzy aggregator	0.1511	0.0222

Table 2 Comparison of results for the elevator

Case	Comparison of results		
	Controller	Mean	Deviation
1	Type-1 fuzzy aggregator [18]	0.4100	0.1020
2	Type-2 fuzzy aggregator [18]	0.2053	0.0510
3	Interval type-3 fuzzy aggregator	0.1522	0.0211

Table 3 Comparison of results for the rudder

Case	Comparison of results		
	Controller	Mean	Deviation
1	Type-1 fuzzy aggregator [18]	0.4070	0.1310
2	Type-2 fuzzy aggregator [18]	0.2037	0.0655
3	Interval type-3 fuzzy aggregator	0.1507	0.0202

fact that type-3 should be able to handle in a better way the inherent uncertainty in controlling the nonlinear plant, in this case, the airplane. Of course, we also believe that type-3 fuzzy should be able to handle other types of uncertainty arising in other kinds of problems (different than control), such as in decision making, prediction, diagnosis, monitoring, just to mention a few.

6 Conclusions

In this article, the main differences among the concepts of type-3, type-2 and type-1 fuzzy sets and systems are discussed and then an account of the existing applications of type-2 were highlighted and future areas of research were outlined. The basic theoretical constructs of interval type-3 were outlined, as well as ways to design and build interval type-3 fuzzy systems were put forward. The advantages of interval type-3 fuzzy logic have been illustrated with a control application. The case of airplane flight control that was considered in previous works with type-1 and type-2 fuzzy control [18] has now been considered with interval type-3 fuzzy logic, and a comparison has been performed. The conclusion was that interval type-3 fuzzy control is able to outperform type-2 and type-1 in the control errors, and we believe this is due to the fact that type-3 is able to manage higher levels of uncertainty, which of course this problem has. In future work, we plan to consider working on constructing better algorithms for inference, type-reduction and defuzzification in type-3 fuzzy, as well other approaches for fuzzy operations. In addition, we plan to consider a more extensive analysis of this problem, including performing optimization of the membership functions with a metaheuristic algorithm. In addition, we plan to apply interval type-3 fuzzy logic in other control problems, as well as in other kinds of application problems to test the possible advantages of this kind of fuzzy systems. It is also possible to develop the theory of general type-3 fuzzy to extend the concepts and operations of interval type-3 and then consider possible applications and comparison of results. Finally, we have to say that we believe that interval type-3 could enhance the quality of the solutions in other areas, such as in time series prediction [19], fuzzy clustering [20] and diagnosis [21], by enhancing the management of the uncertainty in the corresponding areas, and we envision working on these application areas in the near future for the benefit of society and economy [22–27].

References

1. L.A. Zadeh, Knowledge representation in Fuzzy Logic. IEEE Trans. Knowl. Data Eng. **1**, 89 (1989)
2. L.A. Zadeh, Fuzzy logic. Computer **1**(4), 83–93 (1998)
3. J.M. Mendel, *Uncertain Rule-Based Fuzzy Logic Systems: Introduction and New Directions* (Prentice-Hall, Upper-Saddle River, NJ, 2001)
4. J.M. Mendel, *Uncertain Rule-Based Fuzzy Logic Systems: Introduction and New Directions*, 2nd edn. (Springer, Berlin, 2017)
5. N.N. Karnik, J.M. Mendel, Operations on type-2 fuzzy sets. Fuzzy Sets Syst. **122**, 327–348 (2001)
6. J.E. Moreno et al., Design of an interval type-2 fuzzy model with justifiable uncertainty. Inf. Sci. **513**, 206–221 (2020)
7. J.M. Mendel, H. Hagras, W.-W. Tan, W.W. Melek, H. Ying, *Introduction to Type-2 Fuzzy Logic Control* (Wiley and IEEE Press, Hoboken, NJ, 2014)
8. F. Olivas, F. Valdez, O. Castillo, P. Melin, Dynamic parameter adaptation in particle swarm optimization using interval type-2 fuzzy logic. Soft. Comput. **20**(3), 1057–1070 (2016)
9. A. Sakalli, T. Kumbasar, J.M. Mendel, Towards systematic design of general type-2 fuzzy logic controllers: analysis, interpretation, and tuning. IEEE Trans. Fuzzy Syst. **29**(2), 226–239 (2021)
10. E. Ontiveros, P. Melin, O. Castillo, High order α-planes integration: a new approach to computational cost reduction of general type-2 fuzzy systems. Eng. Appl. Artif. Intell. **74**, 186–197 (2018)
11. O. Castillo, L. Amador-Angulo, A generalized type-2 fuzzy logic approach for dynamic parameter adaptation in bee colony optimization applied to fuzzy controller design. Inf. Sci. **460–461**, 476–496 (2018)
12. Y. Cao, A. Raise, A. Mohammadzadeh et al., Deep learned recurrent type-3 fuzzy system: application for renewable energy modeling/prediction. Energy Reports (2021)
13. A. Mohammadzadeh, O. Castillo, S.S. Band et al., A novel fractional-order multiple-model type-3 fuzzy control for nonlinear systems with unmodeled dynamics. Int. J. Fuzzy Syst. (2021). https://doi.org/10.1007/s40815-021-01058-1
14. S.N. Qasem, A. Ahmadian, A. Mohammadzadeh, S. Rathinasamy, B. Pahlevanzadeh, A type-3 logic fuzzy system: optimized by a correntropy based Kalman filter with adaptive fuzzy kernel size. Inform. Sci. **572**, 424–443 (2021)
15. J.T. Rickard, J. Aisbett, G. Gibbon, Fuzzy subsethood for fuzzy sets of type-2 and generalized type-n. IEEE Trans. Fuzzy Syst. **17**(1), 50–60 (2009)
16. A. Mohammadzadeh, M.H. Sabzalian, W. Zhang, An interval type-3 fuzzy system and a new online fractional-order learning algorithm: theory and practice. IEEE Trans. Fuzzy Syst. **28**(9), 1940–1950 (2020)
17. Z. Liu, A. Mohammadzadeh, H. Turabieh, M. Mafarja, S.S. Band, A. Mosavi, A new online learned interval type-3 fuzzy control system for solar energy management systems. IEEE Access **9**, 10498–10508 (2021)
18. L. Cervantes, O. Castillo, Type-2 fuzzy logic aggregation of multiple fuzzy controllers for airplane flight control. Inf. Sci. **324**, 247–256 (2015)
19. O. Castillo, J.R. Castro, P. Melin, A. Rodriguez-Diaz, Application of interval type-2 fuzzy neural networks in non-linear identification and time series prediction. Soft. Comput. **18**(6), 1213–1224 (2014)
20. E. Rubio, O. Castillo, F. Valdez, P. Melin, C.I. Gonzalez, G. Martinez, An extension of the fuzzy possibilistic clustering algorithm using type-2 fuzzy logic techniques. Adv. Fuzzy Syst. (2017). https://doi.org/10.1155/2017/7094046
21. P. Melin, I. Miramontes, G. Prado-Arechiga, A hybrid model based on modular neural networks and fuzzy systems for classification of blood pressure and hypertension risk diagnosis. Expert Syst. Appl. **107**, 146–164 (2018)

22. A. Mancilla, M. García-Valdez, O. Castillo, J.J. Merelo-Guervós, Optimal fuzzy controller design for autonomous robot path tracking using population-based metaheuristics. Symmetry **14**(2), 202 (2022). https://doi.org/10.3390/sym14020202
23. M.W. Tian, A. Mohammadzadeh, J. Tavoosi, S. Mobayen, J.H. Asad, O. Castillo, A.R. Várkonyi-Kóczy, A deep-learned type-3 fuzzy system and its application in modeling problems. Acta Polytech. Hung. **19**(2) (2022)
24. E. Bernal, M.L. Lagunes, O. Castillo, J. Soria, F. Valdez, Optimization of type-2 fuzzy logic controller design using the GSO and FA algorithms. Int. J. Fuzzy Syst. **23**(1), 42–57 (2021). https://doi.org/10.1007/s40815-020-00976-w
25. O. Castillo, J.R. Castro, P. Melin, *Interval Type-3 Fuzzy Systems: Theory and Design* (Springer, Cham, Switzerland, 2022)
26. O. Castillo, P. Melin, A new fuzzy-fractal-genetic method for automated mathematical modelling and simulation of robotic dynamic systems, in *1998 IEEE International Conference on Fuzzy Systems (FUZZ-IEEE 1998) Proceedings*, vol 2, pp. 1182–1187
27. O. Castillo, P. Melin, Intelligent adaptive model-based control of robotic dynamic systems with a hybrid fuzzy-neural approach. Appl. Soft Comput. **3**(4), 363–378 (2003)

A Comparative Study Between Bird Swarm Algorithm and Artificial Gorilla Troops Optimizer

Ivette Miramontes and Patricia Melin

Abstract Nowadays, bio-inspired algorithms are one of the most widely used existing soft computing techniques. Thanks to these methods we can help different problems in improving the results provided. Because new metaheuristics are being proposed more rapidly every day, it is in our interest to meet analysts with these methodologies in order to identify and apply them according to the problem we face. In this research, a new metaheuristic is analyzed, which is called Artificial Gorilla Troops Optimizer, and compare the results obtained with a well-known algorithm that we have used for different optimizations, this is known as Bird Swarm Algorithm. For the present analysis, 10 mathematical benchmak functions were used in order to study the results obtained. After the statistical study carried out, we were able to conclude that according to the dimensions handled, the Artificial Gorilla Troops Optimizer algorithm shows a significant improvement against the Bird Swarm Algorithm.

Keywords Optimization · Benchmark functions · Bio-inspired algorithms · Bird Swarm Algorithm (BSA) · Artificial Gorilla Troops Optimizer (GTO)

1 Introduction

Today there are processes within the industry that allow a very small or null margin of error, due to the delicate nature of the products that are manufactured, which is why more and more different optimization methods are used that help to the improvement and accuracy of these processes. In the area of soft computing, there are metaheuristics or also known as bio-inspired algorithms, which take this inspiration of different aspects, such as genetic-based [1], nature-based [2], swarm intelligence

I. Miramontes · P. Melin (✉)
Tijuana Institute of Technology, TecNM, Tijuana, Baja California, Mexico
e-mail: pmelin@tectijuana.mx

O. Castillo and P. Melin (eds.), *Fuzzy Logic and Neural Networks for Hybrid Intelligent System Design*, Studies in Computational Intelligence 1061,
https://doi.org/10.1007/978-3-031-22042-5_13

[3], physic-based [4] and human-based [5] designing these mathematical models that help us to improve in different aspects. In the different studies carried out, it has been of interest to analyze bio-inspired algorithms, where the Flower Pollination Algorithm has been used for the optimization of neural networks, Chicken Swarm Optimization has been used for the optimization of fuzzy systems, among others [6–9].

Although it has well-known algorithms such as Genetic Algorithms [10] and Particle Swarm Optimization [11], which have been shown to provide excellent results when used in solving different optimization problems, it is in our interest to give way to new metaheuristics, this in order to see in what area they specialize and to know when or what problems can be solved in a better way to it.

The main contribution of this research is to make a comparison of the results of a well-known algorithm such as the Bird Swarm Algorithm (BSA) and a new metaheuristic, in this case the Artificial Gorilla Troops Optimizer (GTO), to test its performance and analyze the results in order to determine if it can be used in the future and even make some modification in order to obtain better results.

This paper has been organized as follows: the literature review is presented in Sect. 2, in Sect. 3 the mathematical benchmark functions are presented, the results are analyzed in Sect. 4 and, the conclusions for this work are presented in Sect. 5.

2 Literature Review

This section gives a brief description of the different methodologies to be used, as well as their application to solve different problems.

2.1 Optimization

Understanding the concept of optimization is of the utmost importance, since it clarifies what you want to achieve by performing it. Formally it can be said that optimization is the search for the best solution to a given problem, and these problems can be focused on different areas [12].

2.2 Bird Swarm Algorithm

Bird Swarm Algorithm [13], is a methodology which imitates the behavior of birds within their habitat. This algorithm based on swarm intelligence was proposed by Meng in 2015, imitating the behavior and social interaction in the swarm. The behavior of the birds is imitated in different actions, which correspond to foraging and flight. This algorithm is based on 5 rules that generally cover how to search for food and escape from predators.

2.3 Artificial Gorilla Troops Optimizer

The Artificial Gorilla Troops Optimizer [14] is proposed by Abdollahzadeh in 2021, which uses five rules to imitate the behavior of gorillas within the swarm and thus be able to carry out exploration and exploitation. The behaviors used to carry out the exploration are the migration to an unknown place, that helps to increase the exploration, the movement of the gorillas in order to have a balance between exploration and exploitation. The third behavior taken into account for the exploration phase is the movement of the gorillas towards known places, this in order to have the ability to search in different spaces within the environment. In exploitation, it is divided into two phases, which helps to significantly increase its search space in which significantly increases the search performance in exploitation.

2.4 Optimization Problems Using BSA

The BSA algorithm has been used in different areas in order to obtain better results in the problems being solved, among which are:

Signal processing for wind speed prediction [15], for the detection of unmanned aerial vehicle objects [16], in the identification of polycyclic aromatic hydrocarbons [17], for the joint planning of substations and lines in urban distribution systems [18] and in obtaining a medical diagnosis of hypertension [19]. It should be noted that several of the works mentioned have applied improvements to the original method.

2.5 Optimization Problems Using GTO

The GTO algorithm is relatively new, but by no means unused. This has been proposed in applications such as:

In the search for the positions and classification of renewable distributed generators (RDG) [20], for power system stabilizer unit adjustment [21], the effective extraction of the parameters of different photovoltaic models, in this work it is also proposed a better one to the original method [22], in the prediction of critical genes obtained from genomic data [23], in the optimal adjustment of power system stabilizer control parameters [14].

3 Mathematical Benchmark Functions

When experimenting with a new metaheuristic, it is common to use mathematical benchmark functions, this with the aim of studying results produced by said metaheuristics and analyzing their performance, an example of these experiments can be found in: Aquila Optimizer [24], Honey Badger Algorithm [25], Tunicate Swarm Algorithm [26], Coronavirus Herd Immunity Optimizer [5], Dingo Optimizer [27], among others.

To carry out the corresponding experimentation of this work, 10 classical benchmark mathematical functions are used, which are divided into unimodal (F1–F7) and multimodal (F8–F10).

Unimodal functions are those that, in their search space, have only one optimal value. This type of functions can be continuous or discontinuous, in addition to being functions of simple analysis.

Multimodal functions are those that have numerous optima (either local or global). In general, these types of functions are difficult in the sense that a global optimal solution cannot be guaranteed in a finite number of steps.

The mathematical representation of the functions used, as well as their range and minimum value, are listed as follows:

- ***F1: Sphere***

$$f(x) = \sum_{i=1}^{n_x} x_i^2$$

Being $x_i \in [-100, 100]$ and $f_{\min}(x^*) = 0.0$

- ***F2: Schwefel 2.22***

$$f(x) = \sum_{i=1}^{n} |x_i| + \prod_{i=0}^{n} x_i$$

Being $x_i \in [-10, 10]$ and $f_{\min}(x^*) = 0.0$

- ***F3: Schwefel 1.2***

$$f(x) = \sum_{i=1}^{n} \left(\sum_{j=1}^{i} x_j \right)^2$$

Being $x_i \in [-100, 100]$ and $f_{\min}(x^*) = 0.0$

- ***F4: Schwefel 2.21***

$$f(x) = \max\{|x_i|, 1 \leq i \leq n\}$$

Being $x_i \in [-100, 100]$ and $f_{\min}(x^*) = 0.0$

- ***F5: Rosenbrock***

$$f(x) = \sum_{i=1}^{n}\left[100\left(x_{i+1} - x_i^2\right)^2 + (x_i - 1)^2\right]$$

Being $x_i \in [-30, 30]$ and $f_{\min}(x^*) = 0.0$

- ***F6: Step***

$$f(x) = \sum_{i=1}^{n}\lfloor |x_i| \rfloor$$

Being $x_i \in [-100, 100]$ and $f_{\min}(x^*) = 0.0$

- ***F7: Quartic***

$$f(x) = \sum_{i=1}^{n} i x_i^4 + rand[0, 1)$$

Being $x_i \in [-1.28, 1.28]$ and $f_{\min}(x^*) = 0.0$

- ***F8: Schwefel***

$$f(x) = -\sum_{i=1}^{n}\left[x_i \sin(\sqrt{|x_i|})\right]$$

Being $x_i \in [-500, 500]$ and $f_{\min}(x^*) = -418.982887272433799n$

- ***F9: Rastrigin***

$$f(x) = \sum_{i=1}^{n}[x_i^2 - 10\cos(2\pi x_i) + 10]$$

Being $x_i \in [-5.12, 5.12]$ and $f_{\min}(x^*) = 0.0$

- ***F:10 Ackley***

$$f(x) = -20e^{\left(-2\times\sqrt{\frac{1}{n}\sum_{i=1}^{n} x_i^2}\right)} - e^{\left[\frac{1}{n}\sum_{i=1}^{n}\cos(2\pi x_i)\right]} + 20 + e^{(1)}$$

Being $x_i \in [-32, 32]$ and $f_{\min}(x^*) = 0.0$.

In Fig. 1 the mathematical functions used are presented graphically.

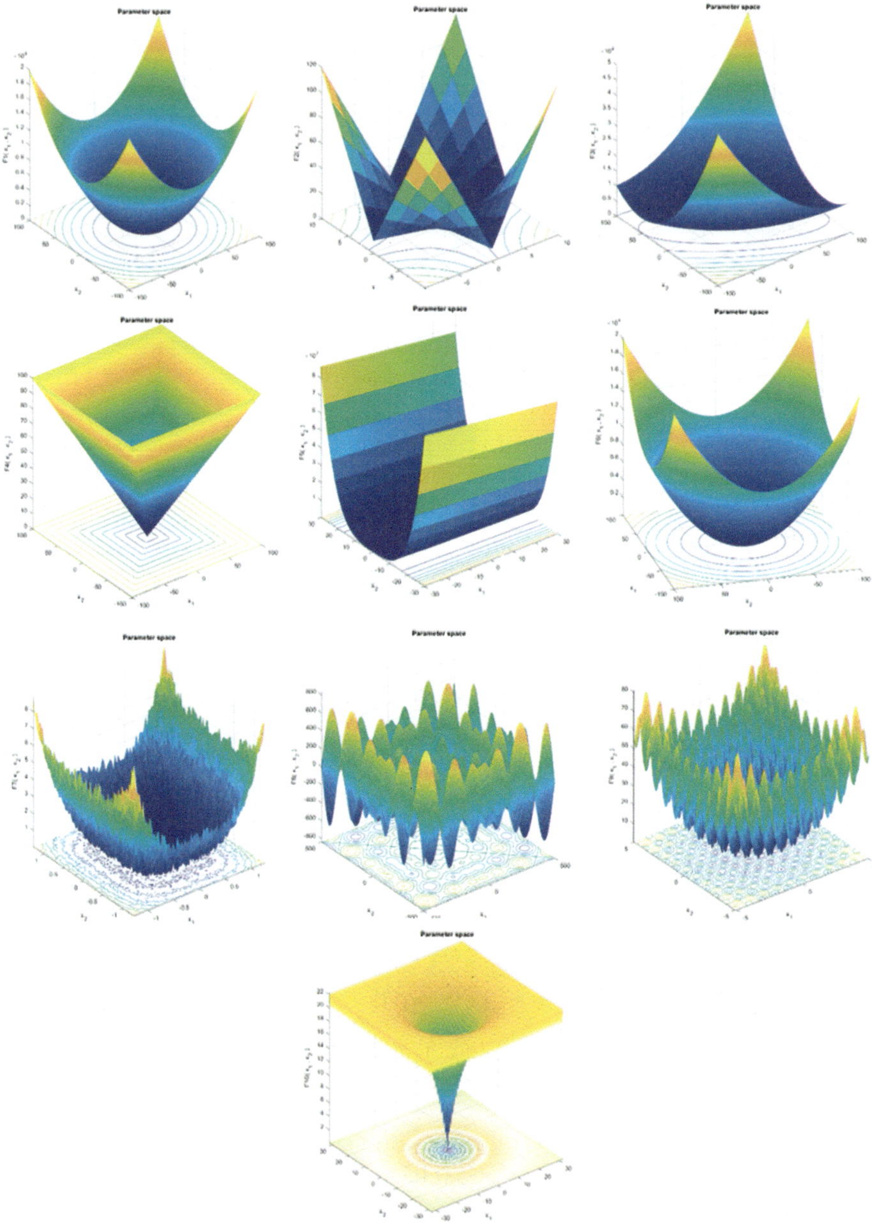

Fig. 1 Graphical representation of benchmark functions

4 Results

In this section, a comparison between the BSA and GTO algorithms is presented in order to analyze the performance of both methods. The experimentation is carried out using 10 mathematical functions, which implement 30 independent runs using a population of 30 individuals and 500 iterations, in addition, the number of dimensions is varied, using 30, 100, 500 and 1000 dimensions, respectively.

The parameters used in the GTO algorithm are:

- Beta: 3
- W: 0.8
- P: 0.03.

The parameters used in the GTO algorithm are:

- a1: 1
- a2: 1
- c1: 1.5
- c2: 1.5
- FQ: 3.

It is worth mentioning that the parameters listed above were taken from the original methods.

In Table 1, the results obtained from the different runs carried out with each of the mathematical functions are presented, and from which the best and worst values, the average of the 30 experiments and the standard deviation are listed. For this study, different dimensions are analyzed. It can be observed that with this algorithm 3 of the 10 functions reached their minimum value.

The results obtained from the experimentation carried out with the BSA algorithm are presented in Table 2. In a similar way to the previous experimentation, the best and worst value obtained are placed, in addition to the average of the 30 runs and the standard deviation. It can be analyzed that with this algorithm larger values were obtained and even only in one of the functions was the minimum value reached.

In order to clarify the comparison made, the results obtained are statistically analyzed, for this, the Z test is used.

The parameters used are:

Ho: $\mu 1 \leq \mu 2$.
Ha: $\mu 1 > \mu 2$.
Critical Value: -1.645.
Alpha: 0.05.
Confidence interval: 95%.

Table 3 shows the results of the Z test, where it can be seen that in 5 of the 10 mathematical functions there is significant evidence to conclude that the GTO method provides better results when tested with 30 dimensions.

Table 1 Results obtained with GTO algorithms

Function	Dimensions	GTO			
		30	100	500	1000
F1	Best	0.00E00	0.00E00	0.00E00	0.00E00
	Worst	0.00E00	0.00E00	0.00E00	0.00E00
	Average	0.00E00	0.00E00	0.00E00	0.00E00
	STD	0.00E00	0.00E00	0.00E00	0.00E00
F2	Best	8.67E−207	1.21E−205	3.03E−200	5.53E−203
	Worst	6.71E−192	4.87E−190	3.43E−187	1.79E−187
	Average	2.44E−193	1.84E−191	1.14E−188	8.29E−189
	STD	0.00E00	0.00E00	0.00E00	0.00E00
F3	Best	0.00E00	0.00E00	0.00E00	0.00E00
	Worst	0.00E00	0.00E00	0.00E00	0.00E00
	Average	0.00E00	0.00E00	0.00E00	0.00E00
	STD	0.00E00	0.00E00	0.00E00	0.00E00
F4	Best	7.31E−208	4.32E−205	3.92E−206	7.13E−202
	Worst	6.75E−192	3.29E−188	4.06E−183	5.21E−188
	Average	2.57E−193	1.68E−189	1.35E−184	1.90E−189
	STD	0.00E00	0.00E00	0.00E00	0.00E00
F5	Best	4.24E−06	3.01E−07	1.51E−05	3.20E−03
	Worst	2.45E+01	9.49E+01	1.26E+01	1.60E+01
	Average	1.58E+00	3.24E+00	1.69E+00	1.85E+00
	STD	6.02E+00	1.73E+01	2.58E+00	3.09E+00
F6	Best	4.79E−10	2.06E−04	2.40E−04	3.08E−05
	Worst	1.04E−06	2.82E−02	2.30E+00	3.11E+00
	Average	1.48E−07	6.46E−03	3.87E−01	6.77E−01
	STD	2.38E−07	6.92E−03	5.11E−01	7.02E−01
F7	Best	6.15E−06	1.53E−05	2.05E−05	6.62E−06
	Worst	3.35E−04	5.11E−04	2.57E−04	3.56E−04
	Average	7.55E−05	1.47E−04	1.01E−04	1.04E−04
	STD	7.09E−05	1.17E−04	7.58E−05	8.01E−05
F8	Best	−1.26E+04	−4.19E+04	−2.09E+05	−4.19E+05
	Worst	−1.26E+04	−4.19E+04	−2.09E+05	−4.18E+05
	Average	−1.26E+04	−4.19E+04	−2.09E+05	−4.19E+05
	STD	1.11E−04	1.67E+00	3.10E+01	2.45E+02
F9	Best	0.00E00	0.00E00	0.00E00	0.00E00
	Worst	0.00E00	0.00E00	0.00E00	0.00E00
	Average	0.00E00	0.00E00	0.00E00	0.00E00

(continued)

Table 1 (continued)

Function	Dimensions	GTO			
		30	100	500	1000
	STD	0.00E00	0.00E00	0.00E00	0.00E00
F10	Best	8.88E−16	8.88E−16	8.88E−16	8.88E−16
	Worst	8.88E−16	8.88E−16	8.88E−16	8.88E−16
	Average	8.88E−16	8.88E−16	8.88E−16	8.88E−16
	STD	0.00E00	0.00E00	0.00E00	0.00E00

The results of the Z test when experimenting with 100 dimensions are presented in Table 4, for this case 4 of the 10 mathematical functions used have significant evidence to conclude that the GTO method presents better results than BSA.

Analyzing the mathematical functions when tested with 500 and 1000 dimensions, the mathematical functions F5, F6 and F8 present significant evidence to determine that GTO provides better results than BSA. This results are presented in Tables 5 and 6 respectively.

5 Conclusions and Future Work

It is quite interesting and very helpful to carry out this type of comparison, since it broadens the panorama of the performance of the metaheuristics, as well as knowing which area they are more inclined towards in solving problems. In this case, a comparison was made of a well-known algorithm in our work, such as the BSA, and a new proposal for a bio-inspired algorithm was taken, which is the Artificial Gorilla Troops Optimizer, which is based on the behavior of gorillas within their habitat. In the study carried out, changing the dimensions of each of the mathematical functions used is experimented with, in order to analyze the results obtained, when doing this, it was observed that in certain functions the results were very similar and we can say that the GTO was significantly better in half of the benchmark functions.

As future work, it is intended to continue working with the GTO algorithm, performing a dynamic adjustment of parameters using fuzzy logic, with which we can make other comparisons and even try other optimization problems to analyze the performance of the method. We can also consider other optimization approaches, as in [28–30].

Table 2 Results obtained with BSA algorithm

Function	Dimensions	BSA			
		30	100	500	1000
F1	Best	1.17E−232	9.16E−232	3.57E−234	1.09E−235
	Worst	8.87E−211	5.10E−212	2.33E−203	1.11E−214
	Average	2.97E−212	1.72E−213	7.78E−205	4.22E−216
	STD	0.00E00	0.00E00	0.00E00	0.00E00
F2	Best	1.67E−116	7.62E−117	3.36E−115	1.07E−16
	Worst	3.58E−108	7.62E−117	1.83E−103	1.47E−01
	Average	2.56E−109	2.13E−105	6.17E−105	5.93E−03
	STD	7.11E−109	1.17E−104	3.33E−104	2.70E−02
F3	Best	4.49E−233	1.50E−231	2.76E−234	2.76E−240
	Worst	3.28E−214	9.72E−210	2.76E−234	2.55E−209
	Average	1.13E−215	6.43E−211	4.91E−210	8.61E−211
	STD	0.00E00	0.00E00	0.00E00	0.00E00
F4	Best	1.65E−115	1.37E−116	7.80E−119	6.52E−114
	Worst	3.12E−106	3.89E−108	1.41E−104	2.55E−105
	Average	1.66E−107	3.66E−109	5.49E−106	8.89E−107
	STD	5.95E−107	1.09E−108	2.60E−105	4.66E−106
F5	Best	2.89E+01	9.88E+01	4.99E+02	9.99E+02
	Worst	2.90E+01	9.90E+01	4.99E+02	9.99E+02
	Average	2.89E+01	9.89E+01	4.99E+02	9.99E+02
	STD	2.59E−02	3.74E−02	3.81E−02	3.16E−02
F6	Best	5.00E+00	2.09E+01	1.22E+02	2.46E+02
	Worst	7.08E+00	2.46E+01	1.25E+02	2.49E+02
	Average	6.15E+00	2.35E+01	1.24E+02	2.48E+02
	STD	5.16E−01	8.48E−01	4.90E−01	7.72E−01
F7	Best	1.12E−05	9.54E−06	1.32E−05	6.94E−06
	Worst	9.50E−04	4.92E−04	6.95E−04	3.36E−04
	Average	1.92E−04	1.32E−04	1.37E−04	1.36E−04
	STD	1.84E−04	1.13E−04	1.37E−04	9.23E−05
F8	Best	−7.24E+03	−1.57E+04	−4.43E+04	−5.68E+04
	Worst	−3.66E+03	−7.22E+03	−2.07E+04	−2.78E+04
	Average	−5.83E+03	−1.19E+04	−2.77E+04	−3.83E+04
	STD	7.81E+02	2.07E+03	5.64E+03	7.43E+03
F9	Best	0.00E00	0.00E00	0.00E00	0.00E00
	Worst	0.00E00	0.00E00	0.00E00	0.00E00
	Average	0.00E00	0.00E00	0.00E00	0.00E00

(continued)

Table 2 (continued)

Function	Dimensions	BSA			
		30	100	500	1000
	STD	0.00E00	0.00E00	0.00E00	0.00E00
F10	Best	8.88E−16	8.88E−16	8.88E−16	8.88E−16
	Worst	8.88E−16	8.88E−16	8.88E−16	8.88E−16
	Average	8.88E−16	8.88E−16	8.88E−16	8.88E−16
	STD	0.00E00	0.00E00	0.00E00	0.00E00

Table 3 Results of Z-test using 30 dimensions

Function	GTO		BSA		Ztest	Result
	Average	StdDev	Average	StdDev		
F1	0.00E00	0.00E00	2.97E−212	0.00E00	0	NS
F2	2.44E−193	0.00E00	2.56E−109	7.11E−109	−1.9721	S
F3	0.00E00	0.00E00	1.13E−215	0.00E00	0	NS
F4	2.57E−193	0.00E00	1.66E−107	5.95E−107	−1.5281	NS
F5	1.58E+00	6.02E+00	2.89E+01	2.59E−02	−24.8565	S
F6	1.48E−07	2.38E−07	6.15E+00	5.16E−01	−65.2809	S
F7	7.55E−05	7.09E−05	1.92E−04	1.84E−04	−3.236	S
F8	−1.26E+04	1.11E−04	−5.83E+03	7.81E+02	−47.4786	S
F9	0.00E00	0.00E00	0.00E00	0.00E00	0	NS
F10	8.88E−16	0.00E00	8.88E−16	0.00E00	0	NS

Table 4 Results of Z-test using 10 dimensions

Function	GTO		BSA		Ztest	Result
	Average	StdDev	Average	StdDev		
F1	0.00E00	0.00E00	1.72E−213	0.00E00	0	NS
F2	1.84E−191	0.00E00	2.13E−105	1.17E−104	−0.9971	NS
F3	0.00E+00	0.00E00	6.43E−211	0.00E+00	0	NS
F4	1.68E−189	0.00E00	3.66E−109	1.09E−108	−1.8391	S
F5	3.24E+00	3.24E+00	9.89E+01	3.74E−02	−161.7026	S
F6	6.46E−03	6.92E−03	2.35E+01	8.48E−01	−151.7935	S
F7	1.47E−04	1.17E−04	1.32E−04	1.13E−04	0.5051	NS
F8	−4.19E+04	1.67E+00	−1.19E+04	2.07E+03	−79.3801	S
F9	0.00E00	0.00E00	0.00E00	0.00E00	0	NS
F10	8.88E−16	0.00E00	8.88E−16	0.00E00	0	NS

Table 5 Results of Z-test using 500 dimensions

Function	GTO		BSA		Ztest	Result
	Average	StdDev	Average	StdDev		
F1	0.00E00	0.00E00	7.78E−205	0.00E00	0	NS
F2	1.14E−188	0.00E00	6.17E−105	3.33E−104	−1.0148	NS
F3	0.00E+00	0.00E00	4.91E−210	0.00E00	0	NS
F4	1.35E−184	0.00E00	5.49E−106	2.60E−105	−1.1565	NS
F5	1.69E+00	2.58E+00	4.99E+02	3.81E−02	−1.06E+03	S
F6	3.87E−01	5.11E−01	1.24E+02	4.90E−01	−956.3352	S
F7	1.01E−04	7.58E−05	1.37E−04	1.37E−04	−1.259	NS
F8	−2.09E+05	3.10E+01	−2.77E+04	5.64E+03	−176.065	S
F9	0.00E00	0.00E00	0.00E00	0.00E00	0	NS
F10	8.88E−16	0.00E+00	8.88E−16	0.00E+00	0	NS

Table 6 Results of Z-test using 1000 dimensions

Function	GTO		BSA		Ztest	Result
	Average	StdDev	Average	StdDev		
F1	0.00E00	0.00E00	4.22E−216	0.00E00	0	NS
F2	8.29E−189	0.00E00	5.93E−03	2.70E−02	−1.203	NS
F3	0.00E00	0.00E00	8.61E−211	0.00E00	0	NS
F4	1.90E−189	0.00E00	8.89E−107	4.66E−106	−1.0449	NS
F5	1.85E+00	3.09E+00	9.99E+02	3.16E−02	−1767.421	S
F6	6.77E−01	7.02E−01	2.48E+02	7.72E−01	−1298.235	S
F7	1.04E−04	8.01E−05	1.36E−04	9.23E−05	−1.434	NS
F8	−4.19E+05	2.45E+02	−3.83E+04	7.43E+03	−280.491	S
F9	0.00E00	0.00E00	0.00E00	0.00E00	0	NS
F10	8.88E−16	0.00E00	8.88E−16	0.00E00	0	NS

References

1. X.S. Yang, Genetic algorithms, in *Nature-Inspired Optimization Algorithms* (2021), pp. 91–100
2. L. Cheng, X.H. Wu, Y. Wang, Artificial flora (AF) optimization algorithm. Appl. Sci. **8**(3), 329 (2018)
3. J. Pierezan, L. dos Santos Coelho, Coyote optimization algorithm: a new metaheuristic for global optimization problems, in *2018 IEEE Congress on Evolutionary Computation, CEC 2018—Proceedings* (2018), pp. 1–8
4. F.A. Hashim, K. Hussain, E.H. Houssein, M.S. Mabrouk, W. Al-Atabany, Archimedes optimization algorithm: a new metaheuristic algorithm for solving optimization problems. Appl. Intell. **51**(3), 1531–1551 (2020)
5. M.A. Al-Betar, Z.A.A. Alyasseri, M.A. Awadallah, I. Abu Doush, Coronavirus herd immunity optimizer (CHIO). Neural Comput. Appl. **33**(10), 5011–5042 (2021)

6. I. Miramontes, P. Melin, G. Prado-Arechiga, Fuzzy system for classification of nocturnal blood pressure profile and its optimization with the crow search algorithm. Soft Comput. Appl. 23–34 (2021)
7. I. Miramontes, P. Melin, G. Prado-Arechiga, Comparative study of bio-inspired algorithms applied in the optimization of fuzzy systems, in *Hybrid Intelligent Systems in Control, Pattern Recognition and Medicine* (Springer International Publishing, Cham, 2020), pp. 219–231
8. O. Carvajal, P. Melin, I. Miramontes, G. Prado-Arechiga, Optimal design of a general type-2 fuzzy classifier for the pulse level and its hardware implementation. Eng. Appl. Artif. Intell. **97**, 104069 (2021)
9. I. Miramontes, C.J. Guzman, P. Melin, G. Prado-Arechiga, Optimal design of interval type-2 fuzzy heart rate level classification systems using the bird swarm algorithm. Algorithms **11**(12), 1–35 (2018)
10. J.C. Guzman, P. Melin, G. Prado-Arechiga, Design of an optimized fuzzy classifier for the diagnosis of blood pressure with a new computational method for expert rule optimization. Algorithms **10**(3), 1–27 (2017)
11. I. Miramontes, P. Melin, G. Prado-Arechiga, Particle swarm optimization of modular neural networks for obtaining the trend of blood pressure, in *Intuitionistic and Type-2 Fuzzy Logic Enhancements in Neural and Optimization Algorithms: Theory and Applications* (Springer International Publishing, Cham, 2020), pp. 225–236
12. E. Bernal, O. Castillo, J. Soria, F. Valdez, Fuzzy galactic swarm optimization with dynamic adjustment of parameters based on fuzzy logic. SN Comput. Sci. **1**(1), 59 (2020)
13. X.-B. Meng, X.Z. Gao, L. Lu, Y. Liu, H. Zhang, A new bio-inspired optimisation algorithm: bird swarm algorithm. J. Exp. Theor. Artif. Intell. **28**(4), 673–687 (2016)
14. M.K. Gude, U. Salma, Artificial gorilla troops optimizer for tuning power system stabilizer control parameters, in *2021 IEEE 2nd International Conference on Electrical Power and Energy Systems* (2021), pp. 1–5
15. L. Xiang, Z. Deng, A. Hu, Forecasting short-term wind speed based on IEWT-LSSVM model optimized by bird swarm algorithm. IEEE Access **7**, 59333–59345 (2019)
16. X. Wang, Y. Deng, H. Duan, Edge-based target detection for unmanned aerial vehicles using competitive bird swarm algorithm. Aerosp. Sci. Technol. **78**, 708–720 (2018)
17. S. Wang, S. Liu, X. Che, Z. Wang, J. Zhang, D. Kong, Recognition of polycyclic aromatic hydrocarbons using fluorescence spectroscopy combined with bird swarm algorithm optimization support vector machine. Spectrochim. Acta Part A Mol. Biomol. Spectrosc. **224**, 117404 (2020)
18. K. Wu et al., A joint planning method for substations and lines in distribution systems based on the parallel bird swarm algorithm. Energies **11**(10), 2669 (2018)
19. P. Melin, I. Miramontes, O. Carvajal, G. Prado-Arechiga, Fuzzy dynamic parameter adaptation in the bird swarm algorithm for neural network optimization. Soft Comput. 1–18 (2022)
20. A. Ramadan, M. Ebeed, S. Kamel, A.M. Agwa, M. Tostado-véliz, The probabilistic optimal integration of renewable distributed generators considering the time-varying load based on an artificial gorilla troops optimizer. Energies **15**(4), 1302 (2022)
21. M.A. El-Dabah, S. Kamel, M. Khamies, H. Shahinzadeh, G.B. Gharehpetian, Artificial gorilla troops optimizer for optimum tuning of TID based power system stabilizer, in *9th Iranian Joint Congress on Fuzzy and Intelligent Systems (CFIS)* (2022), pp. 1–5
22. M. Abdel-Basset, D. El-Shahat, K.M. Sallam, K. Munasinghe, Parameter extraction of photovoltaic models using a memory-based improved gorilla troops optimizer. Energy Convers. Manage. **252**, 115134 (2022)
23. R. Ara Parvin, B. Jana, S. Acharyya, Predicting critical genes from genomic data using artificial gorilla troops optimizer. Easy chair preprints (2022). https://easychair-www.easychair.org/publications/preprint/zd2B. Accessed 23 Apr 2022
24. L. Abualigah, D. Yousri, M. Abd Elaziz, A.A. Ewees, M.A.A. Al-qaness, A.H. Gandomi, Aquila optimizer: a novel meta-heuristic optimization algorithm. Comput. Ind. Eng. **157**, 1–37 (2021)

25. F.A. Hashim, E.H. Houssein, K. Hussain, M.S. Mabrouk, W. Al-Atabany, Honey badger algorithm: new metaheuristic algorithm for solving optimization problems. Math. Comput. Simul. **192**, 84–110 (2022)
26. S. Kaur, L.K. Awasthi, A.L. Sangal, G. Dhiman, Tunicate swarm algorithm: a new bio-inspired based metaheuristic paradigm for global optimization. Eng. Appl. Artif. Intell. **90**, 1–29 (2020)
27. A.K. Bairwa, S. Joshi, D. Singh, Dingo optimizer: a nature-inspired metaheuristic approach for engineering problems. Math. Probl. Eng. **2021**, 1–12 (2021)
28. P. Melin, D. Sánchez, O. Castillo, Genetic optimization of modular neural networks with fuzzy response integration for human recognition. Inf. Sci. **197**, 1–19 (2012)
29. P. Melin, D. Sánchez, Multi-objective optimization for modular granular neural networks applied to pattern recognition. Inf. Sci. **460–461**, 594–610 (2018)
30. D. Sánchez, P. Melin, O. Castillo, Optimization of modular granular neural networks using a firefly algorithm for human recognition. Eng. Appl. AI **64**, 172–186 (2017)

Particle Swarm Optimization of Convolutional Neural Networks for Diabetic Retinopathy Classification

Patricia Melin, Daniela Sánchez, and Rodrigo Cordero-Martínez

Abstract This work proposes convolutional neural networks (CNNs) and particle swarm optimization (PSO) for diabetic retinopathy classification. Particle swarm optimization seeks to minimize the classification error, designing the convolutional neural network using different preprocessing methods to compare results. The parameters optimized to design the CNN are the number of convolutional layers, filters with their filters size, pool size, algorithm for the learning process, number of fully connecter layers with their number of neurons, batch size, and finally the number of epochs. Among the preprocessing applied are: extraction of the retina and applying a histogram equalization to the red, green, and blue channels. The database used to test the proposed method is APTOS 2019, where the best result achieved is 96.59%, with an average of 95.33%.

Keywords Diabetic retinopathy · Convolutional neural network · Particle swarm optimization · Classification

1 Introduction

In recent years, diabetes has become a public health disease. There are millions of people in the world who suffer from it. Some research has been carried out seeking its relationship with other conditions and how it can complicate them, as is the case for heart failure, hypertension, obesity, pancreatic cancer [1]. Type 2 diabetes is also related to retinopathy, nephropathy, and neuropathy [2]. The leading cause of blindness and visual impairment worldwide is diabetic retinopathy. It is classification is into non-proliferative and proliferative. A patient with a non-proliferative stage has

P. Melin (✉) · D. Sánchez · R. Cordero-Martínez
Tijuana Institute of Technology, Tijuana, Mexico
e-mail: pmelin@tectijuana.mx

D. Sánchez
e-mail: daniela.sanchez@tectijuana.edu.mx

R. Cordero-Martínez
e-mail: rodrigo.cordero201@tectijuana.edu.mx

O. Castillo and P. Melin (eds.), *Fuzzy Logic and Neural Networks for Hybrid Intelligent System Design*, Studies in Computational Intelligence 1061,
https://doi.org/10.1007/978-3-031-22042-5_14

retinal hemorrhages and hard and soft exudates. If the severity increases, the proliferative stage occurs with ischemic with neovascularization and traction elevation of the retina. Both stages are associated with diabetic macular edema, the most common cause of vision loss [3]. Nowadays, intelligence techniques have tried to detect or predict if a person has diabetes. In [4], novel diabetes classifying model based on Convolutional Long Short-term Memory (Conv-LSTM) is proposed using the Pima Indians Diabetes Database (PIDD). In [5], the author proposed a features extraction using stacked autoencoders and performing classification with a Deep Neural Network framework. A virtual doctor using artificial intelligence, which can interact with patients using speech recognition to predict type 2 diabetes mellitus, is proposed [6]. Concerning diabetic retinopathy (DR), different works have also been developed to perform its classification and association with other diseases. A study of the association of different degrees of damages of DR and hearing loss is presented in [7]. In [8], authors proposed the automatic DR detection using a deep learning hybrid on pre-trained Inception-ResNet-v2. In [9], a design of the Source-Free Transfer Learning (SFTL) method for diabetic retinopathy detection is proposed. The method consists of 2 modules; the first one is the target generation module, and the second one is a collaborative consistency module. Most of the proposed methods are based on pre-trained models. For this reason, in this work, the use of an optimization method that allows obtaining the optimal convolutional neural networks (CNNs) architecture is proposed. In the literature, there are different optimization methods based on different approaches [10], such as probabilistic, music, mathematical, human [11] or swarm intelligence [12–14]. The main inspiration in the swarm intelligence is based on the nature of fish schools, bird flocks, ant colonies, and other animals with self-organization capabilities [15–17]. A hybrid system is proposed in this work, where CNNs and particle swarm optimization are combined for diabetic retinopathy classification. The proposed PSO designs the convolutional neural networks architectures to perform binary and multiclass diabetic retinopathy classification.

This work is structured as below: Sect. 2 contains the basic concepts applied to perform the method, in Sect. 3, the general description of the method is shown, Sect. 4 contains experiments results obtained with the proposed method. Finally, Conclusions and References are presented.

2 Basic Concepts

In this section, the intelligence techniques used in the proposed method are presented.

2.1 Convolutional Neural Networks

Artificial neural networks (ANNs) are widely known as intelligence techniques that simulate the learning process [18, 19]. An artificial neural network has 2 main attributes: its parallel distributed structure and its ability to learn what allows the generalized producing outputs to inputs not learned by the artificial neural networks [20, 21]. Convolutional neural networks (CNNs) are based on conventional neural networks [22]. This kind of neural network uses a convolution process, where inputs are multiplied using a filter (also known as kernel) with a size m × m, where *m* is an integer value, usually between 3 and 5 [23], producing an output map. The pooling layer allows reducing information. There are different pooling layers such as max, min, and average layers. Finally, the fully connected layers learn the information [24, 25].

2.2 Particle Swarm Optimization

Particle swarm optimization (PSO) was proposed in [26, 27], and it is based on swarm behavior, where a swarm is formed of particles, and each one has a certain number of dimensions. These dimensions represent parameters that allow a possible solution to a problem that moves stochastically in the search space [15, 28]. The next position of each particle is calculated by:

$$x_{id}(t+1) = x_i(t) + v_i(t+1) \tag{1}$$

where $x_{id}(t)$ denotes the actual position of the particle *i*, dimension *d*, at time *t*. A velocity $v_i(t + 1)$ is set to determine the next position. The original algorithm was improved using an inertia weight (*w*). With a big inertia weight value, a global exploration is allowed. Meanwhile, a small value allows a local exploration [29].

3 General Description of the Proposed Method

The proposed method designs CNN architectures for diabetic retinopathy classification. The proposed method can be applied to multiclass (healthy retina or with a stage of damage) or binary classification (healthy retina or not). The optimal parameters are found using particle swarm optimization. After each convolution layer follows a max-pooling layer to reduce image size. The proposed optimization also determines the pool size. The proposed method applied to multiclass classification is shown in Fig. 1.

3.1 Description of PSO

The particle swarm optimization allows the designing of the CNN architectures: the number of convolutional layers, filters with their filters size, pool size, algorithm for

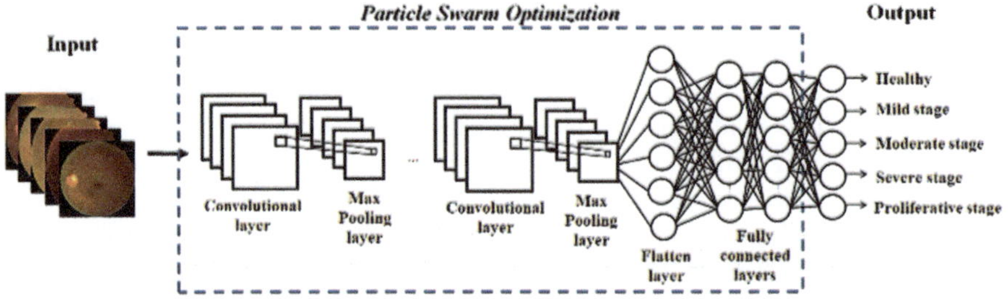

Fig. 1 Proposed method for multiclass classification

the learning process, number of fully connecter, number of neurons, batch size, and the number of epochs. The configuration of the PSO parameters can be observed in Table 1. These values are based on a previous work where modular granular neural networks were optimized [30].

The accuracy equation used in this work is given by the expression:

$$Accuracy = (TP + TN)/(TP + FP + TN + FN) \tag{2}$$

where TP is True Positive, TN is True Negative, FP is False Positive, and FN is False Negative. The proposed optimization seeks to minimize the error of classification, and the objective function is given by:

$$f = 1 - (TP + TN)/(TP + FP + TN + FN) \tag{3}$$

The rectified linear activation function (ReLU) is used before the output layer. The values shown in Table 2 determined the search of the algorithm. For the batch size, the possible values are from 1 up to 4, which means 8, 16, 32, or 64. For the algorithm used to training phase, the possible values are from 1 up to 6, to select one of the six possible algorithms:

1. Adaptive Moment Estimation (Adam).
2. Adamax based on Adam and the infinity norm.
3. Nesterov-accelerated Adam (Nadam).

Table 1 Parameter values for the PSO

Parameter	Value
Particles	10
Maximum iterations	30
Dimensions	13
C_1	1.5
C_2	2
w	0.8

Table 2 Definition of the search space

Parameter	Value	
	Minimum	Maximum
Convolutional layers (CL)	1	3
Number of filters CL #1	8	16
Number of filters CL #2	16	32
Number of filters CL #3	32	64
Size filter	3	5
Pool size Max-pooling layer 1	3	5
Pool size Max-pooling layer 2	3	5
Pool size Max-pooling layer 3	3	5
Fully connected layers	1	3
Neurons of the fully connected layers	10	300
Epoch	5	10
Batch size	1	4
Learning algorithm	1	6

4. Stochastic gradient descent method based on adaptive learning rate per dimension (Adadelta).
5. Adaptative Gradient Algorithm (Adagrad).
6. Stochastic gradient descent (SGD).

Each particle has 17 dimensions with the information needed to design a CNN. The structure of the particle is shown in Fig. 2. In Fig. 3, the flowchart of the proposed method is shown.

3.2 Database

The APTOS 2019 database [31] contains 5590 images with noise and different sizes, where only 3662 have a tag of its classification. These images are used to prove the proposed method (training and testing phase), with a size of 200 × 200 pixels. A sample of this database is shown in Fig. 4. The database contains 5 classes:

- Healthy Retina
- No Proliferative Mild Stage
- No Proliferative Moderate Stage
- No Proliferative Severe Stage
- Proliferative Stage.

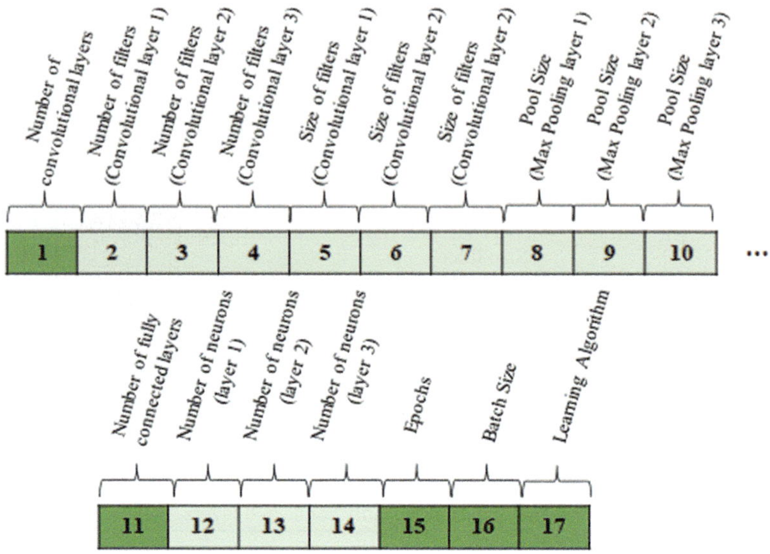

Fig. 2 Representation of a particle

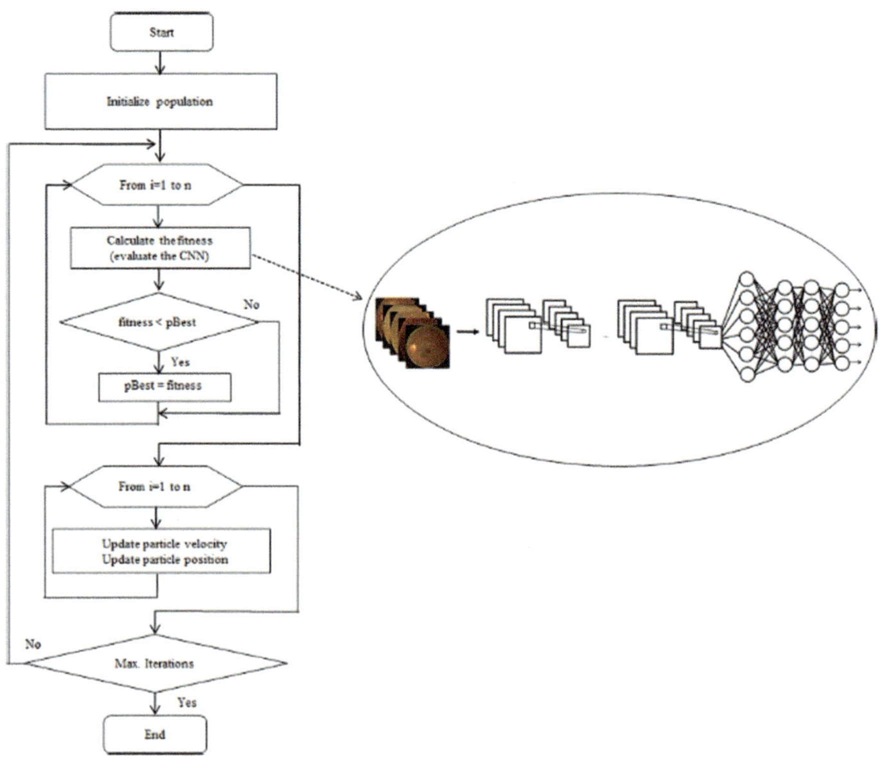

Fig. 3 Flowchart of the proposed method

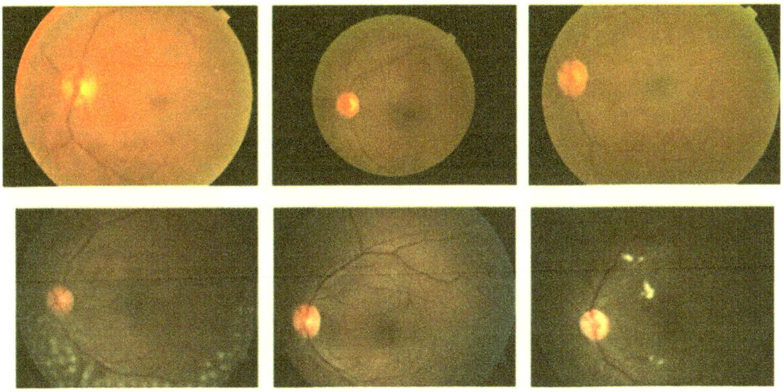

Fig. 4 Sample of APTOS 2019 database

Table 3 Images per class (APTOS 2019 database)

Class	1	2	3	4	5
Images	1805	370	999	193	295

For both classifications: binary and multiclass, 72% of the images are for training, 8% for validation, and 20% for testing. The number of images for each class is shown in Table 3.

Pre-processing #1: This pre-processing has been used before in [8]. In this pre-processing, the image has integer values in its pixels between 0 and 255. A new image is created using the original image; if a pixel value is less than 20, the new pixel value equals 0. On the contrary, it is equal to 1, generating a new binary image. This image allows detecting the position of the largest object to obtain the central pixel of the retina with its height and width. The original image is recut to delete unnecessary information or noise with this information. A sample of this pre-processing is shown in Fig. 5.

Pre-processing #2: This pre-processing is based on the first one to extract the retina image. After the red, green, and blue channels are extracted, histogram equalization is applied to each one to be joined to generate a single image finally. A sample of this pre-processing is shown in Fig. 6.

4 Experimental Results

This section shows the results achieved using the two pre-processing previously described. For each pre-processing and type of classification (binary and multiclass), 30 runs are performed.

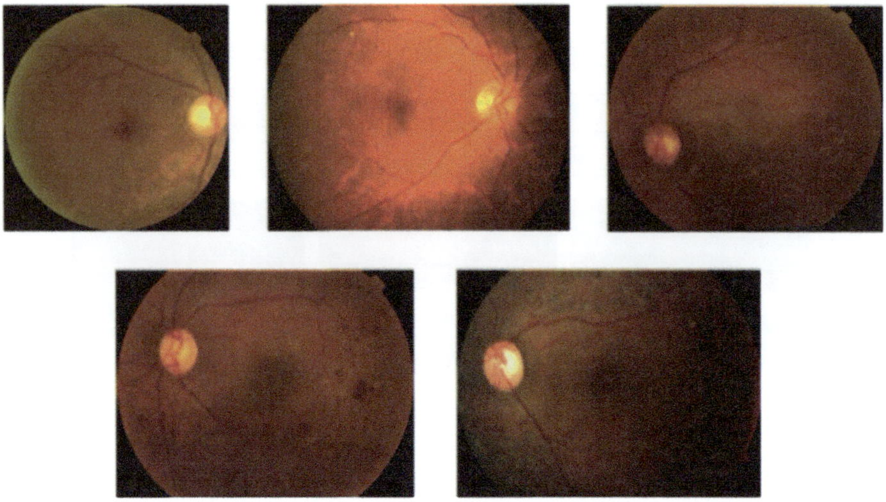

Fig. 5 Sample of pre-processing #1

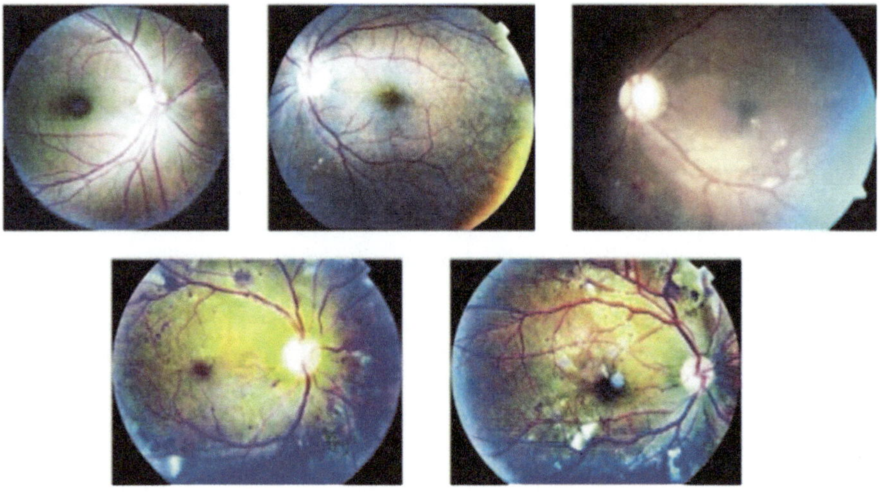

Fig. 6 Sample of pre-processing #2

4.1 Pre-processing #1

In this section, the results achieved using pre-processing #1 are shown.

Table 4 Best architecture (pre-processing #1, binary)

Parameters	Values
Convolutional layers	3
Number of filters and size per layer	8 (5 × 5) 16 (3 × 3) 64 (3 × 3)
Max-pooling filter size per layer	2 × 2 4 × 4 5 × 5
Fully connected layers (neurons)	1 (300)
Epoch	10
Batch size	8
Algorithm	Nadam
Error	0.0341
Accuracy (%)	96.59

Table 5 Results (pre-processing #1)

Best	96.59%
Average	95.33%
Worst	94.68%

4.1.1 Binary Classification

In Table 4, the best architecture for the binary classification using pre-processing #1 is shown. The best accuracy achieved is 96.59%, using 3 convolutional layers, with their respectively max-pooling layers and one fully connected layer.

A summary of the results achieved with pre-processing # 1 for binary classification is shown in Table 5, where the best, average, and worst accuracy values are shown. The average of convergence is shown in Fig. 7.

4.1.2 Multiclass Classification

Table 6 shows the best architecture for the multiclass classification using pre-processing #1. The best accuracy achieved is 77.35%, also using 3 convolutional layers, with their respectively max-pooling layers, but this architecture uses 3 fully connected layers.

A summary of the results achieved with pre-processing #1 for multiclass classification is shown in Table 7. The average of convergence using pre-processing #1 is shown in Fig. 8.

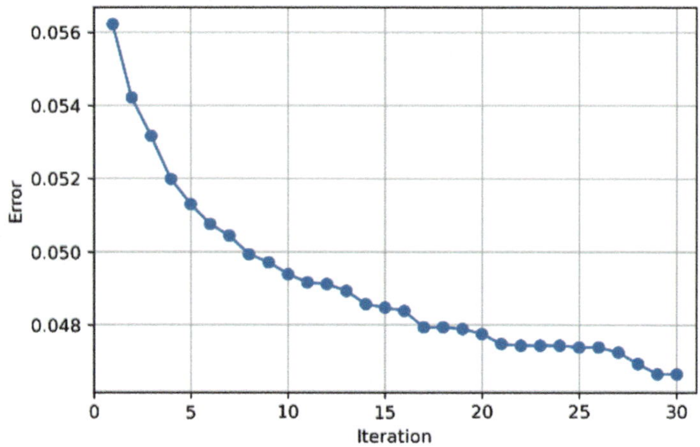

Fig. 7 Average of convergences (pre-processing #1, binary classification)

Table 6 Best architecture (pre-processing #1, multiclass)

Parameters	Values
Convolutional layers	3
Number of filters and size per layer	8 (4 × 4) 22 (2 × 2) 64 (4 × 4)
Max-pooling filter size per layer	4 × 4 3 × 3 4 × 4
Fully connected layers (neurons)	3 (276, 159, 179)
Epoch	8
Batch size	8
Algorithm	Adam
Error	0.2265
Accuracy (%)	77.35

Table 7 Results for multiclass classification (pre-processing #1)

Best	77.35%
Average	76.03%
Worst	74.90%

4.2 Pre-processing #2

In this section, the results achieved using pre-processing #2 are shown.

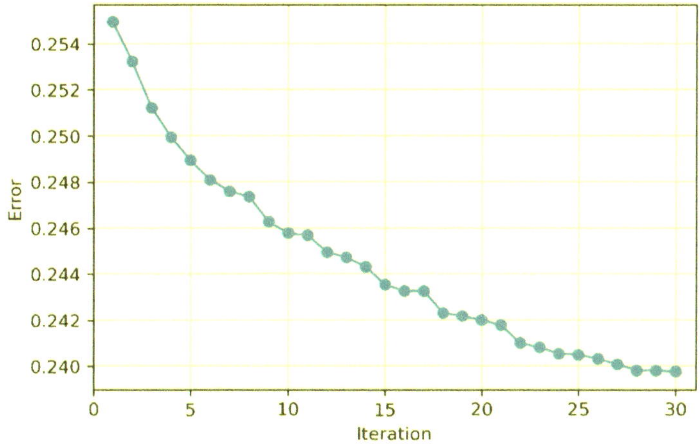

Fig. 8 Average of convergences (pre-processing #1, multiclass classification)

4.2.1 Binary Classification

In Table 8, the best architecture for the binary classification using pre-processing #2 is shown. The best accuracy achieved is 95.36%, using 2 convolutional layers with their max-pooling layers and one fully connected layer.

A summary of the results achieved with this pre-processing is shown in Table 9. The average of convergence is shown in Fig. 9.

Table 8 Best architecture (pre-processing #2, binary)

Parameters	Values
Convolutional layers	2
Number of filters and size per layer	16 (3 × 3) 31 (2 × 2)
Max-pooling filter size per layer	4 × 4 5 × 5
Fully connected layers (neurons)	1 (288)
Epoch	10
Batch size	16
Algorithm	Adamax
Error	0.0464
Accuracy (%)	95.36

Table 9 Results (pre-processing # 2)

Best	95.36%
Average	93.58%
Worst	93.04%

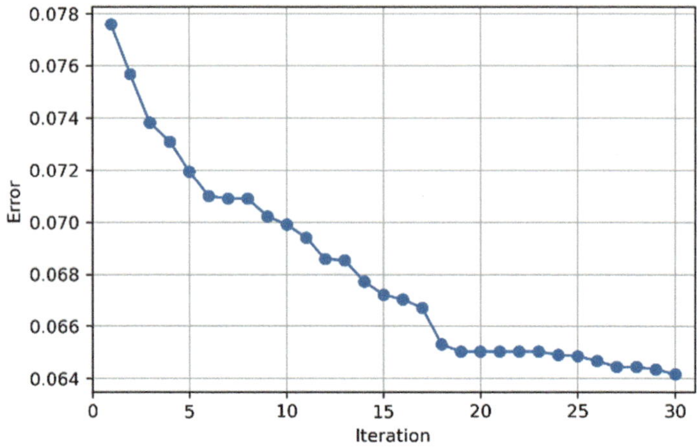

Fig. 9 Average of convergences (pre-processing #2, binary classification)

4.2.2 Multiclass Classification

Table 10 shows the best architecture for the multiclass classification using pre-processing #2. The best accuracy achieved is 74.76%, also using 2 convolutional layers, with their respectively max-pooling layers and 3 fully connected layers.

A summary of the results achieved with this pre-processing is shown in Table 11. The average of convergence is shown in Fig. 10.

Table 10 Best architecture (pre-processing #2, multiclass)

Parameters	Values
Convolutional layers	2
Number of filters and size per layer	8 (2 × 2) 32 (2 × 2)
Max-pooling filter size per layer per layer	3 × 3 4 × 2
Fully connected layers (neurons)	3 (249, 300, 76)
Epoch	6
Batch size	16
Algorithm	Adam
Error	0.2524
Accuracy (%)	74.76

Table 11 Summary of results (pre-processing #2)

Best	74.76%
Average	73.57%
Worst	72.58%

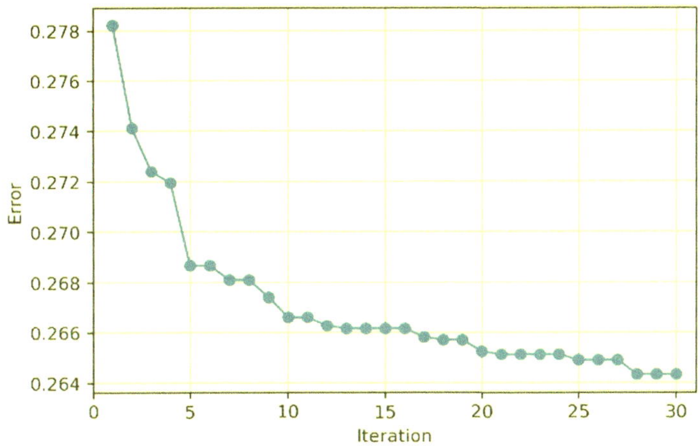

Fig. 10 Average of convergences (pre-processing #2, multiclass classification)

Table 12 Comparison of results (binary classification)

	Best (%)	Average (%)	Worst (%)
Pre-processing #1	96.59	95.33	94.68
Pre-processing #2	95.36	93.58	93.04

4.2.3 Summary of Results

Table 12 shows a summary of results, where the results achieved for the binary classification can be observed. The values are better using the first pre-processing.

A graphical comparison of the convergences for binary classification is shown in Fig. 11, where a better convergence for the first pre-processing is observed.

Table 13 shows a summary of results, where the results achieved for the multiclass classification can be observed. The values are better using the first pre-processing with a wider difference than in the binary classification.

A graphical comparison of the convergences for multiclass classification is shown in Fig. 12, where a better convergence for the first pre-processing is observed, even it can be observed that pre-processing #2 is stagnated in some iterations.

5 Conclusions

The PSO is used to design the convolutional neural networks, where each particle contains information: the number of convolutional layers, filters with their filters size, pool size, algorithm for the learning process, number of fully connecter layers, number of neurons, batch size, and the number of epochs. The proposed method combines CNNs and PSO for retinopathy classification. The results show that the

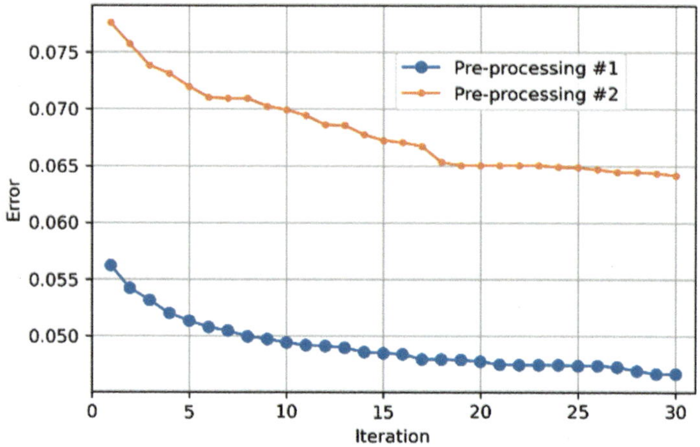

Fig. 11 Average of convergences (binary classification)

Table 13 Comparison of results (multiclass classification)

	Best (%)	Average (%)	Worst (%)
Pre-processing #1	77.35	76.03	74.90
Pre-processing #2	74.76	73.57	72.58

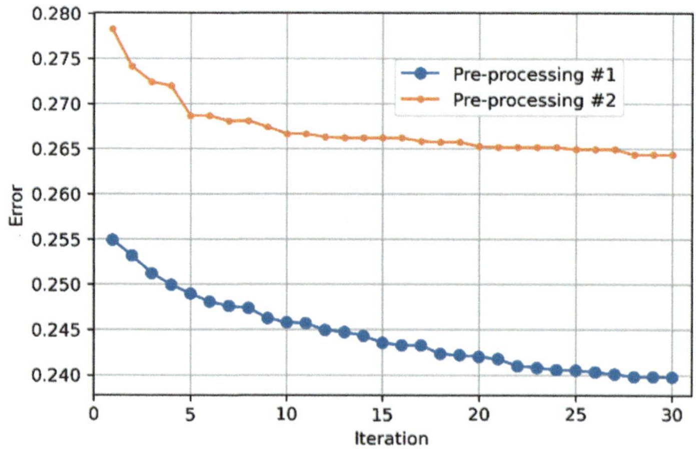

Fig. 12 Average of convergences (multiclass classification)

first pre-processing allowed for better results than the second one, for both kinds of classification: binary and multiclass. The proposed method achieved 96.59% of accuracy as the best results in the binary classification; meanwhile, for multiclass classification, 77.35% of accuracy. It does not mean that the second pre-processing is no good, perhaps the search space must be increased, and better results can be achieved. In future works, other pre-processing, and type of pooling layer will be implemented to compare results and increase the accuracy in the classification.

References

1. R.B. Prasad, E. Ahlqvist, L. Groop, Genetics of diabetes and diabetic complications, in *Diabetes Epidemiology, Genetics, Pathogenesis, Diagnosis, Prevention, and Treatment* (Springer, Berlin, 2018), pp. 81–140
2. A. Amutha, M.R. Anjana, U. Venkatesan, H. Ranjani, R. Unnikrishnan, K.M. Narayan, V. Mohan, M.K. Ali, Incidence of complications in young-onset diabetes: comparing type 2 with type 1 (the young diab study). Diabetes Res. Clin. Pract. **123**, 1–8 (2017)
3. K. Laud, U. Shabto, C. Tello, Diabetic retinopathy, in *Principles of Diabetes Mellitus* (Springer, Berlin, 2017), pp. 407–424
4. M. Rahman, D. Islam, R.J. Mukti, I. Saha, A deep learning approach based on convolutional LSTM for detecting diabetes. Comput. Biol. Chem. **88**, 1–10 (2020)
5. K. Kannadasan, D.R. Edla, V. Kuppili, Type 2 diabetes data classification using stacked autoencoders in deep neural networks. Clin. Epidemiol. Glob. Health **7**(4), 530–535 (2019)
6. S. Spänig, A. Emberger-Klein, J.P. Sowa, A. Canbay, K. Menrad, D. Heider, The virtual doctor: an interactive clinical-decision-support system based on deep learning for non-invasive prediction of diabetes. Artif. Intell. Med. **100**, 1–9 (2019)
7. Y. Alizadeh, M.M. Jalali, A. Sehati, Association of different severity of diabetic retinopathy and hearing loss in type 2 diabetes mellitus. Am. J. Otolaryngol. **43**(2), 1–6 (2022)
8. A.K. Gangwar, V. Ravi, Diabetic retinopathy detection using transfer learning and deep learning, in *Evolution in Computational Intelligence* (Springer, Berlin, 2020), pp. 679–689
9. C. Zhang, T. Lei, P. Chen, Diabetic retinopathy grading by a source-free transfer learning approach. Biomed. Signal Process. Control **73**, 1–10 (2022)
10. M. Dorigo, Optimization, learning and natural algorithms. Ph.D. thesis, Politecnico di Milano, Italian (1992)
11. D.E. Goldberg, *Genetic Algorithms in Search Optimization and Machine Learning* (Addison-Wesley, Boston, 1989)
12. X.S. Yang, Firefly algorithms for multimodal optimization, in *Proceedings of 5th Symposium on Stochastic Algorithms, Foundations and Applications* (2009)
13. R. Rajabioun, Cuckoo optimization algorithm. Appl. Soft Comput. J. **11**, 5508–5518 (2011)
14. E. Rashedi, H. Nezamabadi-Pour, S. Saryazdi, GSA: a gravitational search algorithm. Inf. Sci. **179**, 2232–2248 (2009)
15. K.E. Parsopoulos, M.N. Vrahatis, *Particle Swarm Optimization and Intelligence: Advances and Applications* (IGI Global, USA, 2010)
16. M. Clerc, *Particle Swarm Optimization* (Wiley-ISTE, London, 2013)
17. S. Mirjalili, S.M. Mirjalili, A. Lewis, Grey wolf optimizer. Adv. Eng. Softw. **69**, 46–61 (2014)
18. J. Szoplik, M. Ciuksza, Mixing time prediction with artificial neural network model. Chem. Eng. Sci. **246**, 1–8 (2021)
19. Ç.K. Şimşek, D. Arabacı, Simulation of the climatic changes around the coastal land reclamation areas using artificial neural networks. Urban Clim. **38**, 1–18 (2021)
20. C.C. Aggarwal, *Neural Networks and Deep Learning: A Textbook* (Springer, Berlin, 2018)
21. S. Haykin, *Neural Networks and Learning Machines* (Pearson, London, 2008)
22. U. Michelucci, *Advanced Applied Deep Learning: Convolutional Neural Networks and Object Detection* (Apress, New York, 2019)
23. H.H. Aghdam, E.J. Heravi, *Guide to Convolutional Neural Networks: A Practical Application to Traffic-Sign Detection and Classification* (Springer, Berlin, 2017)

24. E.M. Houby, N.I. Yassin, Malignant and nonmalignant classification of breast lesions in mammograms using convolutional neural networks. Biomed. Signal Process. Control **70**, 1–10 (2021)
25. R. Venkatesan, B. Li, *Convolutional Neural Networks in Visual Computing: A Concise Guide* (CRC Press, Boca Raton, 2017)
26. J. Kennedy, R.C. Eberhart, Particle swarm optimization, in *Proceedings of the IEEE International Joint Conference on Neuronal Networks* (1995)
27. R.C. Eberhart, J. Kennedy, A new optimizer using particle swarm, in *Proceedings of Sixth International Symposium on Micro Machine and Human Science* (1995)
28. R.C. Eberhart, Y. Shi, Comparing inertia weights and constriction factors in particle swarm optimization, in *Proceedings of the IEEE Congress on Evolutionary Computation* (2000)
29. Y. Shi, R. Eberhart, Parameter selection in particle swarm optimization, in *International Conference on Evolutionary Programming* (1998)
30. D. Sánchez, P. Melin, O. Castillo, Comparison of particle swarm optimization variants with fuzzy dynamic parameter adaptation for modular granular neural networks for human recognition. J. Intell. Fuzzy Syst. **38**(3), 3229–3252 (2020)
31. Asia Pacific Tele-Ophthalmology Society (2019) [Online]. Available: https://www.kaggle.com/c/aptos2019-blindness-detection. Accessed: 2 Feb 2022

Printed by Printforce, the Netherlands